# Dublin
MAPPING THE CITY

# Dublin
## MAPPING THE CITY

*Joseph Brady and Paul Ferguson*

*The Publisher would
like to dedicate this book to
Edwin Higel,
in appreciation of his friendship
and in recognition of his love
for his adopted country*

First published in 2023 by
Birlinn Ltd
West Newington House
10 Newington Road
Edinburgh
EH9 1QS

*www.birlinn.co.uk*

Copyright © Joseph Brady and Paul Ferguson 2023

ISBN 978 1 78027 751 6

The right of Joseph Brady and Paul Ferguson to be identified as authors of this work has been asserted by them in accordance with the Copyright, Designs and Patents Act 1988.

All rights reserved. No part of this publication may be reproduced, stored, or transmitted in any form, or by any means, electronic, mechanical or photocopying, recording or otherwise, without the express written permission of the publisher.

British Library Cataloguing in Publication Data
A catalogue record for this book is available
from the British Library.

Designed and typeset by Mark Blackadder

Printed and bound by Replika Press Pvt. Ltd, India

# Contents

| | | |
|---|---|---|
| | Acknowledgements and map sources | ix |
| | Introduction | xi |
| 1561 | Dublin's first mentions | 1 |
| 1610 | The first map of Dublin | 5 |
| 1728 | Improving on Speed: Charles Brooking | 9 |
| 1730 | A chart of Dublin Bay | 15 |
| 1756 | A detailed look at the city: John Rocque | 19 |
| 1757 | The environs of Dublin: John Rocque | 25 |
| 1762 | Merrion Square | 29 |
| 1791 | *A picturesque and descriptive view of the city of Dublin* | 33 |
| 1797 | William Faden's map | 37 |
| 1802 | Wide Streets Commission: reflections on progress | 41 |

| | | |
|---|---|---|
| 1817 | Finding your way about town: Dublin directory maps | 45 |
| 1818 | The Botanic Gardens, Glasnevin | 49 |
| 1821 | Duncan's map of the county | 53 |
| 1830 | The first railway: Westland Row to Kingstown | 59 |
| 1836 | The Society for the Diffusion of Useful Knowledge | 63 |
| 1837 | Reforming municipal government | 67 |
| 1843 | The Castle Sheet | 71 |
| 1846 | Dublin from the air: *Illustrated London News* | 75 |
| 1847 | The Liberator | 81 |
| 1850 | *New City Pictorial Directory* | 87 |
| 1861 | Dublin in three dimensions | 91 |
| 1865 | Suburban living: Rathmines | 95 |
| 1890 | Impressions of Dublin | 99 |
| 1893 | Public health: typhoid fever epidemics | 103 |
| 1900 | The tram service | 107 |
| 1906 | The high death rate | 111 |
| 1907 | Dublin's railway system | 115 |
| 1914a | Dublin of the future: civic projects | 119 |

## CONTENTS

| | | |
|---|---|---|
| 1914b | Dublin of the future: the city view | 123 |
| 1918 | Building the suburbs: north city survey | 127 |
| 1919 | Marino: a model suburb | 131 |
| 1924 | Léar Sgáil conntae Bhaile Átha Cliath | 135 |
| 1925a | Hygiene: the Civic Survey | 139 |
| 1925b | Regional transport network: the Civic Survey | 143 |
| 1925c | Congestion and bridges | 147 |
| 1926 | Destruction and renewal: Sackville/O'Connell Street | 151 |
| 1929 | An Irish-speaking colony, Gaeltacht Park | 155 |
| 1930 | The new suburbs: Bacon's maps | 159 |
| 1932 | Public piety: the Phoenix Park | 163 |
| 1935 | The best shopping and Goad's fire insurance plans | 167 |
| 1936 | Dún Laoghaire civic survey | 171 |
| 1938 | Governing Dublin | 175 |
| 1941a | *Sketch Development Plan* | 179 |
| 1941b | A new city centre | 185 |
| 1957 | Henry and Moore Streets | 191 |
| 1961 | Dublin postal districts | 195 |

| Year | Title | Page |
|------|-------|------|
| 1966 | The monumental landscape | 199 |
| 1967 | Planning the Dublin region | 203 |
| 1969 | Bringing motoring under control | 207 |
| 1971a | Preservation and renewal: the development plan | 211 |
| 1971b | Urban motorways and the city centre | 215 |
| 1972 | The port and the bay | 219 |
| 1980 | Targeting Dublin: USSR General Staff maps | 223 |
| 1992 | A city centre database | 227 |
| 2000 | Life after town gas: Grand Canal Dock | 231 |
| 2006 | Below ground: an underground system for Dublin | 235 |
| 2015 | Dublin from space | 241 |
|  | Further reading | 244 |
|  | Index | 247 |

# Acknowledgements and map sources

Most of the maps and images contained in this book come from our personal collections, which have been assembled over many years. We are very grateful to Charlotte and Andrew Bonar Law for allowing us to use their copy of William Duncan's *Map of the County of Dublin* (1821) and Daniel Heffernan's *Dublin in 1861*. Their two volumes on the maps and prints of Dublin were invaluable sources. Thanks also to the Dublin Port Company and Maria Lopez for permission to use their copy of Charles Brooking's *A Map of the City and Suburbs of Dublin* (1728) and an extract from the 1972 *Port Studies*. The National Archive kindly gave permission to reproduce Jonathan Barker's plans for Merrion Square (1760s). We also acknowledge with thanks the assistance of the Glucksman Map Library, Trinity College Dublin. John Montague generously gave us sage advice about John Rocque and the Wide Streets Commission. Conversations with friends over many years have enhanced our knowledge of Dublin and its maps, and to them we are immensely grateful. We greatly appreciate the help and advice of all at Birlinn in bringing this project to fruition.

## *Map sources*

The source of each map is shown, in square brackets, at the end of the map captions in each chapter. The following abbreviations are used:

| | |
|---|---|
| [ABL] | Andrew Bonar Law |
| [AUTH] | Author's copy |
| [AV] | Anne Vaughan |
| [CC] | Creative Commons |
| [DP] | Dublin Port Archive |
| [ESA] | European Space Agency |
| [JB/CPA] | Joseph Brady and Crampton Photo Archive, UCD |
| [NA] | National Archive, Dublin |
| [PC] | Private collection |
| [TCD] | Trinity College Dublin, Map Library |

# Introduction

Dublin's origins are lost in the mists of time. The city is certainly Viking, but there are those who argue for an earlier settlement on the basis that Dublin was an important nodal point where major routeways met on the way to Tara. The River Liffey can be difficult to cross. It is tidal as far as Islandbridge and responds quickly to rain in the Wicklow mountains. Travellers would have taken time to cross and may often have been delayed. These are the circumstances that have given rise to settlements elsewhere and could, it is argued, have resulted in some form of settlement on the left bank of the Liffey. There is no evidence for this, however, and the idea must remain conjectural. Equally so are the early Ptolemaic maps of Europe which show a settlement called Eblana on the east of the country. The existence of this information from the second century suggests that Ireland was on some trading routes and sufficient knowledge was gleaned from travellers to get a vague idea of what the country looked like. However, to equate Eblana with Dublin is to go too far – it could have been anywhere along the east coast.

Matters become more definite with the Vikings. They found a shallow bay with a good landing spot and a high ridge nearby giving shelter and defence. Over time, their *longphort* (seasonal camp) developed into a successful town that was well connected with the Viking world and which traded far and wide. Unfortunately, there is no surviving map from this period and knowledge of the geography of the city is dependent on archaeology and literary evidence. This is why, although this book looks at some early sources, the first map of the city is by Speed from the early seventeenth century – now one of the most recognisable images of the city. Speed's map of Dublin is just an inset on his map of Leinster, something he did with his regional maps to make them more interesting. Little did he know that it would become, and remain, so important. It is used to try and look backwards in time on the assumption that change was relatively slow and the view of Dublin in 1610 was a reasonable representation of those earlier times. It was also widely copied – nothing else came to replace it for over a century – and the copying was immediate and persistent. Braun and Hogenberg published the first volume of the *Civitates Orbis Terrarum* in Cologne in 1572 with the sixth and final volume appearing in 1617; Speed's map appeared in their sixth volume, having been published just in time. Alain Manesson Mallet was a French cartographer and engineer who published some not very good

---

Greenville Collins, Dublin Bay from *Great Britain's Coasting Pilot* (1693) [AUTH] Replaced in 1730 by Price's more accurate chart but still published to the end of the eighteenth century.

Alain Manesson Mallet, Dublin from *Description de L'Univers* (1683) [AUTH] One of many versions of Speed which appeared during the century.

versions of Speed around 1683. It was, in fact, 1728 before there was an updated map.

The bay had however received some attention during the intervening period. Although the Vikings had found a good place to berth their longships and a good location to defend their settlement, the bay was not really suitable. It was (and is) very shallow, with significant sandbanks and only a narrow channel. As Dublin turned more to trade and trading ships had deeper draughts, it was found that the bay was treacherous to navigate. Not only was the channel narrow, but there was a large sandbank across it which made it very difficult to get into the harbour at low tide. The 'Dublin Bar' became a significant barrier to commerce – ships had to wait for wind and tide to be in perfect accord in order to make a safe passage, and many did not make it. There was interest when Greenville (also Greenvil) Collins published his chart of the bay in 1693. It was not completely accurate, but *Great Britain's Coasting Pilot* was an important resource and provided tide tables and coastal views as well as the geography of the bay. Though Collins continued to be published until the end of the eighteenth century, a better chart was that produced by Price in 1730, and is discussed here.

Two years previously, a new map of the city had been produced by Charles Brooking. In fact, it was more than a map, there was also a perspective of the city and small engravings of important buildings and spaces. The map has often been criticised for its omissions and errors, but it was the first look at the city since 1610 and a great deal had happened. In the improved political and economic climate of post-Restoration Dublin, the city had expanded to the east on both sides of the river, with the better-off beginning to abandon the older city in the south-west. Brooking showed these new developments and the striking contrast between these new regular streets and the organic street pattern of the medieval city. He also described the infrastructural works of Dublin Corporation as they promoted the development of the quays as an aid to business. Just as valuable perhaps are the vignettes that surrounded his plan – many were of buildings that would change over the coming years or disappear.

Brooking did not meet the need for a detailed map of the city, and that had to wait until 1756, two years after John Rocque's arrival. By that time Rocque had a reputation as a skilled cartographer and was a successful London publisher. Within two years he had surveyed the city and began an ambitious publishing project that involved maps of the city and suburbs at different sizes and scales. He upstaged the efforts of the official city surveyor Robert Kendrick, who had plans of his own for a detailed map of the city. For the first time there was a map of the city that showed the landscape in great detail. This involved four sheets and was a bit unwieldly but there was even a pocket edition of the Dublin sheets which showed the city in outline and named the principal streets (a numbered index on the left margin pointed out buildings of interest). Another reason for Rocque's importance was that his first sheets were published in 1756. The following year, the work of the Wide Streets Commission began and were set to transform the city centre over the next 50 years.

Much of the city described by Rocque was set to disappear, and his maps are immensely valuable as a result. Less detailed maps of the city became more generally available following Rocque, and were a great asset in recording the changing city as Dublin emerged as one of the finest capital cities in Europe. Such was the level of pride in the city that it was felt appropriate to produce histories and guides. Walter Harris published his *History and Antiquities of the City of Dublin* in 1766. In addition to engravings of the cathedrals and other important buildings, there was a copy of Speed and a version of Rocque's pocket map, which Harris claimed was updated to 1765. In 1780, John Poole and Robert Cash published their *Views of the most remarkable public buildings, monuments and other edifices in the City of Dublin*. This was a book of engravings but which also included a map of the city. The built-up area was shown in blocks but with the main streets named, and the arrival of the Grand Canal was signalled, as were some 'intended streets' around the docklands. Sackville Mall still enjoyed its exclusivity, but maps would soon show its extension to the river. The detailed engravings of important public buildings included those involved with public administration, education and the hospitals, but also Newgate prison. While they noted that the construction of the new jail was superior to others in the kingdoms, they set out the flaws very clearly, noting that the chapel, being located on an upper floor, was difficult for the inmates to reach, they being in chains. Poole and Cash included descriptions and engravings of the two cathedrals and three churches of note, as well as details of some of the monuments in the cathedrals. The

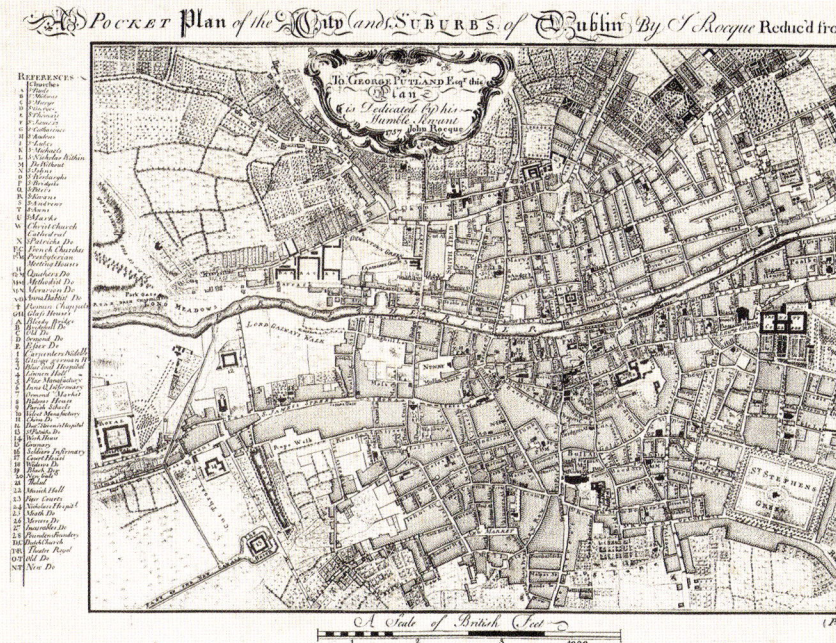

John Rocque, *A pocket plan of the city of Dublin* (1757) [AUTH]
One of many versions of the Dublin map produced by Roque.

engravings of Christ's Cathedral (Christ Church) and St Patrick's Cathedral added to the small number of views of both buildings before the extensive and romantic restoration/reconstructions which took place in the middle of the nineteenth century. Pen pictures of four impressive houses were also provided and, to no surprise, these were Leinster, Powerscourt, Charlemont and Tyrone houses.

Poole and Cash were soon supplanted by the high-quality engravings of James Malton, which he produced during the last decade of the eighteenth century. The engravings proved popular because of their quality and their size – suitable for framing. They showed Dublin at the height of its glory with its new streets, fine public buildings and impressive regular houses. Included in his bound volumes was a large-format map by William Faden, dated 1797.

An innovation from the mid-eighteenth century was the directory, the internet of the day. In an increasingly urbanised society with greater population densities, it was no longer

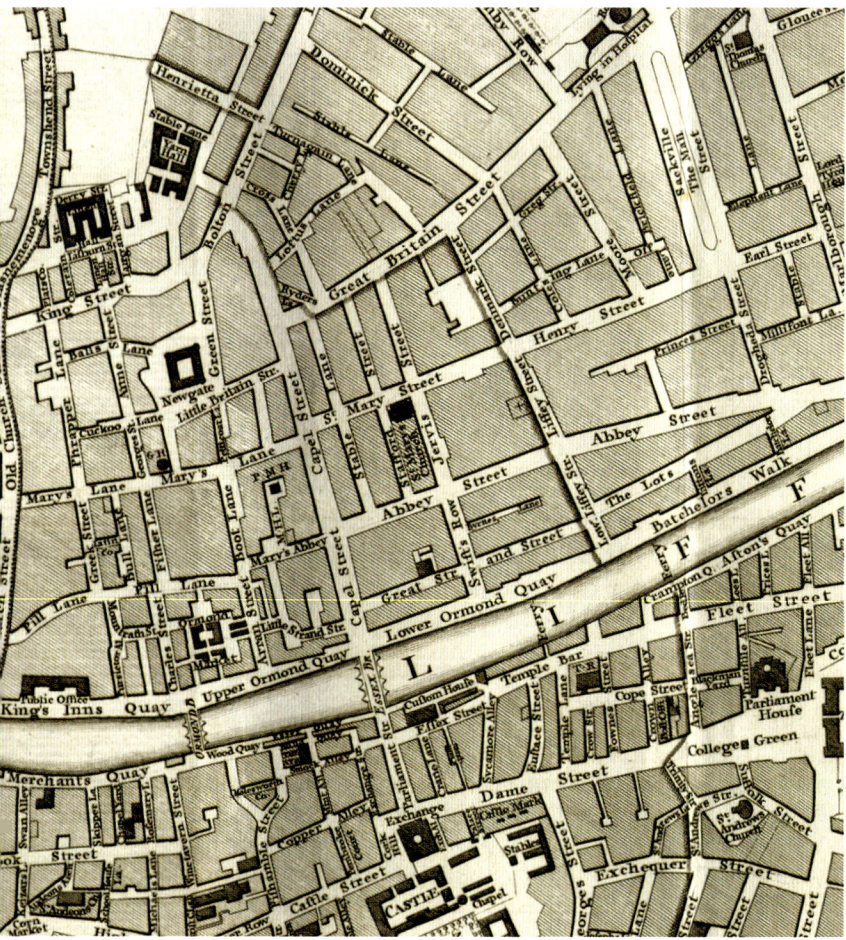

John Poole and Robert Cash, central Dublin showing the new Parliament Street and the developing Dame Street (1780) [AUTH]

possible to know everyone. The directories came to fill that gap by providing information on who was doing what role across the full spectrum of society. To this was added information on services such as the times and costs of postage, or lists of people in various trades and occupations. This soon developed into a street directory where it is possible to find the occupant of any premises in the city. Directories in various forms continued almost to the present day and many of them contained a map. These were updated reasonably regularly and filled in the gaps between the issue of the larger format and more 'official' publications.

Rocque had provided a map of the suburbs and environs of Dublin but by the early years of the nineteenth century there was need for a replacement. Two arrived in quick succession. In 1814, the Grand Jury for Dublin County commissioned William Duncan to produce a replacement but, while he was working, another map appeared on the market which seemed to meet the same need. [John] *Taylor's map of the environs of Dublin* appeared in 1816 on two sheets, though with an unequal coverage to ensure that the city was not divided on the sheets. His 'environs' omitted a large amount of north County Dublin but included parts of Meath, Kildare and Wicklow. It seems that this was done to make a more convenient shape rather than an attempt to delimit some form of hinterland for the city. William Duncan persevered and his map, which needed eight sheets, was ready in 1821. Both maps were decorative and useful, but Duncan's map covered the entire county and was generally preferred over the two. The example discussed here was laid down on linen and dissected into 64 panels.

The problem of access to the bay did not go away. Rocque recorded the first stage in the solution: the building of a seawall from Ringsend midway out into the bay with a small harbour close by to facilitate the early unloading of mail from the packet ships. A lighthouse was built on the seaward side of the wall and the Poolbeg became another of the iconic images of the city. This was only a partial solution, however, and other plans were investigated, including a detailed survey undertaken by William Bligh, of *Bounty* fame, in 1800, which among other things provided the soundings information for Taylor. The solution ultimately chosen was a second wall built southwards from a point in Dollymount. After 1825, maps of the city now included the distinctive arms of the Bull and South walls, and very soon the Bull Island would become a permanent feature.

The directories ensured that there were now regular maps of the city, as did the increasing number of guides and histories of Dublin. G.N. Wright's *An Historical Guide to Ancient and Modern Dublin* contained a map of the city centre together with a set of engravings of significant buildings by George

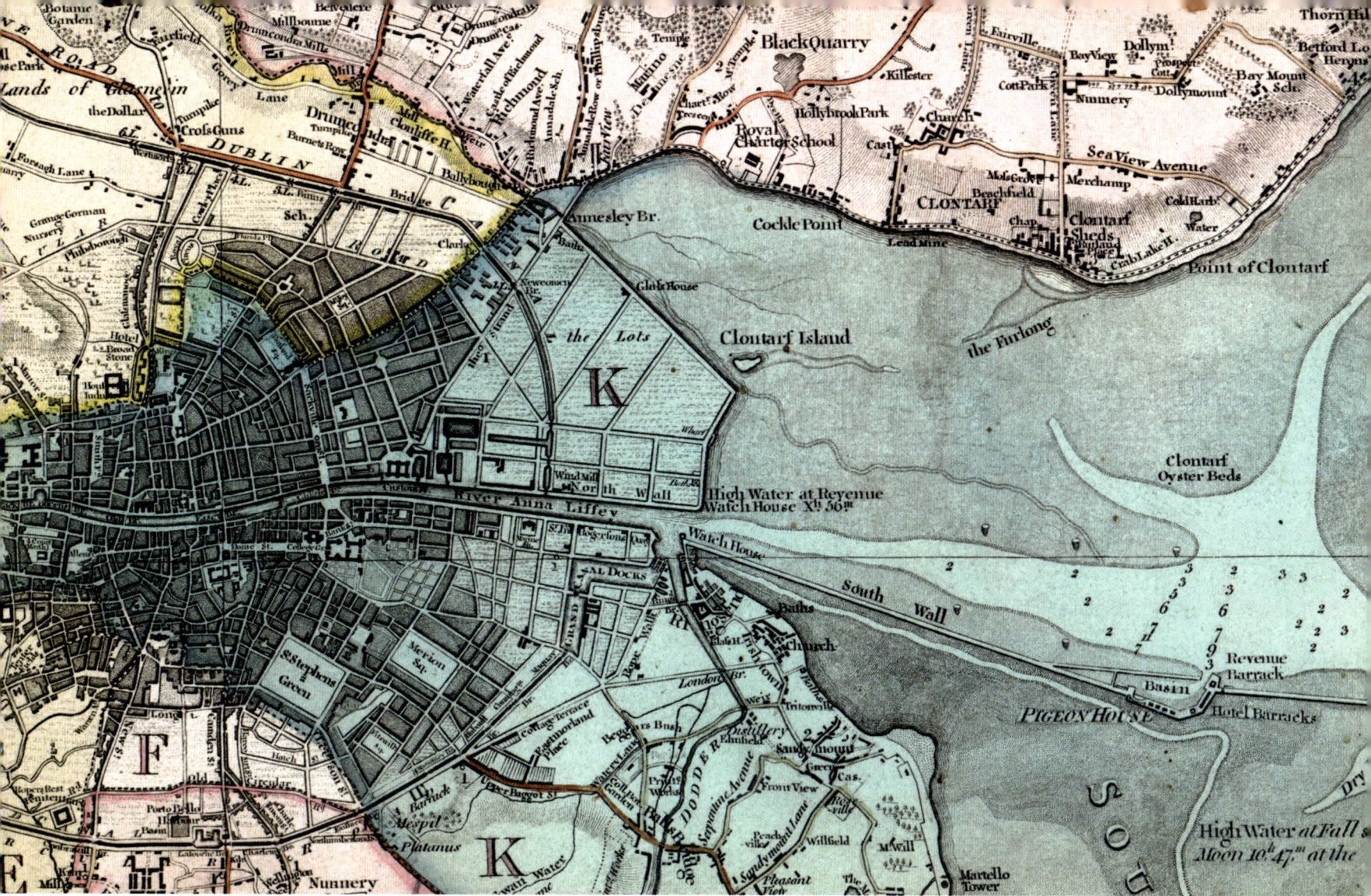

John Taylor, Dublin City, extract from *Taylor's map of the environs of Dublin* (1816) [AUTH]
Taylor's map was very quickly in competition with Duncan's larger format map.

Petrie and was first published in 1821 with a second edition in 1825. However, it was the arrival of the Ordnance Survey that changed matters forever. From 1825 to 1846 teams of surveyors traversed Ireland and produced maps to a high degree of accuracy at 6 inches to the mile. Maps at more detailed scales were produced for Dublin and other important areas, and sheets appeared at 25 inches to the mile and 5ft to 1 mile. This allowed individual buildings and their outbuildings to be shown precisely, and the maps became an important legal and surveying tool. Their accuracy was unparalleled and would not be equalled until the modern era. Unfortunately, they were not updated as regularly as users might have hoped and this provided opportunities for others to fill the gap.

Self-improvement was one of the desirable virtues that Victorians promoted, and could be assisted by profitable reading. Companies were therefore created to produce useful texts and encyclopaedias, especially for the working classes. One such was the Society for the Diffusion of Useful Knowledge (SDUK) and among its publications was an atlas of the world which included detailed maps of important cities. Dublin was included in this range and the SDUK map of the city (1836) and its environs (1837) were reprinted many times

George Petrie, 'A view of Dublin from the North', frontispiece to G.N. Wright, *An Historical Guide to Ancient and Modern Dublin* (1825) [AUTH] Petrie's views of the city were widely copied.

over the following decades. Another mass circulation publication arrived in 1842, and continued in one form or another until 2003. The *Illustrated London News* (*ILN*) reported regularly on Dublin, but its biggest contribution was a large-format bird's-eye view of the city. The quality was good, and it added greatly to the existing collection of maps since it offered a three-dimensional view of the landscape. One of the *ILN*'s rivals was *The Graphic*, which first appeared in 1869. It professed to be a higher quality production with better artists employed and high-quality paper. While it too produced features on Dublin from time to time, its bird's-eye view, which appeared in December 1890, might have given a valuable update on the *ILN*'s perspective of 40 years previously, but unfortunately it viewed the city from a different perspective. Shaw's *New City Pictorial Directory* (1850) was another publication that would have been useful had it succeeded since, in addition to being a street directory, it provided line drawings of the buildings. Heffernan also adopted a visual approach for his unique map of the city in 1861. He added detailed drawings in three dimensions of important buildings to his street map, placing them in their appropriate location. The effect was impressive and very decorative. Perhaps the most decorative map from this period was that produced by Tallis in 1851 as part of his world atlas (and used as the cover to this book). The map was very similar to that produced by the SDUK, but it was enclosed within a decorative border and there were vignettes of some scenes in the city. He decided to include a view of St Patrick's Cathedral (prior to the 'Guinness' restoration between 1860 and 1865), the Custom House, the Four Courts and King's Bridge. The Poolbeg lighthouse was included in a prominent location at the top margin. The lighthouse, shown with a bulbous shape and octagonal lantern, had been replaced by 1820 with the current structure. The image of the Four Courts suggested a large esplanade outside it, suitable for promenade, whereas the reality was much narrower. The memorial to Lord Nelson was incorrectly named Nelson's Column, and it is difficult to gauge the viewpoint of the observer since the perspective offered is impossible – the Tallis company had an office in Dublin so it would not have been difficult to check, but it seems that they relied on existing images.

From the 1860s onwards it was usual to find two maps of Dublin in the variety of atlases which were on the market, such as those produced by Letts, Cassells, and Cram in the United States. There would be a map of the city centre, usually the area between the canals, which would name the principal streets and show some important buildings, especially those of particular relevance to tourists. The buildings would not be shown in detail and were usually just blocks between the roads. They were updated from time to time, so it was possible to see some changes in the urban landscape. The other map was of the environs and tended to show the city and County Wicklow, sometimes also showing Meath. Another source of maps was the increasing number of guides to the city. Dublin was now a regular tourist destination and guides outlined the city and what was to be seen. The focus was on education and useful information; the concept of shopping while on tour had yet to develop. One such early example was *Black's Guide for Tourists*, which appeared in 1857. The 1875 edition was titled *Black's Picturesque Tourist of Ireland* and contained a number of maps and plans. The map of Dublin followed the usual format – a simple outline plan, naming the principal streets and showing the waterways in a shade of light blue. Ward, Lock and Co. (the name went through a series of variations) would have been known to generations of tourists. The 'red guides' were substantial publications and provided both practical information and detailed accounts of what could be seen for the better part of a century. The introduction to the 1878 edition noted that:

A fifth edition of our Guide to Dublin and Wicklow being called for by the public, we have introduced into it a number of new engravings, which will we trust prove acceptable to our friends, and a plan of the city, corrected up to the latest possible date. The binding too has been improved and strengthened.

The 25th edition, published in 1950, was the final one in the series but it remained available until the 1960s. The map of Dublin was designed by Bartholomew, another atlas and map producer, and was a very clear and detailed depiction of the city.

The high-end hotels such as the Metropole and Shelbourne produced guides from the 1880s onwards, with the Gresham Hotel doing so on an occasional basis. The Shelbourne guide continued into the 1960s, but the range and detail of information declined significantly after the 1930s. Every guide to the city had a map of the city centre. The size and quality varied greatly, with most showing the main streets and the location of important public buildings. Some were too small to be useful and some were absolutely too big to be used in any kind of public setting. The map enclosed with the *Welcome to Dublin* guide, published in 1948 by the Dublin Publishing Company and distributed free of charge by the Irish Tourist Association and leading hotels, was certainly a convenient size (slightly bigger than A5). It also tried to be useful and communicate a significant amount of information, though it is not entirely certain that it was successful in that.

By the middle of the nineteenth century, Dublin had serious social problems. Large numbers of underemployed poor people occupied unsuitable, substandard houses in the central area while the middle and upper classes increasingly lived in the suburbs. The Victorians and the Edwardians were very good at studying the problem, though not as adept at solving it, and this produced occasional detailed and fascinating maps such as that produced by Sir Charles Cameron, the city's chief medical officer for health, showing the typhoid fever epidemics of 1891 and 1893. Another such specialised map was that produced by Flinn in 1906 showing the areas of extreme poverty in the city. These added important detail to the standard issue of directory, atlas and tourist guide maps which continued into the twentieth century.

Other specialised maps were also produced. House insurance in the modern sense was not generally available, but fire insurance was. By the middle of the nineteenth century there were many insurance companies in Dublin offering fire insurance, but evaluating individual risk required detailed knowledge of each building. This business opportunity was seen by Charles E. Goad, where fire insurance plans became an essential item in every office, continuing until the internet changed the nature of data availability. Other specialised maps included one of the funeral route for Daniel O'Connell in 1847 – a simple, single-purpose map, it showed where the various groups should assemble. In a similar vein, maps were produced for the celebrations around the centenary of Catholic Emancipation in 1929 and the International Eucharistic Congress in 1932. The maps were designed to give those attending the necessary information about where they should assemble and/or park their cars. They were also intended as a souvenir, especially in 1932 when the range of souvenirs was vast. Important too was the opportunity to advertise. A lot of people would be in the city and the map would be in use. Here was a great opportunity to draw the public's attention to all sorts of goods and services.

By 1910, Dublin had become aware of a new approach to managing cities called 'town planning'. The work of urban reformers had produced some interesting towns in the UK, such as Saltaire, Bournville and Port Sunlight, and the work of Ebenezer Howard and his ideas on garden cities was becoming hugely influential. People in Dublin were well aware of these new trends and of experts such as Raymond Unwin and Patrick Geddes. In 1914, an international competition was held to produce ideas for a town plan for Dublin. The winning entry by Patrick Abercrombie contained a series of maps which set out his vision for the new Dublin, one in which land was managed rationally or zoned. To this was later added the maps produced as part of the Civic Survey in 1925. These demonstrated how maps could be used not only to show the spatial layout of a planned development but also as a means of cross-referencing data from a variety of sources. Maps became a crucial part of town planning, the means to communicate ideas which had a spatial form, and the 1941 *Sketch Development Plan* contained a map which said much more than was in the printed document. It showed how a new communications network could be created, how it would link with new suburbs and how the various elements of the plan could be knitted together. When Dublin finally got its first official development plan in 1971, the maps were very important in allowing people to understand what was intended.

Sometimes the absence of maps allowed suggestions to get less reaction than they might have deserved. Schaechterle's ideas for transforming the road system in the mid-1960s, for example, would have dramatically changed the urban landscape. Dublin would have constructed urban motorways right into the city centre. These would have been high-capacity roads with multi-level interchanges similar to those seen in many European cities. The reaction to Schaechterle's plans was less intense than might have been expected given what he was suggesting, and that is probably because there were no maps or plans of any sort included in the report. People simply could not see the impact of what was intended, and similar omissions from later plans ensured they got a less robust reaction than they might otherwise have had. It was only the special edition of the *Architectural Review* in November 1974 that brought home how Dublin Corporation's plans for inner tangent roadways, as developed for them by Travers Morgan, would actually look on the ground.

It was not just the big projects that were communicated by maps. As Dublin grew and became more complex, it required greater regulation of day-to-day activities. For example, commuters had to become familiar with the locations of traffic lights. These were first introduced in 1937, following years of discussion and debate, but once initial fears about their impact had abated, they were introduced quite widely by 1940. Dublin also got draconian parking regulations at about the same time as traffic lights were introduced. These limited the locations where cars could be parked and the length of time for which they could stay. On busy streets it was limited to 20 minutes, and infractions meant a prosecution in the District Court. Drivers therefore had to get used to the concept of a control zone in Dublin, an area of the city where the parking regulations were in force. Although they did not initially have to pay for parking, it was within that same control zone that parking meters were located when they arrived in 1969. It was a mixed blessing because although they had to pay, motorists could now stay for two hours in most locations.

The dominance of paper maps began to decline as the impact of the internet began to be felt. Having an electronic version of a map could in some circumstances be more convenient that wrestling with a paper copy. It was only with the advent of widespread Global Positioning Systems (GPS) that the decline of paper accelerated. While the military use of GPS dates back to the 1970s, it was only during the 1980s that they became available for civilian use. These early systems were deliberately downgraded by the US military, and it meant that they were of limited use on a day-to-day basis. Additionally, the lack of satellite coverage reduced their usefulness in places like Dublin. That has all changed, and now a combination of GPS and mobile phone data give users access to accurate locations while on the go. The investment by Google in street mapping and street imaging ensures that location can be easily merged with the maps, permitting effortless route planning. One issue though is that the constantly updated maps means that it has become much more difficult to track change in the landscape. Dublin is now regularly surveyed by helicopter, aeroplane and satellite. It is possible to get high-resolution LiDAR (Light Detection and Ranging) data for Dublin from a variety of sources, both governmental and commercial. This raises the question as to whether there is a future for the paper map beyond the ephemeral for the tourist market. It still has the value of being a snapshot in time, whereas many high-tech approaches are dynamic and constantly changing. Furthermore, paper will be accessible long after today's electronic media are lost and unreadable.

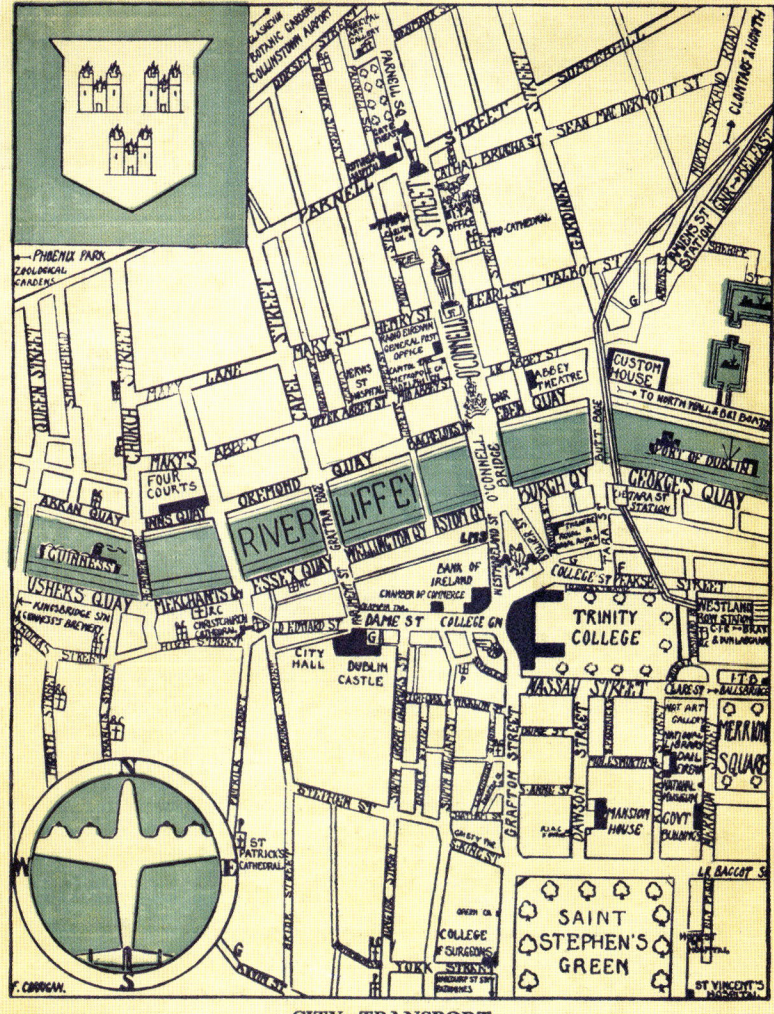

Dublin Publishing Company, Dublin extract from *Welcome to Dublin* (1948) [AUTH] A clear sacrifice of design for information.

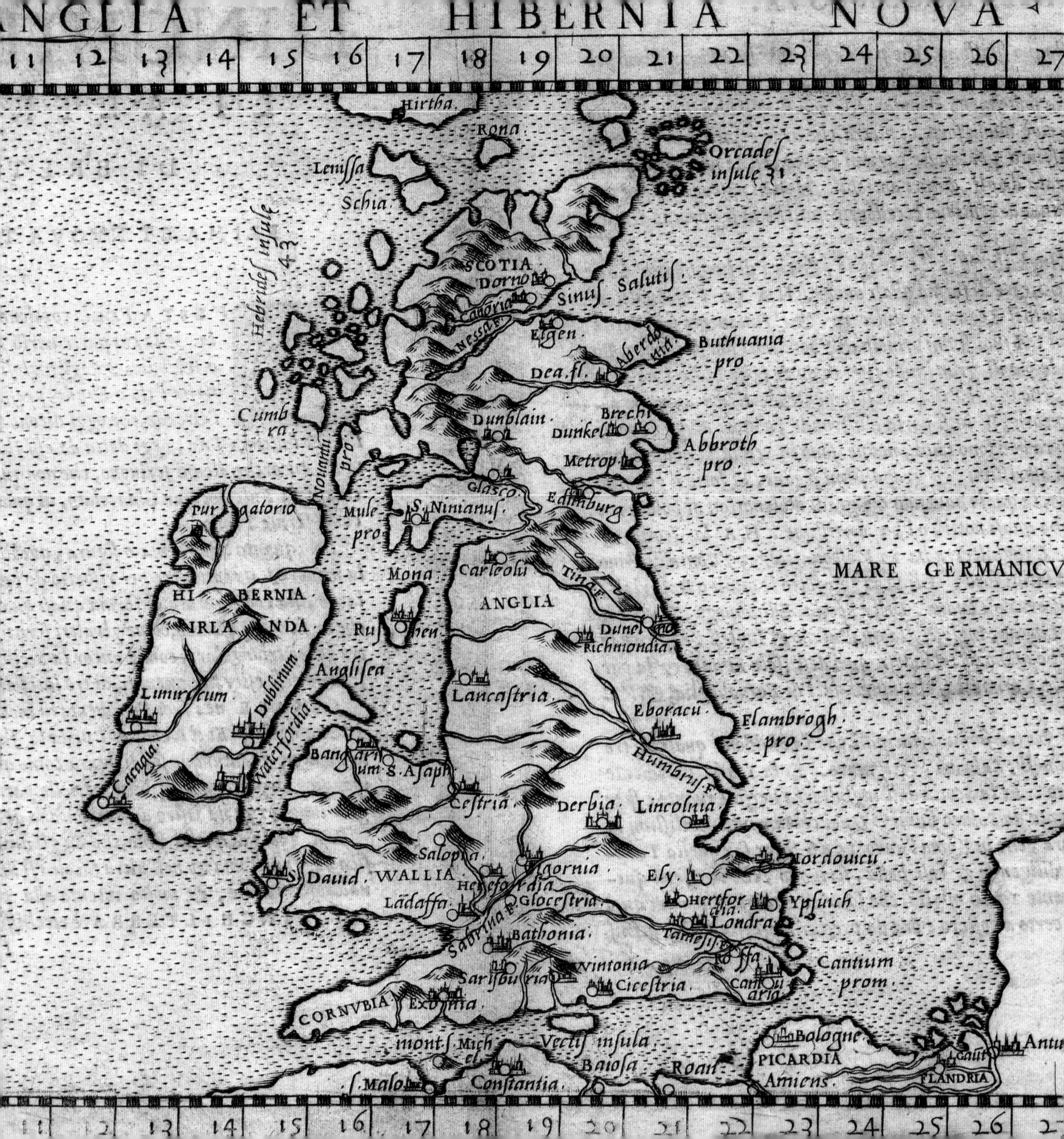

# 1561

## Dublin's first mentions

Dublin claims to be at least 1,000 years old and celebrated its millennium in 1988. That celebration was an economic and social success, and gave a much-needed psychological boost to the city after the doldrums of the decade. The Central Bank of Ireland produced a commemorative 50p coin, one of very few celebratory coins ever produced. A special stamp was issued showing the city in outline. There was a logo for the city, even millennium milk bottles were produced. A regeneration process was beginning to have its impact on Dublin's landscape and the mood of the city was lightened. However, the historical context for the celebration was dubious. The official basis was the commemoration of the subjugation of the Viking city of Dublin in 988 by Máel Sechnaill II, King of Mide and High King of Ireland, following a 20-day siege. Leaving aside the fact that the event took place in 989, it required that there be a city there to subjugate. That it took 20 days to bring about its surrender attests to its substantial nature. Therefore it seems clear that there was a city in Dublin at least a thousand years ago and probably significantly longer.

However, the first detailed map we have of the city dates from the much more recent past; from the *Theatre of the Empire of Great Britain*, dated 1610 and published by John Speed in 1611. If there had been earlier maps, and it seems that there might have been at least one, none have survived. In fact, very few maps survived the medieval period, though they must have been produced. Portolan charts existed in the thirteenth century in the Mediterranean basin and were later expanded to include other regions. These were nautical databases and crucial to safe navigation. It was in the interests of sailors that such charts be as accurate as possible and they were regularly updated, but most survivals date to no earlier than the 1500s. Ireland, however, was still on the periphery of

---

Girolamo Ruscelli, *Anglia et Hibernia Nova* from *La Geografia di Claudio Tolomeo* (1574) [AUTH]
One of many sixteenth-century attempts at mapping Ireland, this version from a Ptolemaic perspective.

Claudius Ptolemy, *Prima Europa Tabula* (1486) [CC] One of the earliest surviving versions of Europe according to Ptolemy.

*Tetrabiblos* was a compendium of astrology and a companion to the *Almagest* that examined the effects of astronomical cycles on earthly matters. However, it is his *Geography* that is most of interest here. The first part was a treatise on cartography but in the second part, the gazetteer, he listed all of the places and geographical features of the known world. Crucially, he developed a system of longitude and latitude which gave precise relative locations for all the places and features he described, and it is these coordinates which allow his maps to be drawn because no maps, if they were ever made, have survived. The earliest surviving Ptolemaic maps are from the fourteenth century, and because he gave very few co-ordinate points for Ireland, the shape of the island was very crudely indicated as a series of sharp angles. He mentioned fifteen rivers, six promontories and ten cities as well as the locations of sixteen tribes. Interestingly, most of the towns mentioned are inland. This is strange because the connection with the world and Hibernia would have been by sea and the source of the information remains a mystery. One of the towns, Eblana, is often suggested to be Dublin and this would not only be the first mention of the city but would push its origins back at least 600 years. However, it is only its relative location on a highly speculative map that suggests that Eblana could be Dublin. The hard evidence that would link Dublin with Eblana is non-existent and no evidence of settlement from this era has emerged.

The other important map from the early centuries of this millennium is the *Mappa Mundi* in Hereford Cathedral. Dating from about 1300, it is the largest surviving world map and it shows Britain and Ireland appropriately at the edge of the known world. Even though the map was produced by a near neighbour, the shape of Ireland is impressionistic and there is no mention of Dublin, or indeed much else in Ireland.

By the middle of the sixteenth century, maps of Great Britain and Ireland were beginning to take on more curved shapes, though Ireland was still quite speculative. Sebastian Münster's *Cosmographia* dates to about 1550 and shows the country as an indefinite shape with Waterford the only place mentioned. Waterford was a major trading centre and would

these charts, and a sea-crossing would have been a dangerous undertaking.

An early source which is suggested to refer to Dublin are the so-called Ptolemaic maps. Claudius Ptolemy was an astronomer and mathematician who lived in the second century of our era. Very little is known about him, though he seems to have worked in Alexandria in the years between AD 127 and AD 148, and was very skilled in mathematics and geometry. Remarkably, much of his work has survived and it proved to be very important during the medieval period and later. The *Almagest* was a multi-volume textbook of astronomy which took a geocentric view of the cosmos, the

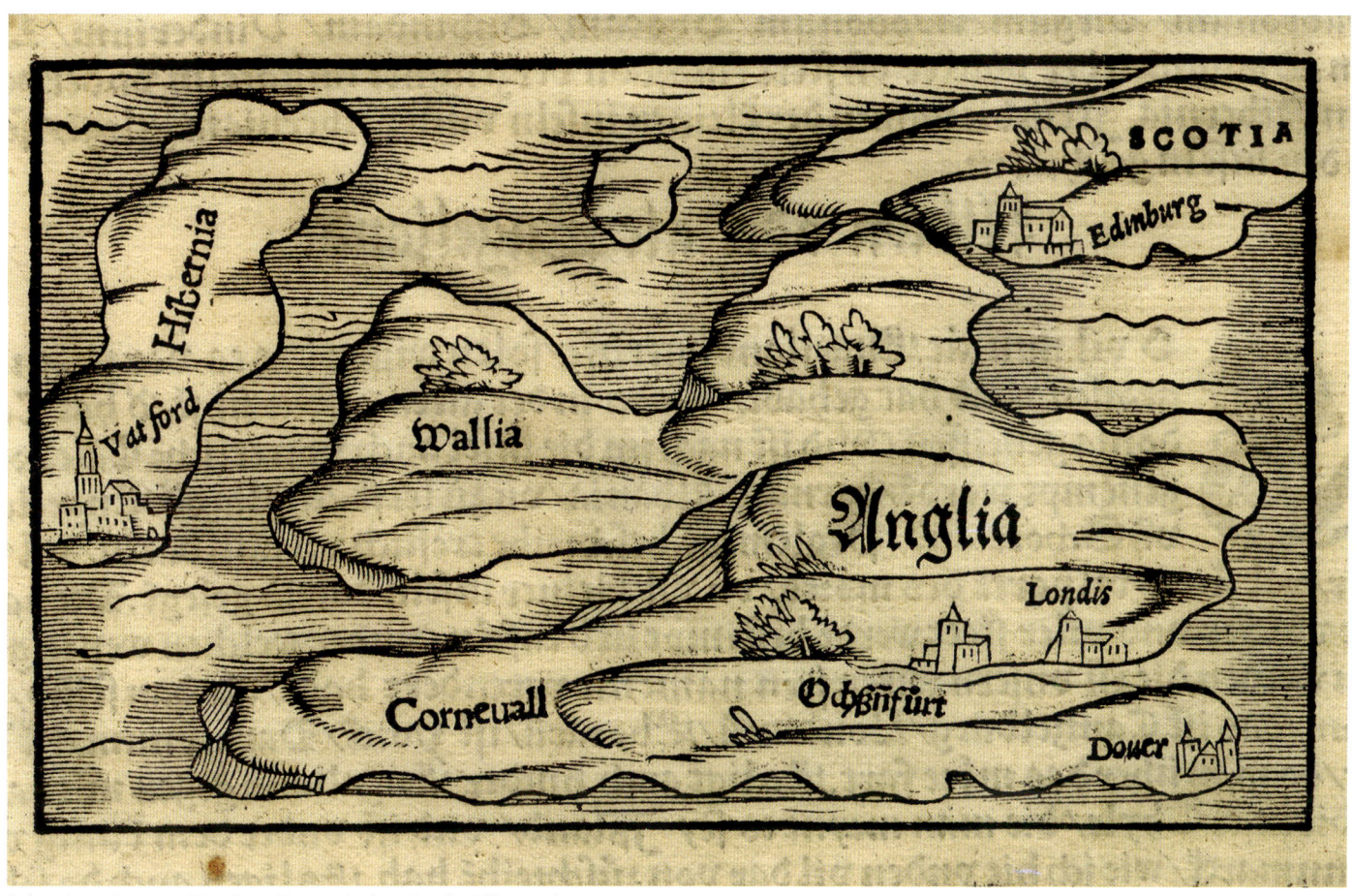

Sebastian Münster, *Von Den Eritannischen Inseln Albione das ist Engelland und Hibernia* from *Cosmographia* (1544) [AUTH]

have been well known on the Continent. Münster drew a much more detailed map of Britain, which appeared in 1540, which mentioned more than 50 rivers and some 70 or more named towns. Ireland appeared only in part and at the edge of the map but Dublin was now shown correctly on the east coast. Thereafter, Dublin was regularly included as the detail of Ireland became better known.

Girolamo Ruscelli's *Anglia et Hibernia Nova* was printed in Venice in 1561 and shows Britain with a much more recognisable shape. Ireland was still a simple rectangle with rounded edges but lines of longitude and latitude were included and Dublin appeared located to the north of a significant river.

Also noted were Waterford, Cork, Limerick and St Patrick's Purgatory. It seems to be a combination of some Ptolemaic geography with more recent knowledge.

By the time Thomaso Porcacchi produced his map of Ireland (*Descrittione dell'Isola d'Irlanda*) as part of his 1572 treatise on islands, the coastline of Ireland was far better known and Dublin was noted together with Howth and the islands Lambay and Ireland's Eye. There was still work to be done on the shape of Ireland, but this was at the beginning of the great age of Dutch cartography that resulted in the hugely detailed and increasingly accurate maps of Gerald Mercator, and those published in the magnificent atlases of Abraham Ortelius.

# 1610

## The first map of Dublin

Despite Dublin's long history the first map of the city is relatively recent. Although John Speed's map, dated 1610, is the first surviving map of the city and served as the most detailed plan for almost a century, it is far from clear when and by whom it was produced. Speed's *Theatre of the Empire of Great Britaine* was an atlas of both islands, based on the counties in Great Britain and the provinces in Ireland. A total of 69 insets showed the principal towns of England and Wales, but none in Scotland and four in Ireland – Dublin, Cork, Galway and Limerick. Some of these maps derived from already existing ones but Speed claimed that he had produced most of them himself from personal observation. The consensus among scholars is that the Dublin map was not produced by Speed simply because it would have been a lot of trouble to go to Ireland to survey Dublin when the other three maps were already at hand. At the same time, the source is unknown. Despite the extensive interest which both Henry VIII and Elizabeth I had in Ireland, it does not seem that this translated into the publication of a map of Dublin.

The map of Dublin appears as an inset to the map of Leinster. It is at a scale of approximately 1:10,500 and about 16 × 18cm. It was issued uncoloured but it is often found with later colouring. The map was produced many times over the next 50 or so years with the final edition appearing in 1676. It is possible to identify the various printings because of small changes to the plate and a general sense of its wear and tear. However, the main feature which identifies the far more common 1676 edition is that a box in the lower left corner says that the map was being sold by Richard Chiswell rather than Sudbury and Humble.

On the verso of his plates, Speed provided a short commentary on the place. Dublin, he noted, was:

---

John Speed, Leinster from *Theatre of the Empire of Great Britaine* (1611) [AUTH]
This sheet is from the 1676 edition.

the royal seat of Ireland, strong in her munition, beautiful in her buildings, and (for the quantity) matchable to many other Cities, frequent for traffick and intercourse of Merchants. In the East Suburbs, Henry the second, King of England (as Hoveden reporteth) caused a royal Palace to be erected: and Henry Loundres, Archbishop of Divelin, built a storehouse about the year of Christ 1220. Not far from it is the beautiful College consecrated unto the name of the Holy Trinity, which Queen Elizabeth of famous memory dignified with the privileges of an University. The Church of S. Patrick being much enlarged by King John, was by John Comin Archbishop of Dublin, born at Evesham in England, first ordained to be a Church of Prebends in the year 1191.

He went on to note that since 1409 the city could choose a mayor and two bailiffs annually, and that the mayor would always be preceded by a gilt sword. With the elevation of the bailiffs to sheriffs under King Edward VI, Speed reckoned that there was nothing further lacking 'that may serve to make the estate of the city most flourishing'.

Doubtless he felt he was being very complimentary when he wrote that 'the people of this County do about the neighbouring parts of Dublin come nearest unto the civil conditions and orderly subjugation of the English'. This was all the more so because in more distant locations, they are 'more tumultuous, being at deadly feuds amongst themselves, committing oftentimes manslaughters one upon another, and working their own mischiefs by mutual wrongs'.

The map showed a town on the right bank of the River Liffey with a channel much wider than the present day. Important buildings and features were numbered and an index was provided in the margin. There was no hint of the topography of the location, but Dublin seemed to occupy a good defensible site with water on three sides. The Black Pool (the Dubh Linn or Dyflin), from which the town's name in English derives, was shown together with the main rivers of the locality, though none are named. Speed showed the city walls, though not the detail of the inner walls of earlier times. As well as naming all of the gates, he suggested that the wall along the river was intact and gated. There was one bridge across the Liffey at Bridge Street. The town itself was quite small with irregular streets but at this scale not much can be gleaned about the nature of the buildings. Churches dominated the public buildings but the Tholsel (the seat of civic governance) was located and named, and the Castle was an obvious and dominant feature. The newly created 'Colledge' was shown in the suburbs to the east, making clear why it was referred to as being 'near Dublin' in its full title. This occupied the lands of the dissolved Priory of All Hallows and had been granted by Dublin Corporation for a university. The city had extensive suburbs to the west, developed in the late medieval period, and James' Street and Francis Street were built up with gates. Development on the other side of the river was more limited, but St Mary's Abbey was identified. Much of this land would have been monastic prior to the Dissolution but by this time it had passed into private hands, with some going to the Corporation's city estate. Similarly, on the southern side, the map identified the former land holdings of Whitefriars. These and other such holdings would become important in the dramatic expansion of the city in the years after the restoration of the monarchy in 1660. No central marketplace existed, as would have been common in mainland European cities of this size, and this would become a problem for the Corporation to manage in the following centuries as the streets became congested by business activities. Some of the street names are suggestive of the location of their economic activities – Sheepe Street (perhaps), Fish Shambles, Wine Tavern, Skinners Row.

Speed's map is a very important snapshot of the city's development and, given the circumstances of the next 50 years, it was probably a reasonable representation of the city at the time of arrival of the Duke of Ormond as the new Viceroy in 1662. Indeed, the city was in better shape in Speed's time than it was when his Lordship made his triumphal arrival on 27 July 1662. However, it gives only glimpses of the city's evolution. Modern research suggests three phases in the city's development. The first phase was a simple cross with a north–

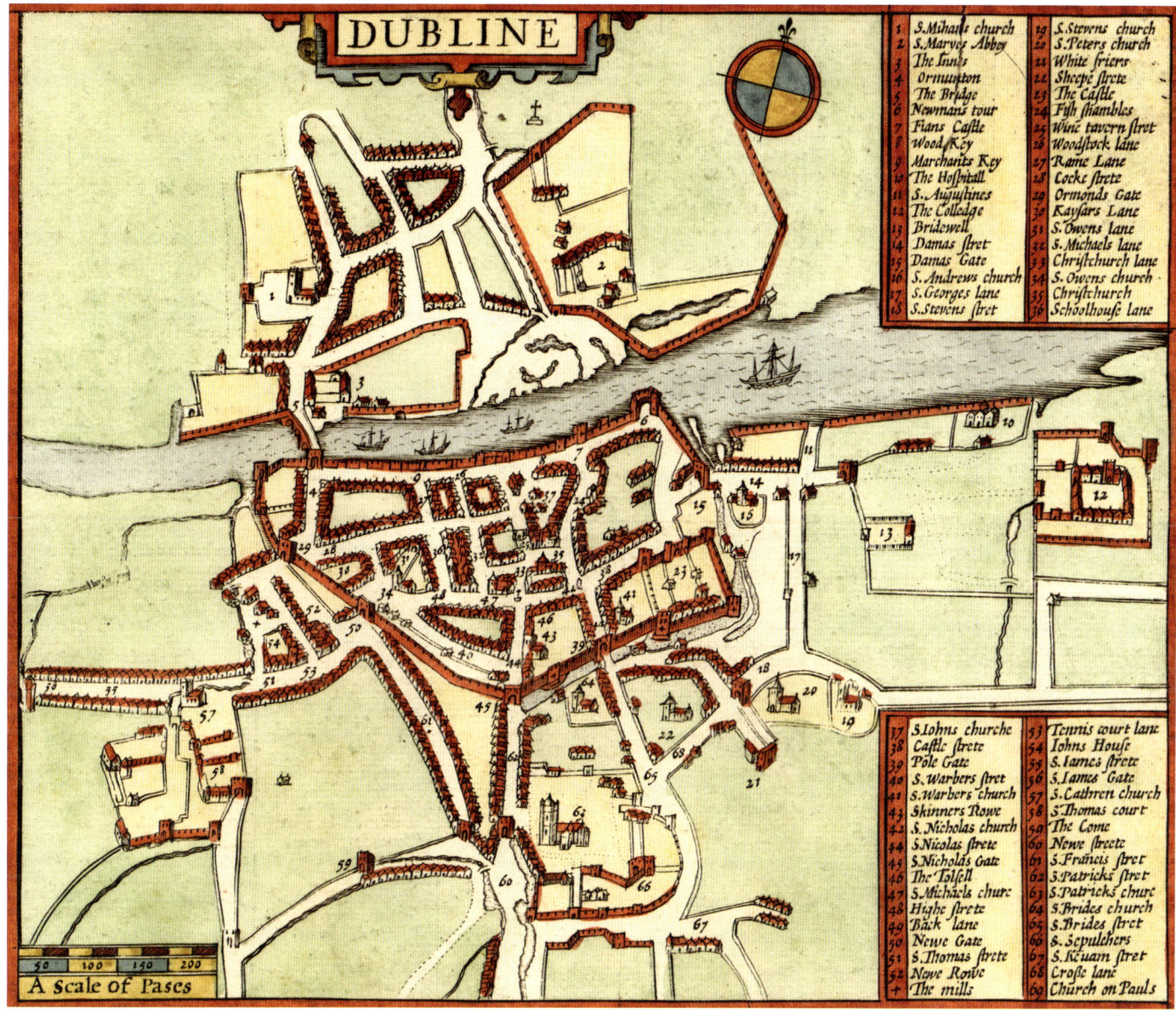

John Speed, Dublin extract from his plan of Leinster (1611) [AUTH]

south and east–west axis centred around Christ Church Cathedral. The second phase saw an expansion to the west along the ridge of the esker which gave the city its defensive advantage. Thirdly, the upper quarter (approximately) of the city, parallel to the river, was a reclamation initiative of the English (Anglo-Normans) who found it increasingly inconvenient to have to use lighters to bring cargo to the shore from larger and larger cargo ships.

# 1728

## Improving on Speed: Charles Brooking

John Speed's map of 1610 was used for many decades but became increasingly out of date, especially when the city began to expand after the Restoration. DeGomme's sketch plan for 1673 had added little and, anyway, development was only beginning to gather pace at that time. So Charles Brooking's map of the city and suburbs was particularly welcome when it appeared in 1728. In fact, it was more than a map. It offered a prospect of the city, giving a sense of the three-dimensional landscape as well as a series of vignettes of important buildings and locations. It was quite a big production at 141 × 58cm, but the amount of information available was impressive, though not necessarily complete. Its size required three plates, and printing on three separate sheets of paper which were then joined.

The map is the dominant feature, and is unusual in that south is at the top of the map, so the reader's eye is first drawn to that part of the city. It depicts a city which had expanded greatly from its medieval origins and the more regular geometry of the new residential areas, particularly north of the Liffey, was visible. One of Dublin Corporation's major projects following the restoration of the monarchy was the creation of continuous quays along the Liffey to aid commerce. This was done largely by means of public–private partnerships and the map shows that this had been successful. The quays were almost continuous on the northern side of the Liffey to Batchelours Walke, except for an obstruction along Arran Quay. South of the river, there was only the section between Anglesea Street and Essex Bridge, and a smaller section between St Georges Key and Astons Key, to be cleared to provide a continuous passage to the developing Sir John Rogersons Key on the far left. (The spelling in this map was not what might be expected – especially 'key' for 'quay'.) Ships could still make the journey up the Liffey as far as Essex Bridge and it was here that the Custom House was located. On the

---

Charles Brooking, the central area of Dublin, extract from *A Map of the City and Suburbs of Dublin* (1728) [DP]

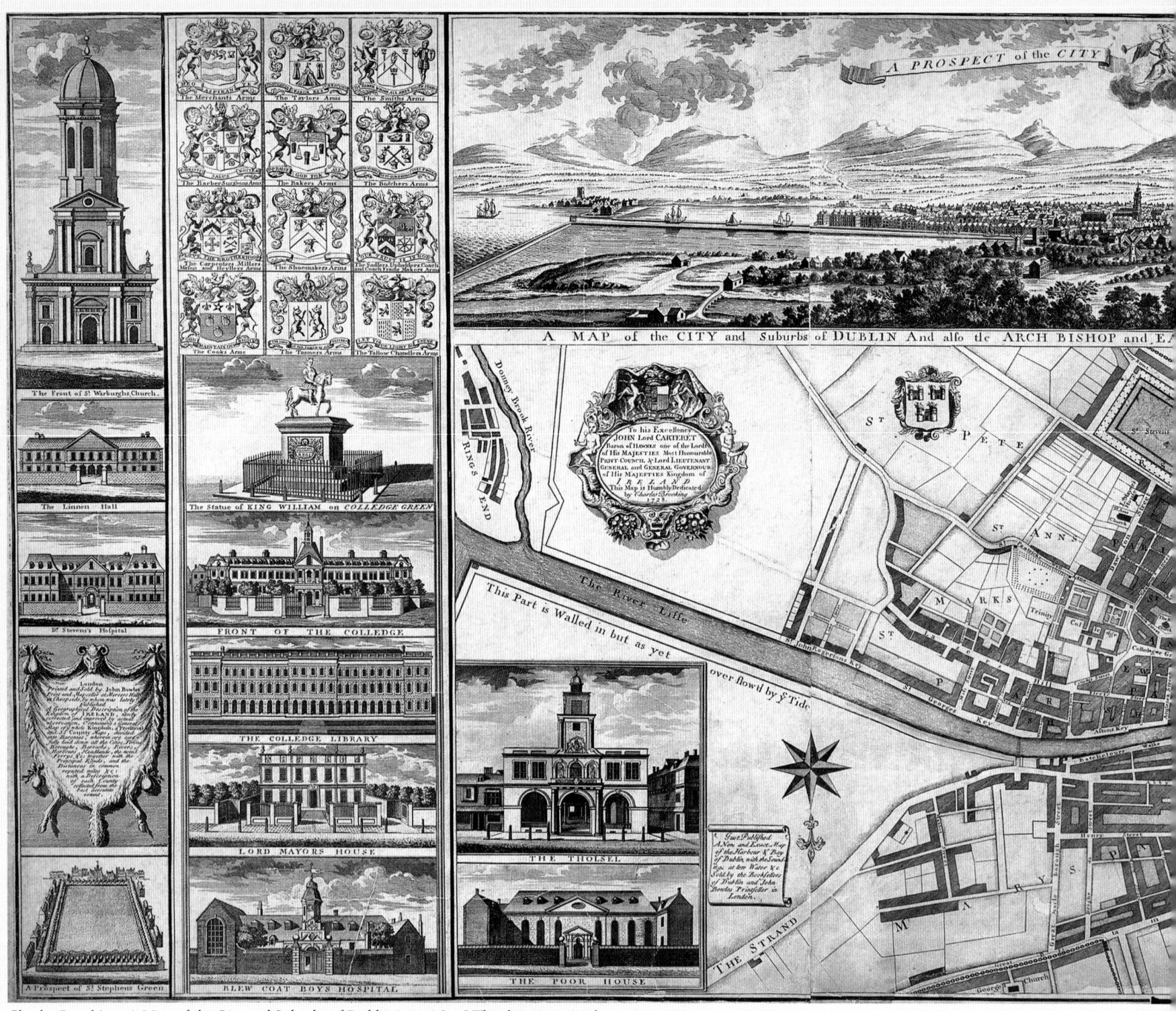

Charles Brooking, *A Map of the City and Suburbs of Dublin* (1728) [DP] The sheet contained a map, a perspective of the city from the north and vignettes of important buildings.

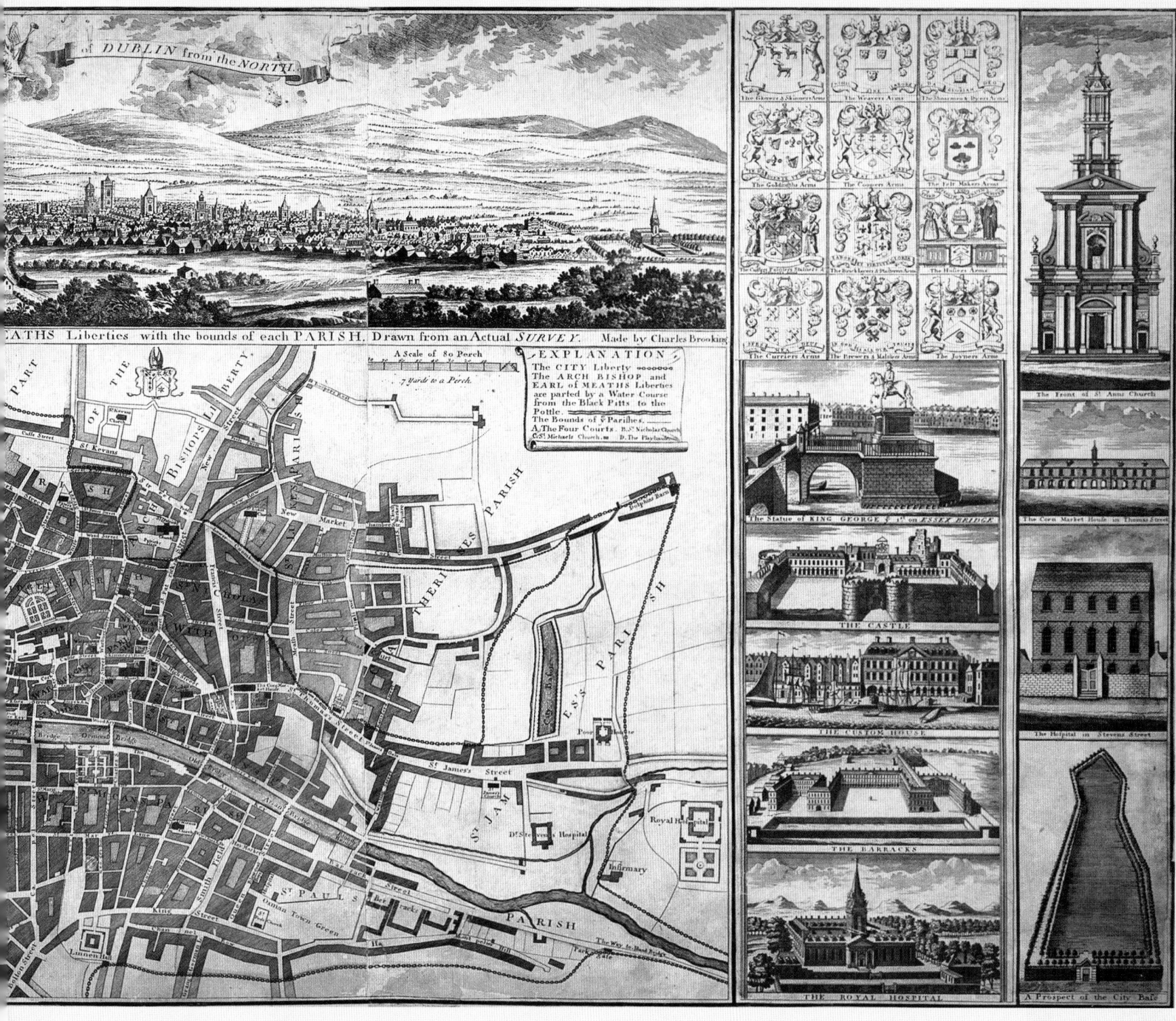

Charles Brooking, An extract from the prospect of the city . . . (1728) [DP]

northside, development had reached as far east as Great Marleborough Street with its adjacent Drogheda Street giving no indication of the major thoroughfare that it was to become. So too Mabbot Street gives no inkling of its role and reputation in the nineteenth century. St Mary's Church was evident on St Mary Street, but Brooking did not include the Roman Catholic St Mary's which had recently been completed on Marleborough Street. Included was Smithfield Market, another of Dublin Corporation's projects intended to bring some order to street trading which had clogged the streets of the medieval city for generations. On the southside of the city, expansion was evident in Trinity College, which now enjoyed a substantial campus and which was the focus of Colledge Green. Dawson Street, with the Lord Mayers House, had taken shape and the streets between it and Merryon Lane had been laid out but seemed not to have yet been built upon.

The problem with Brooking's map is that it was not entirely accurate. One significant omission has already been noted, but he also omitted City Quay by running St George's Quay directly to Sir John Rogerson's Quay. He extended Batchelours Walke as far as Marleborough Street while Aston Quay joined directly to St George's Quay, omitting Burgh Quay. Some other errors were anticipatory. In this he was not unusual, maps often showed buildings or developments which were planned but not actually constructed. In this case, Brooking extended the quay walls as far as the Donnybrook River (Dodder) on the southside and as far as East Wall on the northside, but it would be some time before they would be completed.

The prospect occupied the upper portion of the sheet and gave a view of Dublin from the north with somewhat exaggerated mountains to the south. Given the height of the viewpoint it seems reasonable that the view was from somewhere in Drumcondra on the first major ridge north of the Liffey.

Windmills were evident and still clearly important to the life of the city. The impression given is of a dense urban environment with regular streets and houses with high pitched roofs and tall chimneys with the occasional Dutch Billy façade appearing. There were hints of public buildings but the eye is immediately drawn to the churches in the centre of the city which are given exaggerated towers. The distinctive character of St Werburgh's is particularly noticeable.

Of additional interest is a series of small engravings on either side of the map and prospect, and these are both interesting and important for the view they give of significant buildings as they were at the time. St Werburgh's is shown to have an impressive tower overtopping, and looking somewhat discordant to, a classical temple façade. It was complemented on the right side by a drawing of St Ann's Church on Dawson Street, which is in a more flamboyant baroque style. At least, this is what it might have looked like since, though it was begun in 1720, it never got beyond the first floor and was completed to a new design only in the 1870s. The view of the Castle is important as it shows its transition from a medieval fortress to its Georgian state as a seat of government, though with some medieval elements incorporated. There was a good view of the Tholsel and the Custom House, though the latter would be replaced in the final years of the century by James Gandon's version further down the quays, moving as the port did to the east. Of particular interest is the engraving of St Stephen's Green (Steven on the map). This shows the state of development some 50 years after Dublin Corporation first embarked on its project to turn an unused asset into valuable real estate. Quite a number of substantial houses had been built, but in a variety of designs and there were still opportunities along the southern border. The green itself had been laid out with an enclosing wall and a border of sycamore trees, the planting of which was one of the conditions which the lessors had to meet. The green was a simple pasture at this time, and the later Malton print shows it still being used for grazing.

Compared to the other elements on the sheet, the cartouche was somewhat modest and advertises the sheet as 'Just Published / A New and Exact Map / of the Harbour & Bay /

Charles Brooking, St Stephen's Green, extract from *A Map of the City and Suburbs of Dublin* (1728) [DP]

of Dublin, with the sound- / ings at low Water &c / Sold by the Booksellers / of Dublin and John / Bowles Printseller in / London'. It is not known how well it sold but it was sufficiently popular for an edition in 1732 and for new plates to be made in 1740, the opportunity being taken to bring it up to date. The title cartouche and two views were removed from the map to make room for a plan of the North Lotts. This necessitated changes to the placement of some engravings and the cropping of some. More space was saved by the removal of the coats of arms of the various guilds and this allowed an engraving of the north façade of Christ Church Cathedral – the 'north prospect of the Cathedral Church of ye Holy Trinity in Dublin'. While this version also needed three plates, the arrangement was somewhat different. The map and prospect were printed onto one sheet from their own plates; the engravings were printed on another sheet, which was then cut in two and pasted on either side of the map and perspective.

# 1730

## A chart of Dublin Bay

Greenville Collins' map of Dublin Bay was a very useful navigation aid when it was first published in 1693, but was limited both in accuracy and precision. Matters improved when Charles Price published his *Correct Chart of the City and Harbour of Dublin* in 1730. Price did not enjoy the best of fortune, despite working with many of the leading figures in cartography and the production of sea charts. In 1727, he announced that work would begin on a general atlas for sea and land, which would be global in scope and would comprise about 250 charts. Unfortunately, the project was only partially completed when his personal life took a downturn and he found himself in the Fleet debtors' prison in 1731. By then there were only 31 charts available, but some atlases seem to have been assembled from this total. His economic necessity forced him to take whatever opportunities existed to make ready money and this map was conveniently at hand. His fortunes did not recover however and he died in early 1733.

Bonar Law (2005) suggests that there are two versions of the Dublin chart, both of which are dated to 1730. The one which appeared in the copies of the atlas had a blank dedication cartouche and it was sold by Mr Heath next to the Fountain Tavern in the Strand, Mr Clark at the Royal Exchange, Mr Atkinson at Cherry Garden Stairs and Mrs Penn in Vine Street, Bristol. In the example here the cartouche was reworked to include a dedication to Edmund Halley, the astronomer, and now sold by William Mount and Thomas Page on Tower Hill. This appears separately and not in any of the surviving atlases.

The map included an area from the north of Howth south to Dalkey, and both Howth and Ireland's Eye were given exaggerated mountains. The sea stacks at the north-east of Howth were incorrectly placed between the island and the harbour and made much more dramatic. The map of the city of Dublin was a basic reduction of Brooking, with the port

---

Charles Price, *A Correct Chart of the City and Harbour of Dublin* (1730) [ABL]

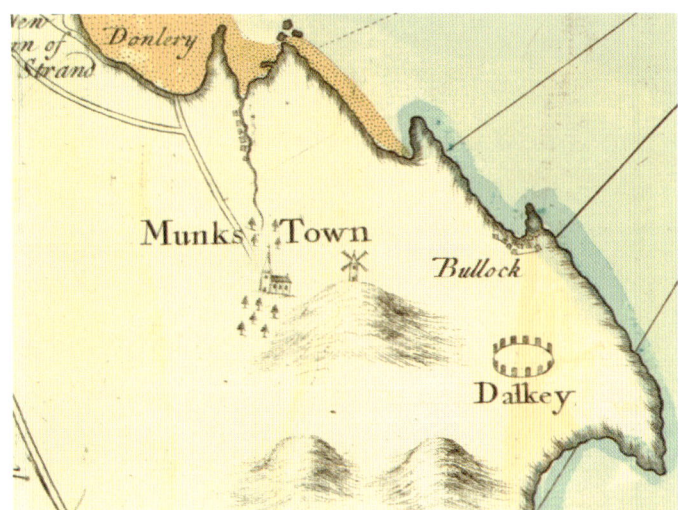

Charles Price, the southern suburbs of Dublin with the new town and stone circle, extract from *A Correct Chart of the City and Harbour of Dublin* (1730) [ABL]

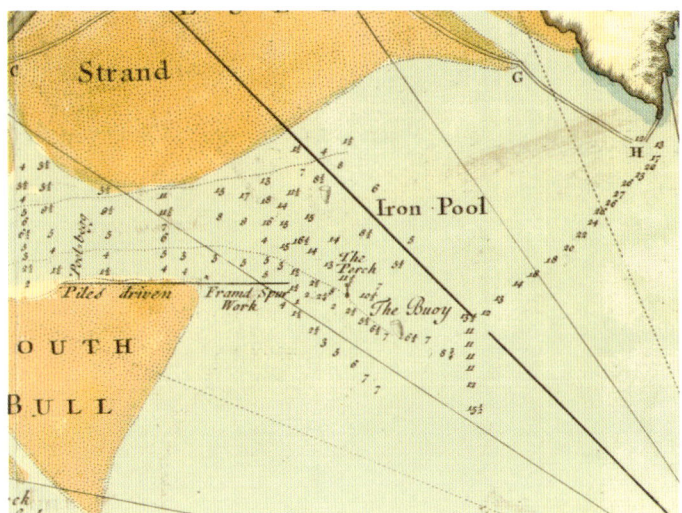

Charles Price, the harbour of Dublin with enhanced soundings, extract from *A Correct Chart of the City and Harbour of Dublin* (1730) [ABL]

surprisingly blank and St Stephen's Green standing out as the most recognisable feature. Ringsend was shown cut off from the city by the 'Donney Brook River', which occupied a much narrower channel than in Collins' map and seems to peter out just to the south of the town. A few suburbs were mentioned, such as Killester and Clandaf (Clontarf) and the Herring Shelds (a small fishing settlement), but these seem to be sprinkled randomly on the landscape. Howth was given more detail, and the harbour and town were included, as was a little drawing of Howth Castle, described as the Lord Hoath's seat. Price recorded a 'new town of the Strand' somewhere around Blackrock, and there was mention of a place called Donlery. He included a drawing of a stone circle at Dalkey, which must have captured his attention. This seems to have been a cromlech enclosed within a stone circle, but it appears that it was destroyed in and around 1797 to provide material for the nearby Martello tower. Much more information was provided about the bay, with detailed soundings on a regular pattern given for the main approach to the harbour. There were far more measurements than in Collins but the latter gave some soundings for the remainder of the bay, which Price omitted.

Price explained that these were taken at low water at spring tide by Thomas Burgh, engineer and Surveyor General, and Captain John Perry, also an engineer with significant European experience. In addition, a detailed description of the bay was provided inside a hanging map frame on the left side of the image. The blank space to the right of the map was filled with an elaborate compass rose and two nice images of sailing ships.

This explained the origin of the map. In 1708 a 'Ballast Board' was established to control and manage the port, and in 1711 began to tackle the problem of the bay in earnest. In 1713 Captain John Perry offered to survey the harbour and suggest methods for its improvement. This was done but Perry's ideas were not taken up. Instead, work began in 1716 on placing a line of wooden piles on the southern side of the channel far out into Dublin Bay, the structure becoming the forerunner of the South Wall. Perry returned in 1725 at the request of the Lord Lieutenant. Colonel Thomas Burgh, the Surveyor General, recruited Dublin surveyor and instrument-maker Gabriel Stokes to assist Perry to take soundings and 'to fix proper marks for the purpose of referring to in future,

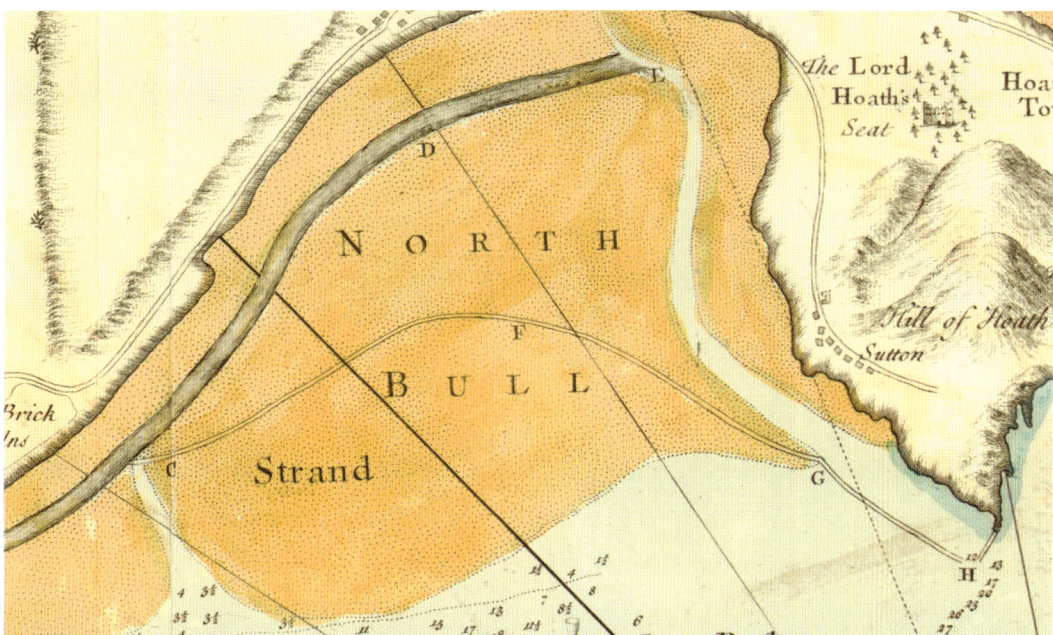

Charles Price, the canal project of Captain John Perry, extract from *A Correct Chart of the City and Harbour of Dublin* (1730) [ABL]

that a true state of the harbour might be ascertained'. Stokes was tasked with producing the survey on which the soundings would be shown. This led in 1728 to Burgh having the Stokes map engraved and printed in London by Emanuel Bowen. It was this map that Price now used.

The text explained that while there were mountains to the south of the bay they were too far away to provide any shelter from SSE or SW winds, which could often be violent. There was good anchorage on the eastern side of the Dublin Bar and ships commonly sat at anchor there when they 'want Tide over the Barr on which a SSE Moon makes full sea on Change Days'. Price also noted that there was a buoy on the deepest part of the Bar where the water was only 6 feet deep at low tide but was 8 or 9 feet at neap tides and 12 feet at Spring tide. The problem with the buoy was that it could be moved by the tides and could prove a 'false friend'. Matters improved when the piles were completed in 1735 and a lightship stationed there. The map also showed the progress that was being made with the laying of the piles. Two disconnected lines of piles had been driven, one from Ringsend and another eastwards from a point called 'Poolbegg' at the end of which the piles gave way to 'framd spur work'. Price differentiated the large north and south Bull sandbanks into their various components and named each of them. The 'Cock Lake' was not a lake in the normal sense of the word but a channel which appeared when the ebbing tide exposed the sandbanks. Before the work on the South Wall, it had been sufficiently deep to permit fishing boats to return to harbour at low water.

The origins of the map explains why it included the 1725 proposal by Captain John Perry to build a harbour at Sutton Creek and to link it back to the Liffey at Ringsend by building a canal. Perry's proposal had been rejected by the Ballast Board in 1726, which decided to continue with its line of piles. The idea seems to have been to avoid the Dublin Bar and its associated sandbanks by building a pier and harbour at Sutton. A connection with Dublin would be provided by a canal which would run parallel to the Clontarf shore from the northern side of the harbour at Ringsend. The River Tolka already followed such a channel and so a good flow of water could be assured. However, among the criticisms of the plan was that the winds were unfavourable for a harbour at Sutton for eight months of the year.

# 1756

## A detailed look at the city: John Rocque

Brooking's map of Dublin was an important update on that of John Speed and, even though it fell short in some ways, it sufficed for the next 30 years. John Rocque was born around 1704, one of four children of a Huguenot family who fled first to Geneva, and then sometime after 1709 to England. He was a surveyor, map-maker, engraver and map seller. His big success was his map of London which was published in 24 sheets in 1746. This was by far the most detailed map of the city ever published up to that point and it did much to make his reputation. He must have felt, though, that there were opportunities in Ireland because he arrived in 1754 and stayed for six years.

Not only did he work on Dublin, but he also produced maps of Armagh County, Kilkenny, Cork and Thurles, but it is his Dublin work that is of concern here. He produced a number of maps of the city and county in different configurations and scales. The most detailed was the four-sheet survey of the city and its immediate environs – so detailed in fact that it was not surpassed until the Ordnance Survey in the nineteenth century. To complete the map, the city was surveyed and while it was not done to the planimetric accuracy of the Ordnance Survey, it was much, much better than Brooking. The map showed the street pattern with its twists and turns and varying widths. Individual buildings were shown and their gardens and back yards delineated. It was not possible that these were absolutely accurate – there was not enough time to complete the survey – but the clearest view ever appeared of the city and its geography.

The map was on four large sheets – each approximately 69 × 49cm – at a scale of 200 feet to an inch (1:2,400), the same scale that he had used in London. The shape of Dublin, an oval, gave Rocque the space to include whatever ancillary

---

John Rocque, the elaborate cartouche and dedication, extract from
*An Exact Survey of the City and Suburbs of Dublin* (1756) [AUTH]

information he wanted within the frame of the map. For the top left-hand sheet, where much of the land was still in agriculture, Rocque chose an elaborate cartouche in which he dedicated his work to a range of notables, including the Earls of Kildare and Bessborough, and which also referred to his recent (1751) appointment as 'chorographer' to Frederick, Prince of Wales.

The scale of the map made it possible to include an impressive range of streets, lanes and alleys, which otherwise would have passed from notice. Buildings were not greatly differentiated. There was a stippled shading for dwelling houses, a darker shade for public buildings and an intermediate one for warehouses, stables, etc. Much more attention was given to churches, and Presbyterian, French, Quaker, Dutch and Roman Catholic churches and chapels are indicated, not by distinctive shading but by letters and symbols overprinted on what was not dissimilar to the shading used for warehouses and stables. Formal gardens were shown with trees and bushes but it would be demanding too much of the map to suggest that the patterns were the result of exact survey.

The north-west of the city was dominated by the mass of the Royal Barracks complex on Barrack Street, together with the artillery ground on Oxmantown Green, between the barracks and the Blue Coat School, founded in 1669 as The Hospital and Free School of King Charles II. The north-east sheet showed the fashionable city with its regular street pattern. The relative wealth of the inhabitants was signalled by the presence of stable lanes where the carriages and servants of the owners were housed; many of which became mews in more recent times. Bartholomew Mosse's brand new (1757) lying-in hospital was a prominent feature on the map. The adjacent pleasure gardens, modelled on London's Vauxhall and Ranelagh gardens, had been opened in 1748 as a fund-raising mechanism for the hospital. Not as noticeable, but just as important in the development of the high status of the north-east city, was Henrietta Street, just off Bolton Street. A comparison of the plot sizes occupied by the houses shows just how different this little street was to the rest of the city. This was the first development of Luke Gardiner, the first of the Gardiners who went on to develop so much of north-east Dublin, and he gambled that there was demand among the rich for large, impressive and modern houses. Building on the street commenced in the 1720s and continued into the 1750s, and it quickly became one of the most prestigious streets in the city. It is also a metaphor for the decline and revival of the city in the nineteenth and twentieth centuries. The street slipped from being a prestigious address to being one of the worst slums in the city before beginning its slow (and yet incomplete) revival. More noticeable was Sackville Mall. Brooking's map shows some development in the area but Luke Gardiner began creating a residential area in the Grand Manner in the 1740s. The street was widened to its present proportions and the building of fine townhouses was encouraged through the nature of leases offered. The status of the development was enhanced by the provision of a central mall, lit by candles in the evening, which became an important location for evening promenade. Rocque's map shows the mall reaching as far as Henry Street, with development almost complete. The work of the Wide Streets Commission, and Rocque's son Charles, would result in the street being extended to the river as part of the city's main axis and would turn it into a major thoroughfare.

The south-east sheet showed that Merrion Street and the streets around Molesworth Street were built up, with Kildare House making a major statement. There were still fields to the east but the Fitzwilliam estate would soon transform that. To the south, development stopped close to St Stephen's Green where the four edges (walks) were largely complete, though with some vacant plots. The sycamore trees were still evident around the perimeter but the inner area had been remodelled to provide a focus of the equestrian statue of George II. Rocque was also up to the minute in including this since the statue would not be erected until 1758, though the plinth was in place at time of publication. The contrast between these new, regular developments and the older, more organic city became more obvious to the west. Here the streets varied in width, the streetscape was irregular and the map showed a warren of lanes and alleys. The main axis of the medieval city

1756

John Rocque, the environs of Sackville Mall, extract from *An Exact Survey of the City and Suburbs of Dublin* (1756)
[AUTH] If there had been an overall planning authority, it might have been possible for the Gardiners and Mosse to co-ordinate their developments so that Sackville Mall and the Lying-in Hospital were better aligned.

can be seen along Skinners Row/Castle Street and Fishamble Street/Werburgh Street, but Christ Church Cathedral was hemmed in by buildings of all sorts, and passage around the area involved the negotiation of irregular routes. The second cathedral, St Patrick's, stood in a little more open space but it too had a complex and crowded environment.

In many cases, it was the smaller details that added character to the image of the city. Some of the more exotic names included Cutt Throat Lane off Mount Brown, or Worlds End Lane near the [North] Strang. There was also Dog and Duck Yard, Indian Alley and Tripilo. Furthermore, Dirty Lane and the adjacent Dunghill Lane left the viewer in no doubt as to what the mounds shown on the map comprised. Although not named, the presence of so many similarly engraved

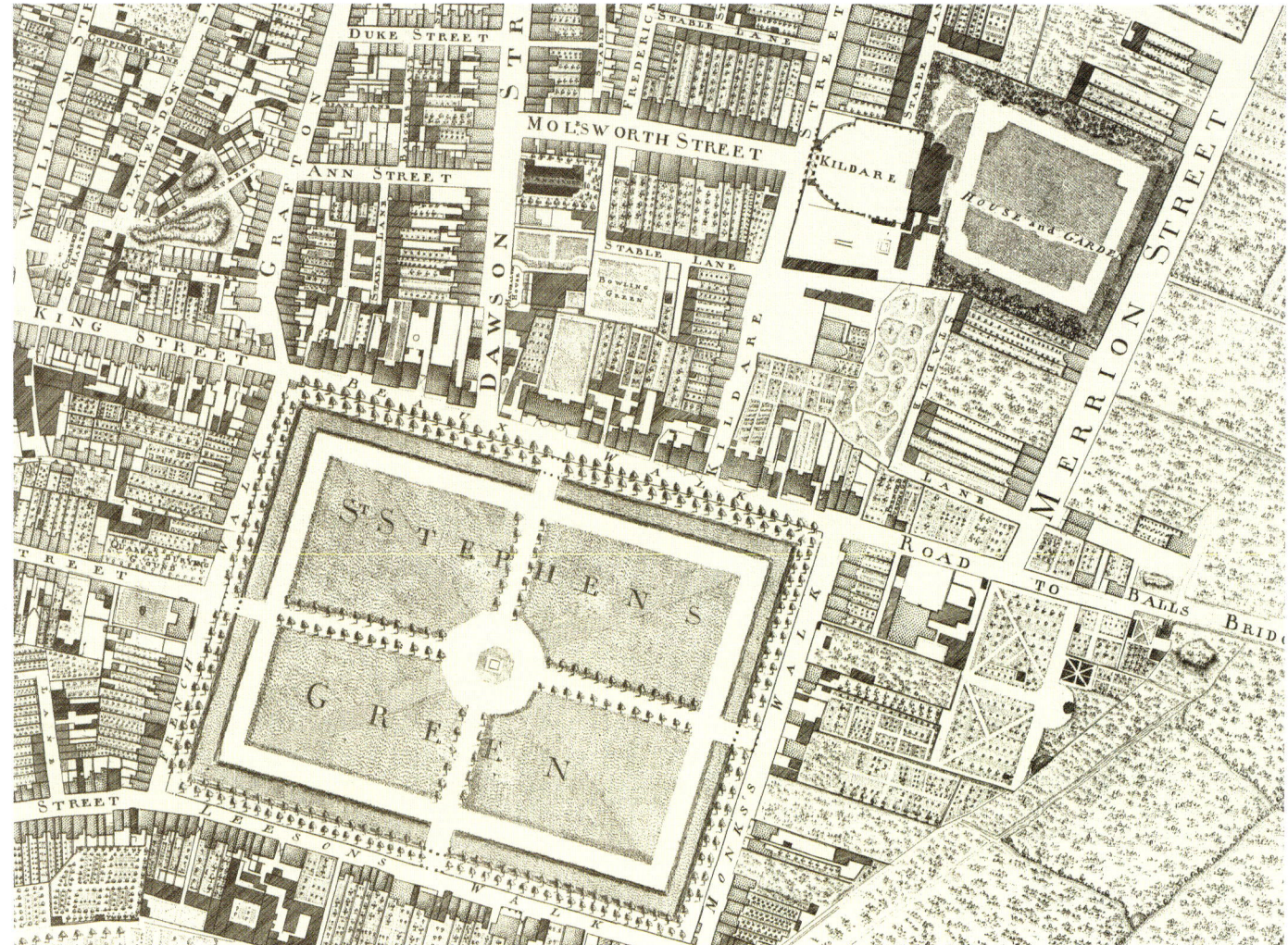

John Rocque, St Stephen's Green and Merrion Square, extract from *An Exact Survey of the City and Suburbs of Dublin* (1756) [AUTH] The development of St Stephen's Green had progressed well but the Pembrokes had yet to begin to develop their property, though they would soon eclipse the Gardiners.

mounds across the city would suggest that they were hard to avoid. For these reasons alone this set of maps is important, but its value is greatly increased because it captured Dublin on the edge of momentous change. In 1757 the Wide Streets Commission began its work, and over the next century it transformed the streetscape, sweeping away many of the streets and features captured in such detail on the maps.

This edition was revised over the next few years with a version published in 1773 described as being with 'additions and improvements by Mr Bernard Scalé to 1773'. Scalé, his brother-in-law, was a surveyor and cartographer who came to Dublin with Rocque. He stayed on in Dublin after Rocque left and established a successful surveying business, but seems to have returned to England following the publication of the 1773 edition.

John Rocque, the contrast between old and new streetscapes, extract from *An Exact Survey of the City and Suburbs of Dublin* (1756) [AUTH]

# 1757

## The environs of Dublin: John Rocque

Rocque's city map provided an unprecedented level of detail, but he was keen to capitalise on his surveying and he produced a range of maps at different scales to meet every possible need. In 1757, he published a four-sheet plan, confusingly entitled *A Survey of the City Harbour Bay and Environs of Dublin*. This covered an area south of a line drawn from the north of Howth Head to Clonskeagh. Each sheet was large, approximately 70 × 50cm, and allowed considerable detail to be shown.

The main roads were shown and named, together with field boundaries. He used a general shading to indicate cultivation and occasionally he indicated more formal planting, which might have been orchards. The gardens associated with large country houses were delimited and sometimes the layout was shown. This was the case with Marino, the suburban home of Lord Charlemont, whose house was new in the 1750s. Formal gardens were shown, laid out in geometric Renaissance style. Similar, but not quite as extensive, gardens were shown around Artane Castle, a little to the north. The landscape was relatively flat, the main feature being the gradual increase in height to the north. It was only at Howth that Rocque was able to use hachuring to effect and it gives the impression of a very rugged and dramatic landscape. The main settlements were named but otherwise placenames were rather sparse, though denser than Duncan's later map. Rocque eschewed the opportunity to place a large cartouche in the north-west as he did for the city sheets, though the one present is impressive enough. Instead he chose to include three vignettes – Kildare (Leinster) House, the Royal Barracks and the Royal Hospital. The barracks was particularly impressive and is viewed from across the river in an ordered and cultivated landscape. Rocque filled the relatively empty space of the bay with an elaborate compass rose in the north-east sheet and a copy of Speed's map of the city, displayed within a scroll, on the south-east sheet.

The south-west sheet encompassed the city that had been

---

John Rocque, *A Plan of the City of Dublin and the Environs . . .* (1757) [AUTH]

John Rocque, the Barracks viewed from the modern Victoria Quay, extract from *A Survey of the City Harbour Bay and Environs of Dublin . . .* (1757) [CC]

described in the four sheets above but extended to take in Glasnevin in the north, in the west to take in the Phoenix Park and in the east to give a view of the port. This was also made available as a single separate sheet and Bonar Law (2005) suggests that this was the first map published by Rocque in 1756, perhaps to ensure that people (and customers) were aware of the project being undertaken. The title ran across the top of the page, outside the neatline of the map, presumably to facilitate joining to the other sheets and it noted in English and French that this was a map of Dublin on the same scale as London and Paris.

In the bay, the space between the parallel walls running eastwards from Ringsend was being filled in to create the roadway that would become Pidgeon House Road. The channel was narrow, though busy with shipping of all types, and the soundings confirmed its relative shallowness. The sandbanks were named and the channel cut by the Ballybough (Tolka) River opened out into the wider Clontarf Pool. The coastal area along modern-day Clontarf is named 'The North Strand' and is, via a dogleg at Ballybough, a continuation of 'The Strand' road leading from the city centre. The land to the east of this road, bounded on the seaward side by the East Quay was shown laid out for future development with three main roads dividing it into four similarly sized sections. Even more dramatically divided and laid out were the North Lotts. In 1674, Jonathan Amory leased this ground from Dublin Corporation – Amory's Ground is noted on the sheet. He decided that the most profitable approach was to subdivide it into the narrow lots running inland from the quays, which he could then lease to other developers.

The extension of the sheet into the Phoenix Park allowed Rocque to show the new house of Nathanial Clements set in its own grounds. This would become the Viceregal Lodge, and ultimately Áras an Uachtaráin. The Magazine Fort, on the higher ground above the river, was also shown, as was the location of the 'saluting battery'. Intriguingly, Rocque shows a star-shaped feature which he named 'The Fortification'. An evolution of the medieval wall, star-shaped fortification systems were to be found around significant towns and cities on the continent of Europe, and were designed to provide greater security for the main wall of the settlement by forcing enemy artillery to stay out of range of the guns placed at the apex of each arm. However, each of these arms needed to be defended too, leading to complex interwoven systems of fortification which ultimately became futile in the face of rapidly improving artillery. The design was also popular for forts and

## 1757

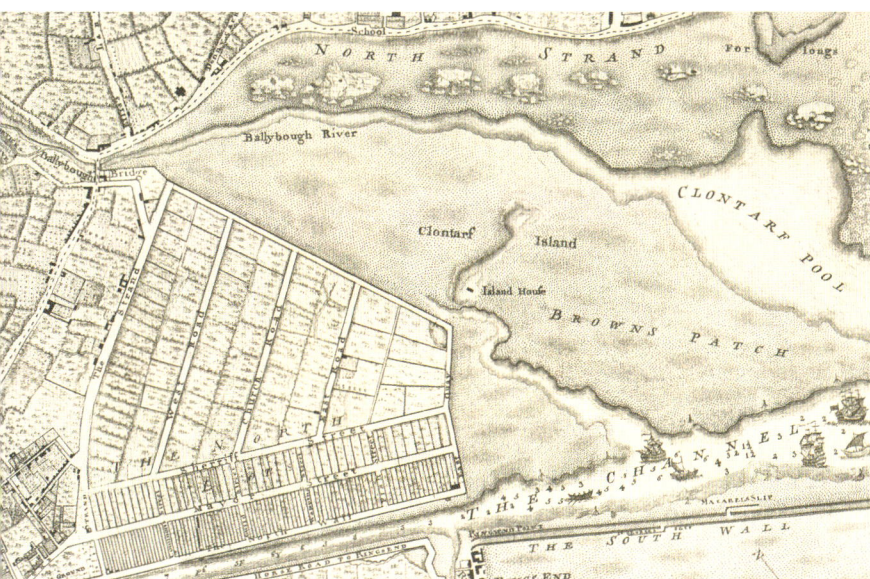

John Rocque, the North Lotts and a view of bay showing the developing South Wall, extract from *A Plan of the City of Dublin and the Environs* (1757) [AUTH]

John Rocque, the Phoenix Park prior to its modern redevelopment showing a fortification of which there is little current trace, extract from *A Plan of the City of Dublin and the Environs* (1757) [AUTH]

Bernard de Gomme suggested such a 'citadel' for Dublin in his plan of 1673. It is unclear, however, how extensive or complex was the 'fortification' that was built in the Phoenix Park. It might have involved no more than ditches and dykes, and certainly does not appear to have caused much difficulty when it was levelled in 1834 as part of the remodelling of the park under Decimus Burton.

Ringsend and Irishtown were isolated from the rest of the city, hemmed-in by the coastal sandbanks and the wide channel of the River Dodder. Sea-bathing was becoming fashionable during Rocque's time and two bathing places were noted between Ringsend and Irishtown. This was a segregated activity – men generally swam naked – with the men's place at Ringsend and the women accommodated at Irishtown. Also shown along the coast was the Conniving House. It is a strange name (even for a pub), but it seems that this small, thatched building was an important location for music and song in the years after 1725. Rocque was parsimonious in his naming of places and its reputation must have deemed it worthy of inclusion.

As with the other maps, these sheets were also improved by Scalé in 1773. The main additions were the completed streets of the Wide Streets Commission, the outline of Merrion Square and the arrival of the Grand Canal.

Increasing his range further, Rocque also published a single-sheet reduction of the city and suburbs in 1757, as well as a 'pocket plan' of the same area. This had an index of features on the left edge. In 1760, he published a four-sheet set for the county, entitled *An actual survey of the county of Dublin on the same scale as those of Middlesex, Oxford, Barks and Buckinghamshire by John Rocque, 1760*. This too was amended and updated over the years with an edition in 1773 and a further one dated 1802 which read 'Printed for Robt Laurie and Jas Whittle, N° 53 in Fleet Street, Novr 1. 1802'. For the sake of consistency, this too was published as a single-sheet reduction in 1762, with new editions in 1775 and 1799. Very few copies of the four-sheet maps survived, probably because they were rather unwieldly in use. The single-sheet reductions are easier to find, but even these are quite scarce.

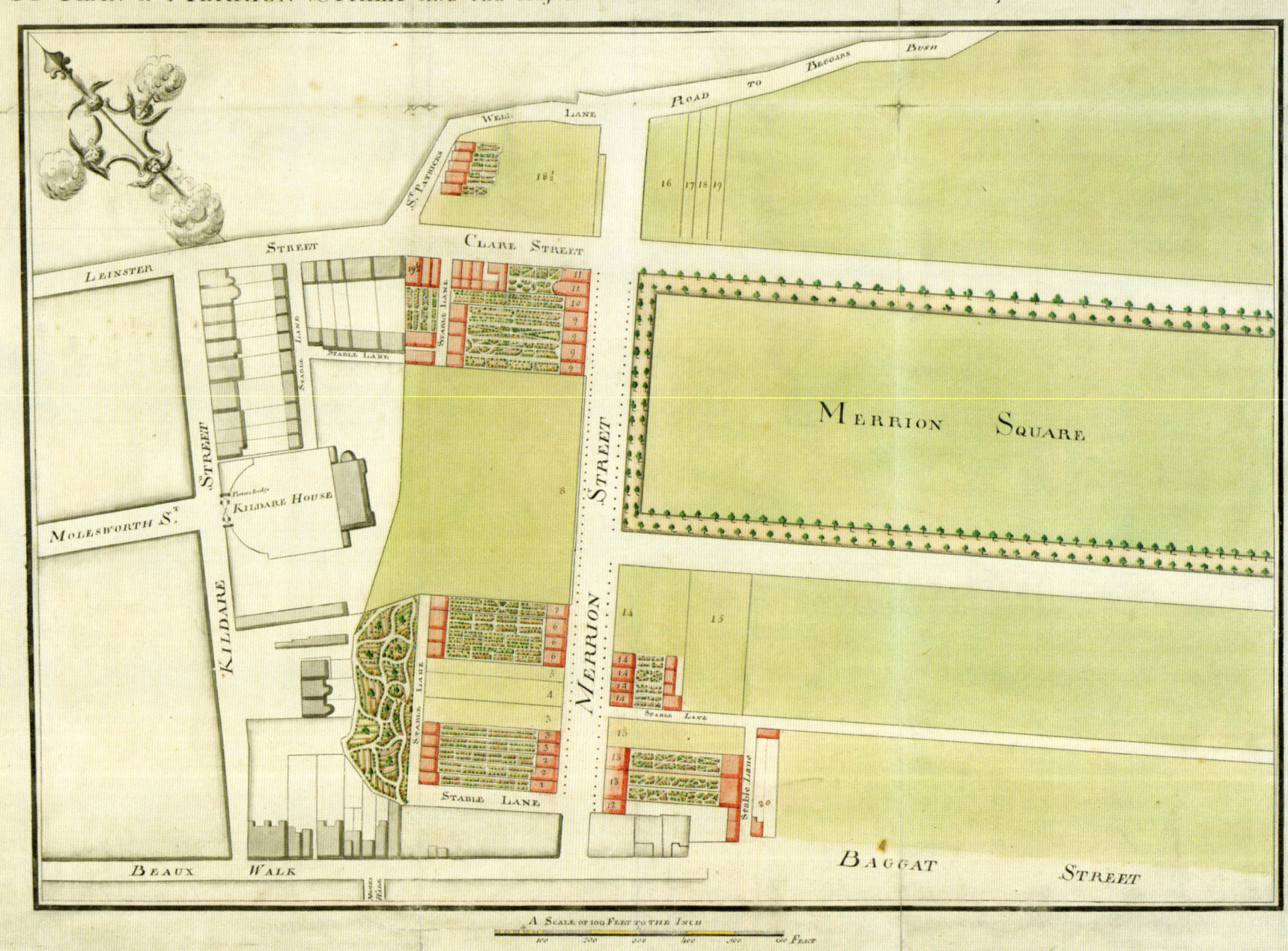

# 1762

## Merrion Square

While the Wide Streets Commission was of enormous importance in developing Dublin's city centre in the 100 years from 1757, it could not have operated without the involvement of the many private developers in the city. These developers are commemorated in the names of the streets that they built, but two families stand out: the Gardiners and the Fitzwilliams. The Gardiners were responsible for the creation of much of the north-east of the city, with Sackville Mall and Mountjoy Square among their signature projects. Unfortunately, due the profligate nature of Charles Gardiner, who died in 1829, the estate began a slow decline during the nineteenth century into tenementation and poverty. This was not however the case with the Fitzwilliams, whose estate was located in the south-east quadrant and which has retained its high status to the present day.

While the family had been amassing a huge property portfolio in south Dublin and beyond since the 1300s, it was the sixth Viscount Fitzwilliam of Merrion who began the residential development of their land in Dublin in 1776. They came to it rather late, the Gardiners had been building since the 1720s and Sackville Mall was by then the most desirable location in the city and an important place for evening promenade along its candle-lit central mall. The Fitzwilliams found themselves on the edge of the developed city with the completion of Leinster House in 1748, which looked out onto their property, but they seemed in no hurry to capitalise upon this arrival.

While Renaissance planning favoured the regularity of a grid pattern for residential areas, it was recognised that this could produce a boring landscape. One means of ameliorating that was to keep a number of grid squares open which could be used as traffic centres, sites for important monuments or

---

Jonathan Barker, *A Plan of Merrion Street and the Adjacent Neighbourhood* (1762) [NA]
The first steps by the Fitzwilliam (later Pembroke) family in developing their extensive property.

residential parks or squares. The Fitzwilliams decided that they would build a residential square in the continental style. No one had done this since Dublin Corporation had laid out St Stephen's Green almost a century previously and it would be some time before the Gardiners would attempt one at Mountjoy Square.

The design task was given to John Ensor, one of the best-known architects of the time, but the plans discussed here were produced by Jonathan Barker. He was a surveyor and cartographer who undertook a variety of projects for the Fitzwilliams in Dublin and Wicklow until his death in 1767. The first plan, at a scale of 1 inch to 100 feet (1:1,200) dates from 1762 and outlined the basic structure of the project. Rocque recorded development along Merrion Street with houses and fine gardens at both the Clare Street and Baggat Street ends. Barker's plan shows some additions but still with some undeveloped plots on the Baggat Street side. The houses had stable lanes to the rear and the buildings seemed to be of considerable size. These not only provided accommodation for carriages and horses, but were often where the servants lived. The presence of stable lanes was a clear indication, if one was needed, of the high status of these residential developments. The stable lane between Kildare Street and Merrion Street opened onto a formal garden.

More detail of what was intended was provided by two further plans, which are both from 1764. These show that the ambition of the Fitzwilliams was limited – there was no intention to plan in a Grand Manner for their entire property portfolio. Such was the size of their land holding that they could have taken the opportunity to lay out an entire suburban development, even if the immediate intention was limited to the square. The Gardiners would have loved such an opportunity, but were hampered by the fragmented nature of their holding.

The earlier of the two plans showed that a new road would be built in the vicinity. This would be 1,460 feet long and would link the road to Donnybrook with the road to Blackrock. The second plan showed more detail about the intended streets. Barker's plan was oriented towards the west with Kildare/Leinster House at the top edge. The 'square' was a rectangle 520 feet wide and 1,273 feet long. Rectangular shapes were often favoured for such developments since true squares can look squat when viewed from one of the edges. However, this was somewhat longer than might have been expected. No internal arrangement was suggested for the square but it was to be edged by trees. Barker had included two lines of trees on his 1762 plan but this was reduced to one here. Merrion Street was already in existence but its more noble width was not repeated for the other streets. The 'new street' which would terminate the square and run parallel to Merrion Street was given a more modest width of 50 feet, that along the southern edge was given 60 feet while the northern edge was given a width of 78 feet. This was probably to reflect its role as a major route into the city.

The 1764 plan shows housing on both sides of the new road from its junction with 'the road to Donnybrook', now Leeson Street. Barker added a few features of interest with a drawing of Dr Barry's house in the bottom left corner and a drawing of a quarry with a windmill at the junction of the Blackrock road and the new street. This was probably a disused quarry, where gallows had been located in previous centuries. The earlier suggestion of another new road seems to have been dropped and its route was shown only by very faint lines. One of the elements of the plan which is particularly interesting is that Barker gives an indication of the streetscape. There seems to have been no intention to create a uniform layout and the houses were shown in a variety of shapes and sizes with different roof lines and some Dutch Billy gables. The Wide Streets Commission would soon come to prefer uniform façades for the redeveloped Dame Street, and the Gardiners would favour a single palatial façade, behind which individual houses would be built, for each of the sides of Mountjoy Square.

When building commenced on the north side of the square, John Ensor was content to have houses of a variety of scale and character, though all within the broad framework of Georgian design. He allowed No. 12 and No. 25 to break the parapet line and the fenestration was not uniform along the

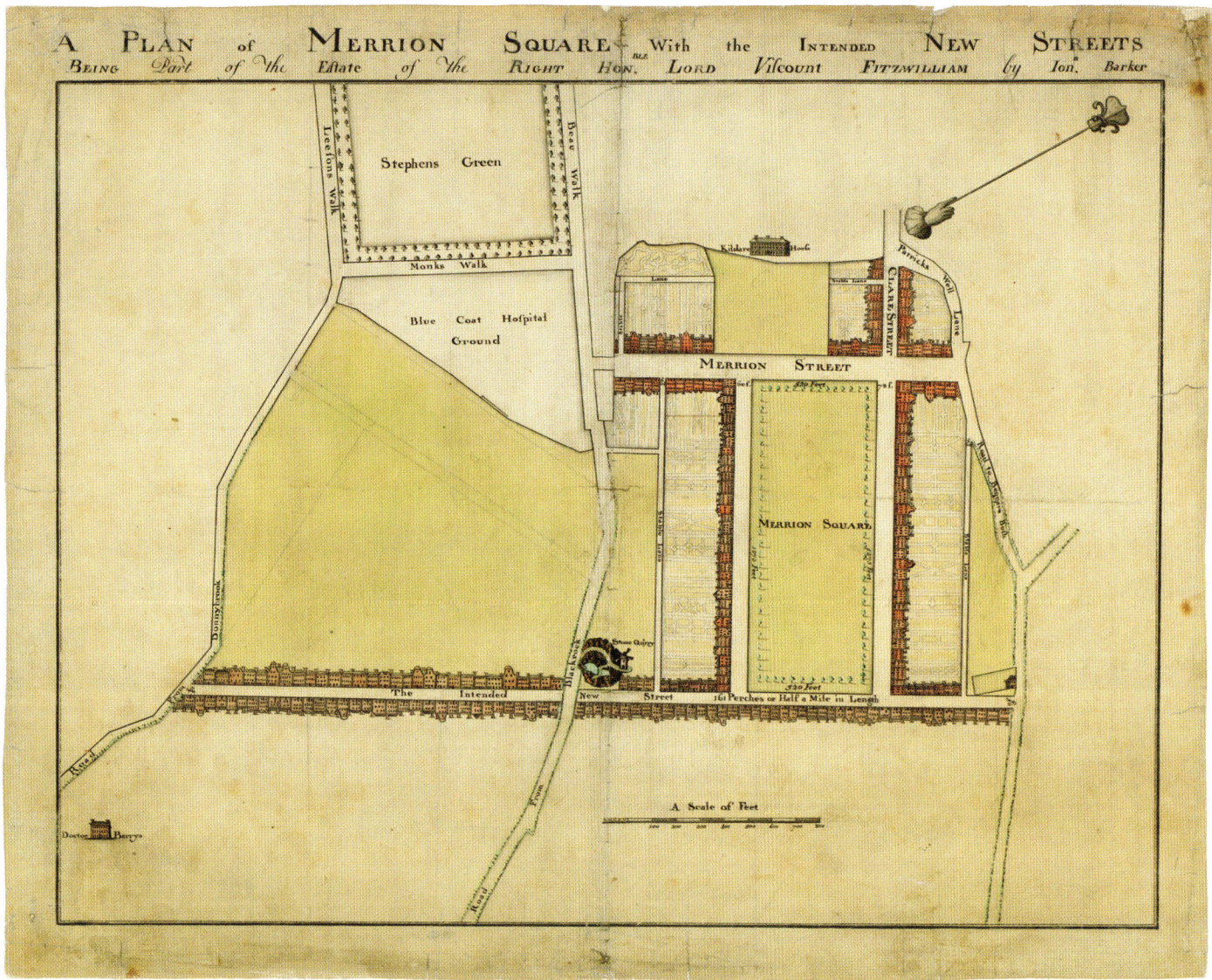

Jonathan Barker, *A Plan of Merrion Square with the Intended New Streets* (1764) [NA]

street. This individuality was continued with the doorcases. In line with the style of the times, they were tall and narrow but it seems that no two were alike. There was greater uniformity on the eastern side of the square, which was leased to Samuel Sproule in 1780 – he was a notable architect, but also a Wide Streets Commissioner. His houses had similar plot sizes and regular three-bay fenestration. Only the existence of earlier and more elaborate houses, such as the five-bay at Nos. 44–45, disrupted the uniformity. Uniformity of layout was also characteristic of the southern side where the houses are mostly three-bay, though the parapet line was not maintained along the entire terrace. These tended to be more modest houses, developed in a piecemeal way by a variety of developers, each building no more than a handful of houses at a time.

# 1791

## *A picturesque and descriptive view of the city of Dublin*

James Malton's images of Dublin have become commonplace that they can sometimes be ignored in favour of more 'interesting' images, in the same way that Vivaldi's *Four Seasons* is often met with groans. Yet, these Malton engravings are hugely important in capturing a city that was at the height of its prominence with a city centre that was new and exciting. By the time that Malton was producing his engravings, the work of Wide Streets Commission had resulted in a new central axis on the grand scale, with a streetscape following a unified design that was in line with approved European thinking. Visitors to the city in the late 1790s were greatly impressed by what they saw and the commentaries suggested that the city could hold its own with any in Europe.

Dublin was now deemed worthy of historical study. At least this was the sentiment expressed in the preface to Walter Harris' history of the city that appeared in 1766. In the writer's view it was timely, given the excellent state of the metropolis, to reflect on its origins. The book

> must prove the source of uncommon satisfaction to the patriot native: when he is informed of and reflects on the many alterations in the face of things and places, the vast improvements and enlargements, most of them the work of less than a single century, we may with exact propriety apply the words of the poet to the amazed and delighted citizen.

Poole and Cash produced their series of engravings of public buildings in 1780, and Malton's work can be seen as part of this growing pride in the new city that was emerging. His work was on a far grander scale than previously and soon his graphics would be taken from their bindings to adorn the

James Malton, St Stephen's Green from
*A picturesque and descriptive view of the city of Dublin* (1791) [PC]

James Malton, Four Courts and River Liffey, from *A picturesque and descriptive view of the city of Dublin* (1799) [AUTH] The view shows that there is not yet a completely clear passage on the right bank. The steep camber on the bridge would soon necessitate replacement.

walls of fine houses and public buildings. The engravings are of different dates, from 1791 to 1799, because although Malton states that they were all 'taken' in 1791 it took time to bring them all to completion, and he updated them as he went along to ensure that they were correct as of the date of publication. It is suggested that they were released in parts but, with the set complete in 1799, they were then sold as a bound volume. Each volume, measuring 42.4 × 57cm, contained a short history of Dublin and three important maps. There was an improved version of Speed (1792), a map of the bay (1795) and a detailed map of the city (1797). This was produced by well-known cartographer William Faden, and Malton makes the point that his version of the Speed map was to the same scale to permit comparison.

The title page was elaborate: 'A picturesque and descriptive view of the city of Dublin displayed in a series of the most interesting scenes taken in the year 1791 by James Malton with a brief authentic history from the earliest accounts to the present time.' There was a dedication

> to the Right Honorable the Lord Mayor, Aldermen, Sheriffs, Common-Council, Freemen, and Citizens of the City of Dublin. This work intended to contain a concise yet complete description of the capital of Ireland; Is humbly Dedicated and given to their Protector by their obedient Humble Servant James Malton. London June 1st, 1794.

Each engraving was accompanied by text which explained the viewpoint and also something about the nature and history of the building. Each also had a dedication to a person of importance and influence; the image of the upper Castle Yard was dedicated to John, Earl of Westmoreland, Lord Lieutenant General and General Governor of Ireland. Malton missed no opportunity!

It seems that the original issues were uncoloured, and that the coloured sets only began to appear in the bound volumes and not until after Malton's death in 1803. Intact bound volumes are rare and usually found only in libraries; most others have long since had the plates removed for individual display but the letterpress can sometimes still be found. Original Malton watercolour drawings exist and can be found in the National Gallery of Ireland, Trinity College Dublin, the British Museum, the Victoria and Albert Museum in London, Mount Stewart in County Down, and the Huntington Library, California.

James Malton, a view over Essex Bridge towards Parliament Street, from *A picturesque and descriptive view of the city of Dublin* (1797) [AUTH]. The escort about to turn onto the bridge suggests a visit to Dublin Castle from the Viceroy. Lotteries were popular and profitable for a time.

The image of St Stephen's Green is dated 1796 and gives an important view of the configuration of the green at this time – its development can be traced from Speed, through Brooking and onto Rocque. It was an open meadow with little adornment except for the equestrian statue of George II, now on its plinth in the middle of the green where horses appear to graze. The green was surrounded by a raised walkway and Malton has captured Dubliners engaged in promenade – essentially showing off – something that was practised in Sackville Mall in the evenings and later in the Rotunda (the indoor location a reflection of the weather). Sycamore trees, which were originally a condition of getting a lease on property around the green, are prominent. In the background, Malton has provided a view of the housing around the green. St Stephen's Green was developed from the late seventeenth century and the housing styles reflected the changes since then. There were a couple of examples of Dutch Billy gables but most reflected the modern Georgian style, though the streetscape was not uniform.

A view of the Law Courts, looking up the Liffey, shows the newly completed Four Courts building. It also shows that the quays were not completed. While there seemed to be unimpeded passage on the left bank of the Liffey, this was not the case on the other. Creating the quays had been a project of Dublin Corporation since the restoration of the monarchy and the progress achieved, largely by public–private partnerships, can be seen in both Brooking and Rocque. Also shown was Dublin Bridge, said to be the site of the original bridge over the Liffey. It was in need of replacement and, in 1818, the new three-arch and flatter Whitworth Bridge took its place.

Another picture of interest is the view from Capel Street looking over Essex Bridge. The first project of the Wide Streets Commission was Parliament Street, and the vista shows the regular architecture of the new street, closed at the end by the Exchange Building. The main motivation for the project was the creation of a convenient passageway for the Lord Lieutenant across Essex Bridge to the Castle – the image shows what might be such a progress about to turn onto the bridge. Other points of interest are the street lights which are shown on both Capel Street and Essex Bridge. The housing on Capel Street was in the new style with shops on the ground floor. In this case, they show the great interest in lotteries, a state lottery having been initiated in 1779, managed by commissioners, which was greeted with enthusiasm by the population.

# 1797

## William Faden's map

Malton's engravings tend to get most notice and the maps are often ignored, but there were three maps included in the bound volume of prints. One was a redrawing of Speed's map of 1610. It is a very attractive engraving and because it was larger than Speed's original it was possible to name the streets and main features of the town on the map, rather than have them in a marginal index. It shows a town with regular streets and very regular buildings, all much the same size, and gives an impression of neatness which must have been far from the reality. As a nice touch it shows a three-masted sailing ship heading out to sea. The map of the bay was described as a 'Correct Survey of the Bay of Dublin 1795'. In fact, there was little or no navigational information, though the two large sandbanks of the North and South Bull were shown, but Rocque's maps gave names to the various elements of the sandbanks and showed the channel of the River Tolka. The Dublin Bar, that most inconvenient of physical features, was indicated and it looked to be a significant barrier. The new South Wall with its lighthouse at its seaward end was impressive but only the Pidgeon House was mentioned as one of the features of the wall. The information on the city was selective, with only the eastern part drawn. Some significant suburban locations were included, though these were mostly coastal, including Black Rock, New Town, Dun Leary, Butter's Town, Ballybough Bridge, the Sheds at Clontarf, Ratheny, Baldoyle and Killbarrack. Howth and the Hills of Dalkey were shown in hachuring and give the impression of being quite dramatic features. Otherwise, the information shown on this map was sparse, even the single three-masted ship in the bay suggested a rather quiet port city.

The third image was a large folded map at a scale of about 1:7,250 and 75.8 × 50.5cm. Malton assured his readers that it was up-to-date and had all the most recent features. He did not draw this map, rather it was produced by William Faden

---

William Faden, *A Plan of the City of Dublin* (1797) [AUTH]

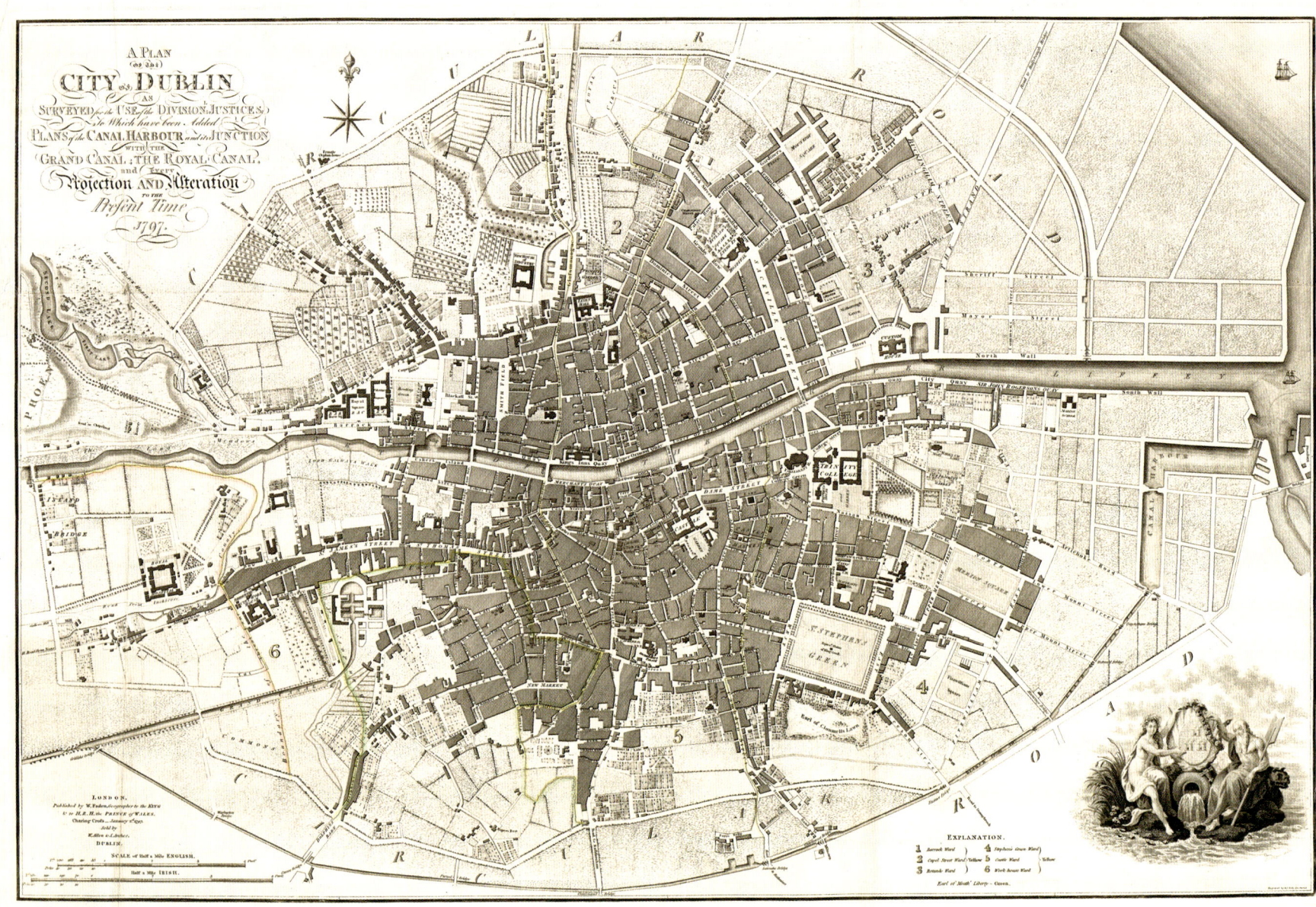

and tends to be known as 'Faden's map'. Faden was one of the most successful cartographer of his time, with a thriving business in London and the title of 'Geographer in Ordinary to His Majesty'. The map was included in all the early editions of Malton and he seems to have had an exclusive deal with Faden because the map is quite rare and there is scant evidence that it was sold separately, though the map itself says it was available from W. Allen and J. Archer in Dublin. Not only was the map a manageable size, but it also contained such new additions to the city as the Royal and Grand canals and the Canal Harbour. These were of sufficient note to merit a mention in the cartouche in the top left corner. Whatever exclusivity it had though was limited because Wilson included a very fine map in his directory for 1798 which captured the same landscape, though not on as large a sheet.

Faden showed a city enclosed between two circular roads. These were so named in large letters, even though they were far from complete at the time. Not only would they improve transport links by joining the main radial routes, they would also give shape and definition to the city, though the canals would soon displace them from that role. Dublin was small and compact, with few examples of squares or other open spaces within the built-up area. The city was set to continue to grow and some of the new development areas had been laid out. The date of the map is useful because it captured the impact of the transformation caused by the Wide Streets Commission during the time of its greatest activity. Sackville Street had been completed and was joined via Carlisle Bridge to the new streets of Westmoreland Street and D'Olier Street, though there was still some work to be done at the College Green end of D'Olier Street to improve its access.

The shape of the Gardiner estate was now clear and their signature square, Mountjoy Square, had been laid out, but there was not yet much building along Gardiner Street. Their ambition to create a unifying focus for their somewhat dispersed development involved the creation of a circus at the top of Blessington Street, Eccles Street and a yet unnamed street. The Gardiners were set to proceed with this project by 1792. Leases had been sold and the size of the plots and their configuration had been fixed with the expectation that all would be complete within 12 years. However, the project never got much further than the outline which Faden included. This Royal Circus was never built, though it remained on maps for the next 20 or more years. The Fitzwilliams had decided to continue the axes of Merrion Square as far as the Grand Canal and Upper and Lower Mount Street had been laid out. Fitzwilliam Street had also been extended almost, but not quite, to Leeson Street. They had decided to create another square, not quite as big as Merrion Square, along that road. Building had begun on Fitzwilliam Square and it would provide another link from Baggot Street to Leeson Street.

Work on the Royal Canal had begun only in 1790, so the project was still in an early stage, and therefore very little of the canal was shown. It was originally intended to terminate at Broadstone, but the map describes the stretch to Broadstone Harbour, adjacent to one of the Houses of Industry, as a 'branch from the Royal canal'. The intention was now to continue the canal via a complex set of locks to the Liffey and to capitalise on the opportunities provided by the port. As the potential for the canals to solve the problem of the city's water supply was recognised, this stretch of the canal provided water for the Blessington Basin, a reservoir that supplied water to the north side of the city. Work would commence on the reservoir within the next five years.

The Grand Canal was more developed by the time of the map. What was described as the main branch terminated at Grand Canal Harbour, which was designed to service the adjacent Guinness brewery. This also provided the water for the 'City Bason' (sic), a vital element in the city's water supply since 1721 and evident on Rocque's map. Both basins remained crucial to the water supply of Dublin until the new Vartry scheme, fed from reservoirs in Wicklow, replaced them in the 1860s. A second branch, opened in April 1796, looped around the built-up area to terminate at a large harbour that joined the Dodder at its confluence with the Liffey, creating a future geographical, political and social border.

The size of the map allowed all of the major and many of the minor streets to be named. Though the streetscape is shown in block form, the shape of the urban landscape is captured together with a lot of minor detail. The linen bleaching fields of the tenters, for example, close to Blackpitts, are shown, as are the brickfields off James' Street. The map was issued uncoloured, but on this copy the ward boundaries were outlined in colour. While they followed roads and geographical features for the most part, around James' Street they ran through buildings, causing all kinds of problems when it came to appointments to the city council.

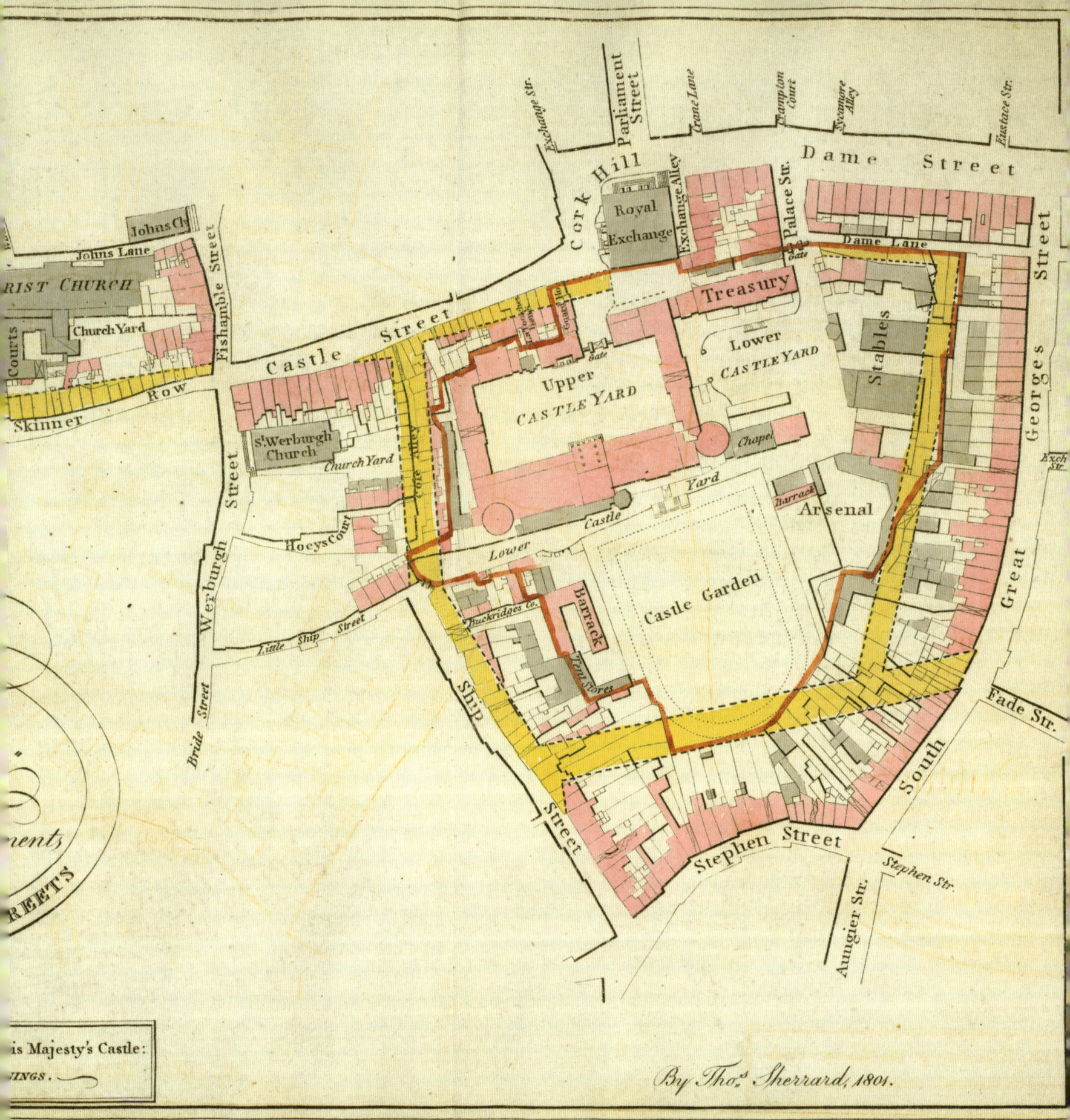

# 1802

## Wide Streets Commission: reflections on progress

Rocque's map showed the organic nature of the street pattern of the medieval city, with meandering streets of varying widths, but this was only part of the story. Dublin lacked a large market space and business consequently spilled onto the streets where the associated detritus both narrowed the passages and made them noxious. This was to the annoyance of both Corporation and shopkeepers, and various attempts were made to control trading but to little avail.

The movement of the city eastwards was well established by the early decades of the eighteenth century, and the problems of congestion in the medieval city might have faded from public concern but for one particular issue. The Viceroy had his state apartments within Dublin Castle but in the 1750s a new and imposing out-of-season house had been obtained in the Phoenix Park – the Viceregal Lodge. Getting to and from this location exposed the Viceroy to a less-than-pleasant journey through narrow and probably foul-smelling streets.

The solution was innovative: establish a commission with a single focus and give them the requisite power to do the job. It needed legislation, and *An Act for Making a Wide and Convenient Way, Street and Passage from Essex-Bridge to the Castle of Dublin, and for other Purposes therein mentioned* was enacted in 1757. It noted that the streets were 'narrow, close, and crooked' and that making the new streets would enhance the safety of passengers and would be of benefit to the public. They got to work quickly and by 1762 the new Parliament Street had been opened, demonstrating a ruthlessness that would be characteristic of their approach. They were influenced by the architectural approaches of their day and they were aware of what was going on elsewhere in Europe. The street was wide and the proportions of both street and buildings followed the approved Renaissance rules. They

---

Wide Streets Commission, 'Improvements to the Environs of Dublin Castle',
extract from *Minutes of the Commissioners . . .* (1802) [AUTH]

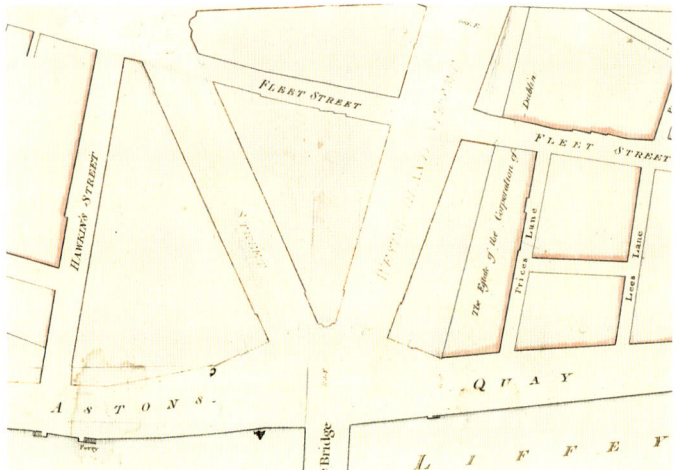

Wide Streets Commission, 'Design for the new streets leading to South Quays' (1793) [TCD] This is an extract from a larger plan showing the area between College Green and the river.

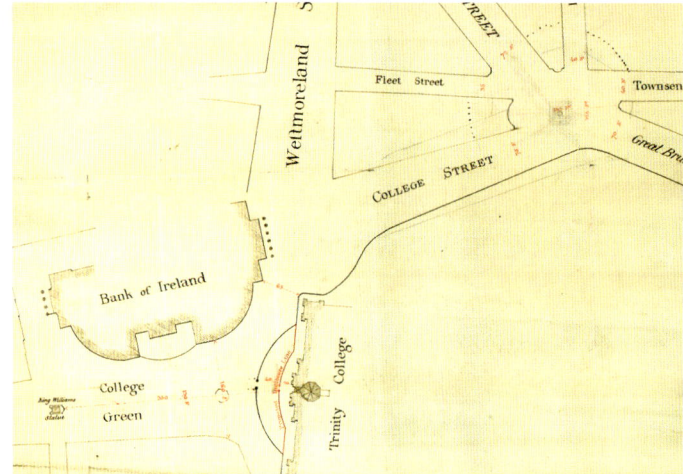

Wide Streets Commission, 'Design to alter the entrance to Trinity College' (1814) [AUTH] An extract from an unexecuted proposal to simplify to the entrance to Trinity College and thus improve traffic flows.

favoured a regular streetscape and when the opportunity presented itself this resulted in a uniform façade for a single block, giving the impression of a single building but with individual ones behind. Elements of this can still be seen on Dame Street but the best example may be seen in the block between Westmoreland and D'Olier Street.

Other projects followed and the remit of the commission was gradually extended to include much of the city, though they expressed no interest in the decaying south-west. This increased remit was no problem for the owners of the large estates in the city since they either were commissioners themselves or closely associated with them. Work on Dame Street and its environs continued during the 1760s and into the 1780s. The opportunity was taken to create a new axis that would link Gardiner's property on the northside of the Liffey with the newly developed Dame Street. In 1782, Parliament approved a plan that would open a route between Sackville Mall and the river, across a new bridge which would split into two avenues – one 'direct to the House of Lords, the other in a South-east direction to communicate with Townsend Street and the Eastern Part of the City'. The commissioners had various designs produced for each of their projects, many of which were drawn by Thomas Sherrard whose surveying business prospered from its association with the Wide Streets Commission. The extract reproduced here from 1793 shows these new streets as regular and wide, and focused on the new Carlisle Bridge. The drawing also shows them taking the opportunity to shave a little from Astons Quay to widen it and make the line more regular. In this plan, one of the streets was tentatively named Westmoreland Street – the name very lightly drawn – but the other street is unnamed, though there was space left to insert it.

The project created a fine axis with impressive vistas along Dame Street and from Carlisle Bridge. The space in front of Trinity College – College Green – was now the focus of a number of major routes but while this proved to be visually impressive, it quickly became snarled with traffic. That the Wide Streets Commission realised this problem early is clear from the second extract, taken from a plan signed Sherrard and dated November 1814. It proposed to shave the hemicycle entrance to Trinity College, thereby simplifying and widening the road linkages. This plan also shows another idea for polyvium at the Trinity College side of D'Olier Street. The idea was to make it into a circus on the European model with a monument as a focus. However, this idea is more crudely drawn and may be no more than a later doodle.

While the axis project was under consideration, attention was also given to improving the route from lower Abbey Street and the quays to the Custom House. Other projects during these decades focused on Rutland Square, the south end of Cavendish Row, a road from Summer Hill to Great Northern Road, Barrack Street to Islandbridge (Parkgate Street and Conyngham Road).

However, despite the significant impact of their work and the widespread praise which the city was now enjoying, the commissioners were never given the necessary long-term resources to plan for the city. They moved from project to project and they found money hard to raise. In 1782 they were granted a tax of one shilling per ton of coal imported, later extended to 1810, against which they could also raise loans. In 1790 they got Parliament to enact a card and club tax but reading their minutes gives the sense of a group always on the lookout for funding. Onc of the outcomes of the Act of Union was that government was now more distant from the commissioners and they found it harder to persuade that government to fund them. This led them in 1802 to prepare a case to the Earl of Hardwicke, then Lord Lieutenant General. This was published as *Extracts from the Minutes of the Commissioners . . . together with a general statement of their proceeding, engagements and funds . . .* It set out in detail the work that the commission had accomplished, and the costs involved, as well as proposals for new projects. The point was made that they were undertaking this work in response to public demand and that the 'commissioners have not, nor can they have, any Interest but public Good'. They had used all the funds available and more was needed to complete the work in hand and to realise these additional projects, so they asked for a local tax or some other form of funding. This made it a fascinating and important statement both for the quality of the information provided and also because of the insight it gave into their thinking.

The commissioners took the opportunity to put forward a small number of proposals. One dealt with the environs of Dublin Castle, and they may have felt that the Viceroy had a particular interest. The plan was to widen and simplify the perimeter of the Castle. Not only would this improve the circulation but it 'would insulate His Majesty's Castle . . . and effectively secure it from all danger, and the apprehensions of Fire from the Forge Manufactories, which are now carried on contiguous and adjoining thereto . . .' The project involved opening a new street from South Great George's Street to Ship Street, with widening along that street to Castle Street. The street level on Ship Street would be raised to prevent flooding. Castle Street itself would be widened and this would be part of a larger project to improve and widen along Skinner Row. This was 'at present very narrow and dangerous' and would 'render a safe and commodious Way to Christ Church'. Dublin lacked formal marketplaces and they proposed another project nearby to enhance the western part of the city. The plan showed a warren of narrow lanes on the south side of Thomas Street, between Francis Street and Vicar Street, which would be cleared. They wanted to build a new market hall which would provide a safe and secure market and storage for corn. They reckoned that the modest taxes proposed would save business worth £1 million for the city.

Legislation was passed in 1807 which allowed the commission to levy a 'Wide Streets' tax of up to one shilling in the pound on the rates. However, this was not an unfettered right and they had to demonstrate the need for the improvement and receive official sanction. They continued to complain that they did not have the money necessary to undertake all the improvements suggested to them and this position continued into the 1840s. By the time of the 1846 Dublin Wide Streets Bill they were looking for an annual support of four pence in the pound but found themselves being opposed by Dublin Corporation. While they did useful work in the years after 1802, it was not as obviously spectacular as that undertaken previously, and the commission was only a shadow of its former self when it was abolished by the Dublin Improvement Act of 1849, with the final meeting of the commission taking place on 2 January 1851.

# 1817

## Finding your way about town: Dublin directory maps

Society became more complex and more urban as the Industrial Revolution progressed, and people had to interact with services on a more formal basis than had previously been the case. Even though in Dublin there was little of the intense industrialisation experienced in other parts of the United Kingdom, the same kind of increasing complexity affected the life of the middle and upper classes. Information was needed, but sources were limited. Directories were developed to fill the gap and from the eighteenth century evolved to provide as much information as possible on government, the military, the gentry, the administration as well as useful data on matters such as communications and transportation. While there were provincial and professional directories, the geographical coverage was variable but it will come as no surprise to find that Dublin had the most comprehensive coverage. These were not yet street directories though, that would have to wait until the nineteenth century.

The *Gentleman and Citizen's Almanack* (also known as *Watson's Almanack*) first appeared in 1729. This comprised a series of lists and tables such as a list of the justices of the peace for each county and the magistrates of towns. The time of sunrise and sunset was given, as well the correction necessary to be applied to sundials, which were still an important element in the regulation of many houses. In 1752 it was joined by *Wilson's Dublin Directory* and it became usual to bind these two publications together with the *English Registry*, becoming formalised after 1801 as the *Treble Almanack*. Someone who purchased the 1796 edition would have obtained a neatly bound volume of about 90 × 160mm; small enough to be carried easily. It offered the 'most Complete Lists published of the present Civil, Military and Naval Establishments of Great Britain and Ireland'.

The *English Registry* must have been a little esoteric for many owners. It provided a list of members of the House of

---

Map of Dublin City from *Wilson's Dublin Directory*, included with the *Treble Almanack* (1817) [AUTH]

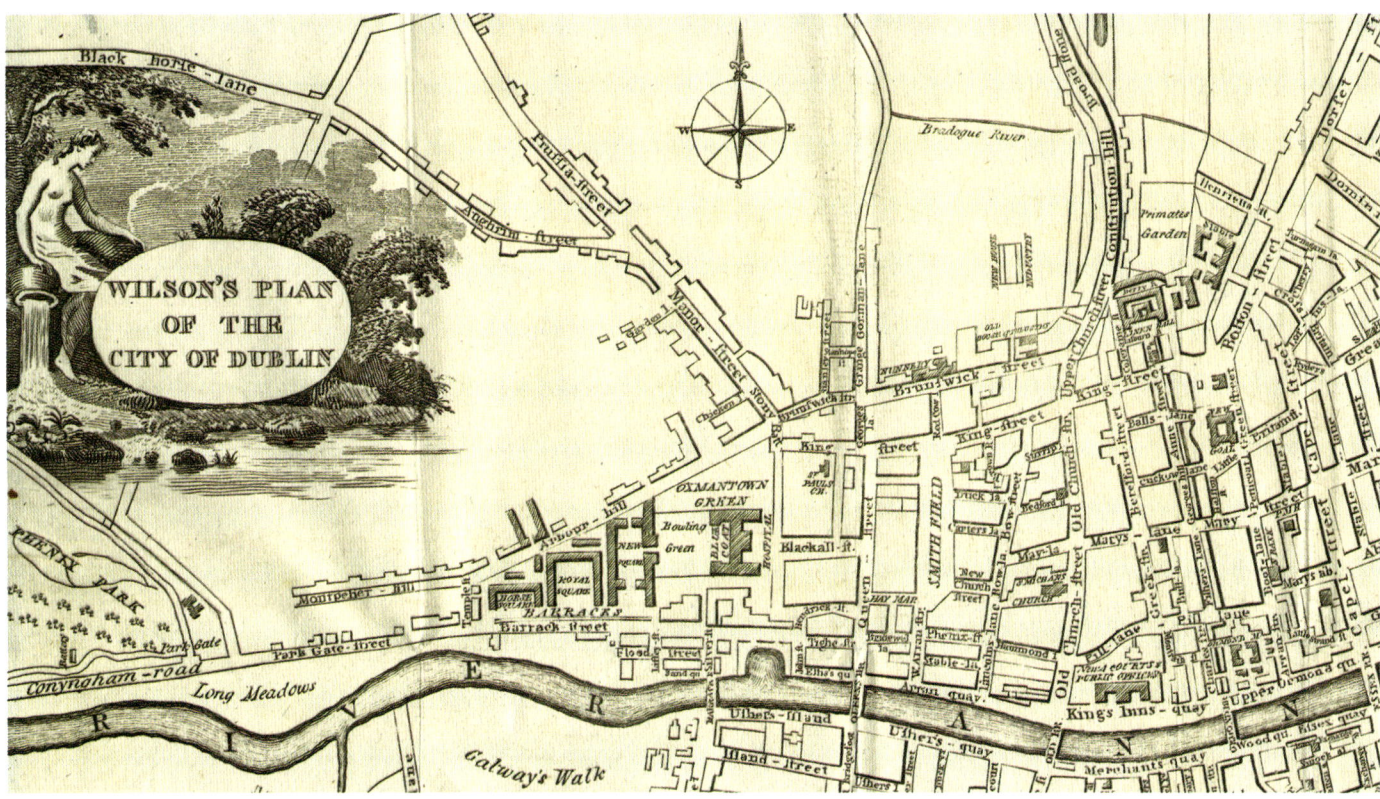

The north-west of the city showing the cartouche, extract from map included with *Wilson's Dublin Directory* (1795) [AUTH]

Lords and Commons, the baronets of Great Britain, as well as a complete list of the Royal Navy of England and the captains and commanders of the navy.

The third element was *Wilson's Dublin Directory*. This listed the streets in the city and the grid square within which the street was located. Thus, Garden Lane East could be found near Manor Street in grid-square Ce. The list of 'Merchant and Traders' provided the name, address and nature of business. So an almanack-owner seeking to buy some tea would find that Catherine Alley was a tea-dealer at 44 Bride Street. The problem with the list was that it was arranged alphabetically and not by the nature of service provided. The only way to find what service was being sought was to proceed down the list from the 'A's (no doubt to the advantage of those with names beginning with the earlier letters of the alphabet).

This was a long list and confirms that the city had a wide range of trades and services available. It must also have been a city where legal matters were of significant importance. The list of barristers and attorneys was astonishingly long and goes on for pages and pages and puts into the shade that for public notaries, physicians and surgeons.

Maps of the city now became a regular feature of directories and remained so until the relatively recent past. A map first appeared in *Wilson's Dublin Directory* in 1760. It was on a single folded sheet, about 330 × 220mm, and described itself in a circular decorative cartouche as 'A New Plan of Dublin'. Most of the built-up area was included from the edge of the Phoenix Park in the west to Merrion Street in the east. The New Gardens (Rotunda Hospital) marked the northern extent while to the south the coverage extended to just beyond

St Stephen's Green. Essentially this was the area that Rocque had surveyed and published. At this size, the amount of detail that could be accommodated was limited and the streetscape was shown as blocks and only the main streets were named. In fact, the need to provide space for the naming of these principal streets necessitated giving the streets an impressive width in many cases. A better and slightly larger plate (339 × 230mm) was made for subsequent years and produced a clearer image. Though the maps were produced in most years, the amount of updating was limited in the years up to 1775. In 1776 it was time for a new design and 'A Plan of Dublin', dropping the 'New' was included which was somewhat larger at 387 × 245mm. The larger size and simplifying how buildings were shown provided more space to include street names and made the grid more useful. The map also included the names of the electoral wards. A further major revision in 1784 kept the same design elements but changed the size slightly. Some further design changes were made in the years up to 1799 but the basic format was retained. The most notable difference was a more elaborate cartouche, especially after 1793 when an unclothed woman against a background of shrubs supported an oval title explaining that this was 'Wilson's Plan of the City of Dublin'. Though the content of the maps did not change greatly from year to year, this was enough to capture the monumental redevelopment of the city centre under the Wide Streets Commission, the arrival of the two canals and the building of the North and South Circular roads, which quickly became a shorthand for the boundary of the city.

By 1800 it was time for a new map because, as Wilson put it in the preliminaries to his 1799 edition, 'the old plate of the City of Dublin heretofore given in the Dublin Directory being quite worn out . . .' The new version returned to showing the built-up area in block format but the quality of engraving was good enough to include many street names and other details on a sheet which was not greatly different in size to that previously (386 × 282mm). This was another 'New Plan of Dublin' and the title appeared in an elaborate cartouche of the arms of the city. The Royal Canal made its appearance and so did

The south-east of the city showing the cartouche, extract from map included with *Wilson's Dublin Directory* (1830) [AUTH]

Gardiner's pipe dream for a Royal Circus. The engraving become more detailed after 1806, as did the cartouche which now filled in the space in the south-east corner. It became a 'map' rather than a 'plan' after 1813. Wilson was now confident enough in the utility of his map that the date was included in the cartouche, the design of which continued to become more and more elaborate in the years up to 1833. It finally disappeared in a new and significantly updated map for 1834, when the space was needed to recognise the extent of development around Baggot Street and Beggars Bush. Big changes, though, were on the horizon. The massive Ordnance Survey project would soon issue detailed and planimetrically correct maps for the city, and the public would soon understand terms such as 'the six-inch sheet' or the 'twenty-five inch sheet' or even the 'five-foot plan'.

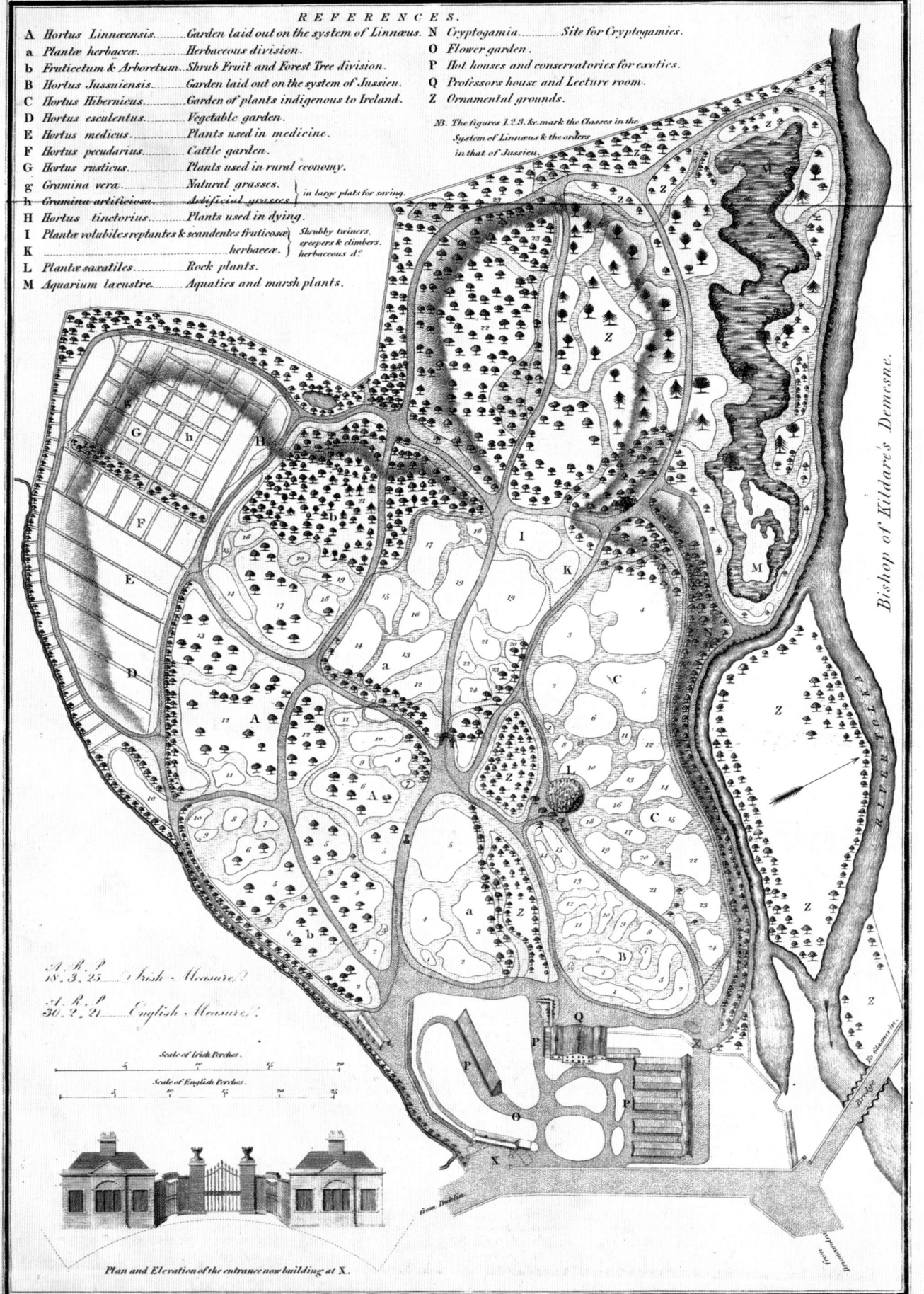

# 1818

## The Botanic Gardens, Glasnevin

The Revd Robert Walsh must have been relieved when he finished the *History of the City of Dublin* in 1818. It had been begun by John Warburton and then taken up by the Revd James Whitelaw when Warburton died. Whitelaw died in 1813, leaving the work unfinished until brought to conclusion by Walsh in 1818. It was a comprehensive work of just under 1,400 pages in two volumes with plates, plans and maps. It was also up to the minute in that it included a section on the new Botanic Gardens which were being developed in Glasnevin.

Modern botanic gardens developed from physic gardens, often attached to monasteries in the medieval period, where plants were grown for their medicinal purposes. However, with the Renaissance came a greater scientific interest in plants, and gardens were often attached to universities and medical facilities. By the seventeenth century, interest had moved from the purely medical application of plants to a wider exploration of botany, but the *History* noted that in Ireland it was not 'till the year 1790, that public attention was drawn to this interesting subject'. In that year a Dr Wade, who was a botanical enthusiast, persuaded the Irish Parliament to support the establishment of a botanical garden. The parliament voted £5,000 to the Dublin Society in 1790, £300 of which was for the garden and Dr Wade was commissioned to produce a plan for such a garden in 1793. This seems to have spurred more interest, and further grants followed. This permitted the purchase of a site, near the village of Glasnevin, though there had been a proposal to establish the garden in the Phoenix Park. Glasnevin was regarded as a good location with a mild climate and well-suited to growing plants.

Thomas Tickell, a minor English poet who had been appointed Under Secretary of State under Joseph Addison in 1717, built and laid out a small demesne in Glasnevin – one

---

J. Warburton, James Whitelaw and Robert Walsh, 'Plan of the Botanic Garden near Glasnevin', from *A History of the City of Dublin* (1818) [AUTH]

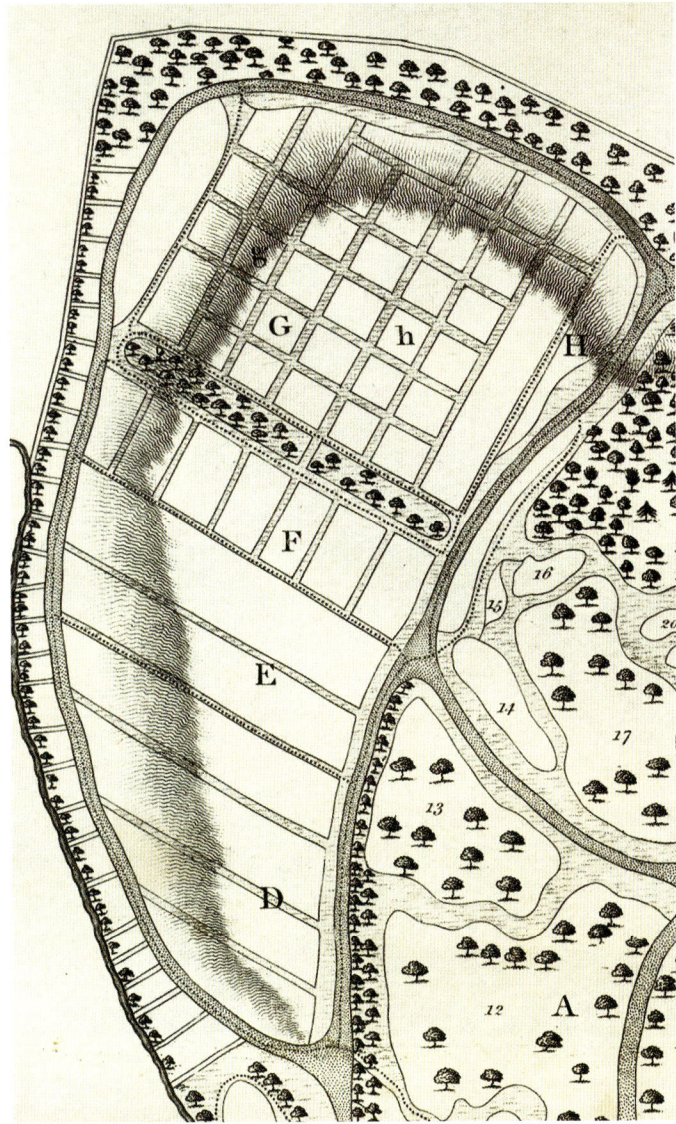

The practical gardens, extract from 'Plan of the Botanic Garden near Glasnevin' (1818) [AUTH]

Walsh included an engraving of the gardens, which is reproduced here, and explained that though the gardens were structured under scientific principles, this was not done in a boring or regimented way. Each of the sections was 'insulated in green swards, and communicating by pathways' with the spaces between filled with shrubs. This ensured that the whole presented an appearance of 'unstudied confusion' while following 'exact regularity' in the classification.

Garden A was designed to provide visitors with an overview of the entire plant world, organised according the Linnaean system. For those with more specialised knowledge the plan showed where the various divisions of the system could be found. A total of 6 acres was devoted to this and the visitor could find 'prefixed to each plant . . . a metal label inscribed with its number in the Glasnevin Catalogue, and its class and order, generic, specific and English names'. Conservatories were provided to protect the exotics which could not take the Glasnevin climate. These were marked P on the plan and were located close to the entrance. The intention was to provide comprehensive coverage of the plant world. Garden B was nearby but smaller, and it was intended for people who were interested in seeing in broad terms how another classification system, the Jussieu, organised plants.

By now the visitor was heading down the slopes towards the river and the Irish garden which aimed to show every plant indigenous to Ireland, arranged on the Linnaean system. This was still a work in progress in 1816 but great strides had been made in recent years and there were now 1,345 different plants available.

Those who wanted to view plants with a more practical purpose needed to head in a different direction, towards the boundary with Glasnevin cemetery. Here the plots were laid out on a regular grid in contrast to the more naturalistic layout of the others. The Kitchen Garden (D) looked at different methods of cultivation and the impact of manures as well as the best and most productive of food crops. Six apprentice gardeners were given the opportunity to develop their skills here and following a period of two years were ready to take up employment. The adjacent garden (E) was devoted to

of the features being a straight avenue of yew trees, called Addison's Walk, that survives to the present. This formed the basis of the garden, which was extended to 30 acres with other land purchases, and gave the Dublin Society a steeply sloping site toward the River Tolka that provided them with appropriate space for 'every botanical purpose'.

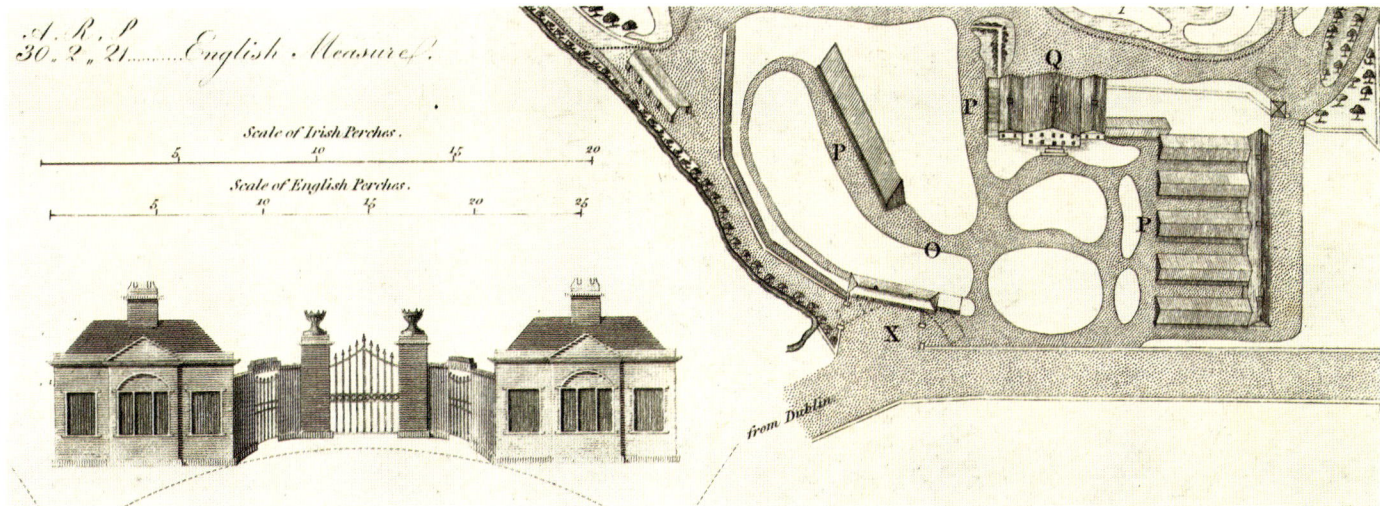

Conservatories, lecture rooms, main entrance and flower garden, extracted from 'Plan of the Botanic Garden near Glasnevin' (1818) [AUTH]

plants which had a medicinal purpose (Walsh lamented the fact that the link between medicine and plants was diminishing and that much was being lost in the process). The visitor next encountered the Cattle Garden (F) and the Hortus Rusticus (G), plants used in the rural economy. In the former, the educational focus was made clear, and plants were divided into those which cattle liked to eat, those which were wholesome but which were not liked, those which would do them harm and those which they did not eat at all. As Walsh put it:

> the great utility of this division is obvious, the farmer sees at once before him the result of long experience, and without the tedious and expensive test of his own practice he may at once adapt his stock to his field, promote the growth of such vegetables as are useful, extirpate such as are injurious, and convert the hitherto despised weed into an useful and wholesome pasturage.

With a similar motivation, the rural division looked at grasses which were good for hay making and providing dry fodder for cattle.

Zones I, K and L were small plots with a very specific focus – plants for dying clothes, climbers and rock plants. The Tolka River had a marshy bank along much of its course in the gardens and it was decided to use this to display aquatics and marsh plants. The river split into two channels towards the main road. Along the narrower of the two, at a base of a steep slope was a site which would be devoted to cryptogamics – fungi, algae and plants without seeds, including ferns and mosses. Walsh makes these sound almost alien and noted that not much was known about them and that this section was not yet complete.

The house in which Tickell lived was devoted to instruction and had been modified to provide a lecture room for almost 150 people. With all this important scientific endeavour evident, there was almost an apology for the provision of a space devoted to flowers, just inside the main gate. In fact, not all of the site was used for scientific purposes and the map showed much of the further reaches of the site as ornamental gardens. The Botanic Gardens were seen as a resource rather than an amenity, but the public were allowed to visit, free of charge, on two days a week as long as they signed in. Visits on other days required the permission of designated members of the Dublin Society, though it was pointed out that those with clear interests in botany would be readily admitted in all seasons.

# 1821

## Duncan's map of the county

William Duncan must have been concerned when John Taylor published his map of Dublin city and its environs in 1816. Duncan had been given the same job by the Grand Jury for County Dublin (an early version of a county council), but Taylor had published his work more quickly. Evidently, Duncan retained the confidence of the Grand Jury and he published his *Map of the County of Dublin* in 1821. Unlike Taylor, who relied on other sources, Duncan claimed that he had 'surveyed, laid down and drawn' the map. He had been, after all, the principal draughtsman in the Quartermaster General's department since 1815. It was a big map, published in eight sheets, each 86 × 57cm, was available in colour and in black and white, and a version was dissected into 64 sections, mounted on linen and presented in a box.

Duncan included all of County Dublin in his map, whereas Taylor omitted a large amount of north County Dublin but included parts of Meath, Kildare and Wicklow. It seems though that this was done to make a more convenient shape, rather than an attempt to delimit some form of hinterland for the city. Since County Dublin is longer than it is wide, it required Duncan to make some important design decisions. Including the entire county from north to south would have left him with a lot of blank space, but he dealt with this by placing the Barony of Uppercross in the top left-hand corner. This should have been to the bottom of the map since it encompassed the area to the south of Blessington. The town of Ballymore Eustace was the most distinctive settlement and Duncan also included the large demesnes of Russboro and Tyrone Lodge.

This still left Duncan with space that he could put to other purposes. In the top right-hand corner he chose to include an elaborate cartouche in the style of the previous century. The dedication was to 'His Royal Highness Field Marshal the Duke of York'. This was a rather unusual choice. Granted, the Field Marshal was the heir presumptive to the Crown after

William Duncan, *Map of the County of Dublin*, extract showing the cartouche and a Martello tower at Glasthule (1821) [ABL]

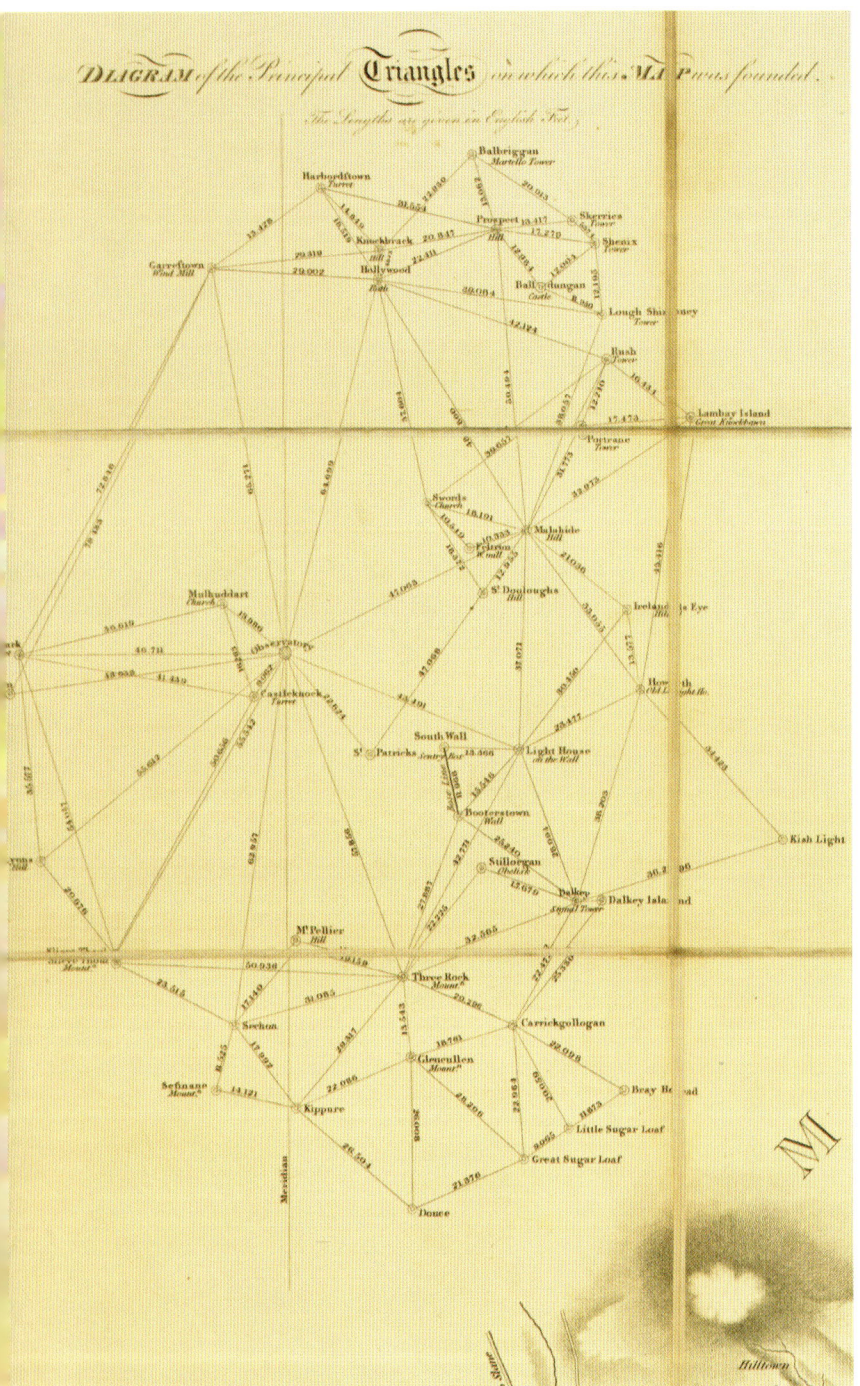

William Duncan, *Map of the County of Dublin*, extract showing the triangulation system (1821) [ABL]

George IV, but he had never shown any interest in Ireland despite being a powerful figure at court. In the space below the dedication, Duncan included an engraving of a Martello tower. The British built such towers in many parts of the empire during the first part of the nineteenth century, but they were sufficiently novel to merit inclusion in place of some of the more usual buildings or views. In Britain and Ireland they were built in response to the perceived threat of invasion by Napoleon, and about 50 were constructed in Ireland. Twenty-six were built on the east coast, concentrated mainly around Dublin Bay, in line of sight of each other, providing the ability to communicate with one another, or warn of any incoming attacks. Duncan drew the tower at Glasthule with a view out over Dunleary Harbour and Howth in the distance.

There was also room at the top of the map to include the main triangulation system that he used to survey the county. This would have been of interest to the specialists who bought the map but it also served to assure readers that the map had been properly constructed. The base line was in Dublin, a line between the sentry box on the South Wall to a point along the seawall at Booterstown. As expected, he made use of the high ground in the county for his points, but he also went to the trouble of travelling to the lightship on the Kish Bank.

Duncan provided a table of distances which occupied the entire lower left side of the map. This gave the distance in miles and furlongs between each town, village and fair location. He felt it was useful to have some sense of distance within the city as well so he included a small table showing how far it was from Dublin Castle to a number of city locations. The distance to the Royal Canal Bridge on the Drumcondra Road was given as a fraction over 1 mile. This must have been of limited use, however, because they are straight-line distances from the Upper Castle Gate.

The map of the city itself was unremarkable; the directory maps had been providing a view of the changes taking place there. It was more valuable for what it said about the suburbs. There was some development beyond the Grand Canal to the south but it was discontinuous and not radial expansion. There was a series of new roads at Ranelagh at right angles

William Duncan, *Map of the County of Dublin*, extract showing the north city and suburbs (1821) [ABL]

to the main road on both sides while in Rathmines some terraces had been built on the main road, just south of the canal as well as close to the original village. (As will be seen later, Rathmines would develop from a number of nodes.) The angular line of the Swan River, so important in explaining the later geography of Rathmines, was striking. On the northside, the Earl of Charlemont's demesne at Marino stood out, but otherwise the landscape was one of small settlements and individual villas. The map shows the completion of the work on the harbour. The South Wall was now complete to the Poolbeg lighthouse and what Duncan described as the 'new wall or breakwater' extended from Clontarf to meet it.

It had taken until the beginning of the century to agree on how to deal with the Dublin Bar, and from the various solutions offered it had been decided that a second wall from Dollymount would be most effective. The new wall would capture a large volume of water within the harbour at high tide resulting in an intense flow at the ebb creating a scouring action on the sandbank. It also had the advantage of preventing sand from the northern part of the bay being washed into the harbour. Work began in 1819 on a design which involved a wooden bridge to connect the embankment with the shore, leaving a channel along the coast. The embankment ran for 5,500 feet from the bridge at a height of 18 feet above low water at ordinary spring tide. To this was joined a 1,500-foot stretch which was at the level of high water on neap tides and a final 1,000 feet (later extended to 2,000 feet) which rose only to half flood level. This was done to limit the force that could be exerted by the tides at spring tide, which it was believed had the potential to damage the main embankment and explains why this element of the wall is underwater for some of the day. The North Wall was completed in 1824

William Duncan, *Map of the County of Dublin*, extract showing the harbour area, 1821 [ABL]

and by the middle of the nineteenth century a depth of 5 metres could be assured in the channel. The wall also altered the dynamics of the bay with the result that the North Bull sandbank accumulated and quickly became an island – and just as quickly became an important amenity for the city, especially once the tramline reached Dollymount.

Although the map had less detail than Rocque, the county map was important in bringing information up to date, showing new settlement, roads, and other developments. The map showed demesnes and villas (gentlemen's seats), churches, both active and in ruins, and Martello towers. It was also an excellent record of the various rivers and streams of Dublin, many of which have now been culverted. As an indication of its importance, Duncan named the Poddle as the 'City Water' close to its source in Templeogue. From there he traced its course from Perrystown to Kimmage and Dolphin's Barn until it reached the 'City Bason' beside the new Grand Canal Harbour. The northside of the city was now served by its own 'Bason' which lay on the spur of the Royal Canal towards Broadstone. Along the south coast, suburban settlement was now evident, even before the arrival of the railway. Blackrock had lines of settlement on all of the roads leading from it and there was another cluster at Booterstown. Kingstown's harbour had developed as a major feature with both distinctive piers now complete. The port had been created because of the need for an asylum port for Dublin, notwithstanding the completion of the two walls there. Work had begun on the east pier only in 1817 and the west pier was under construction in 1820, so the map has to be one of the earliest representations of the new harbour. It also showed that Kingstown was just beginning to develop, with most settlement concentrated along George's Street and the junction with the road to the harbour.

Duncan claimed in a strapline just under the cartouche that the map was published on 19 July 1821. If that is so, then it is all the more remarkable that it named the town Kingstown since the change did not happen until after the King's departure in September 1821. Duncan must have revised the map, but without changing the publication date.

William Duncan, *Map of the County of Dublin*, extract showing the 'city water' (1821) [ABL]

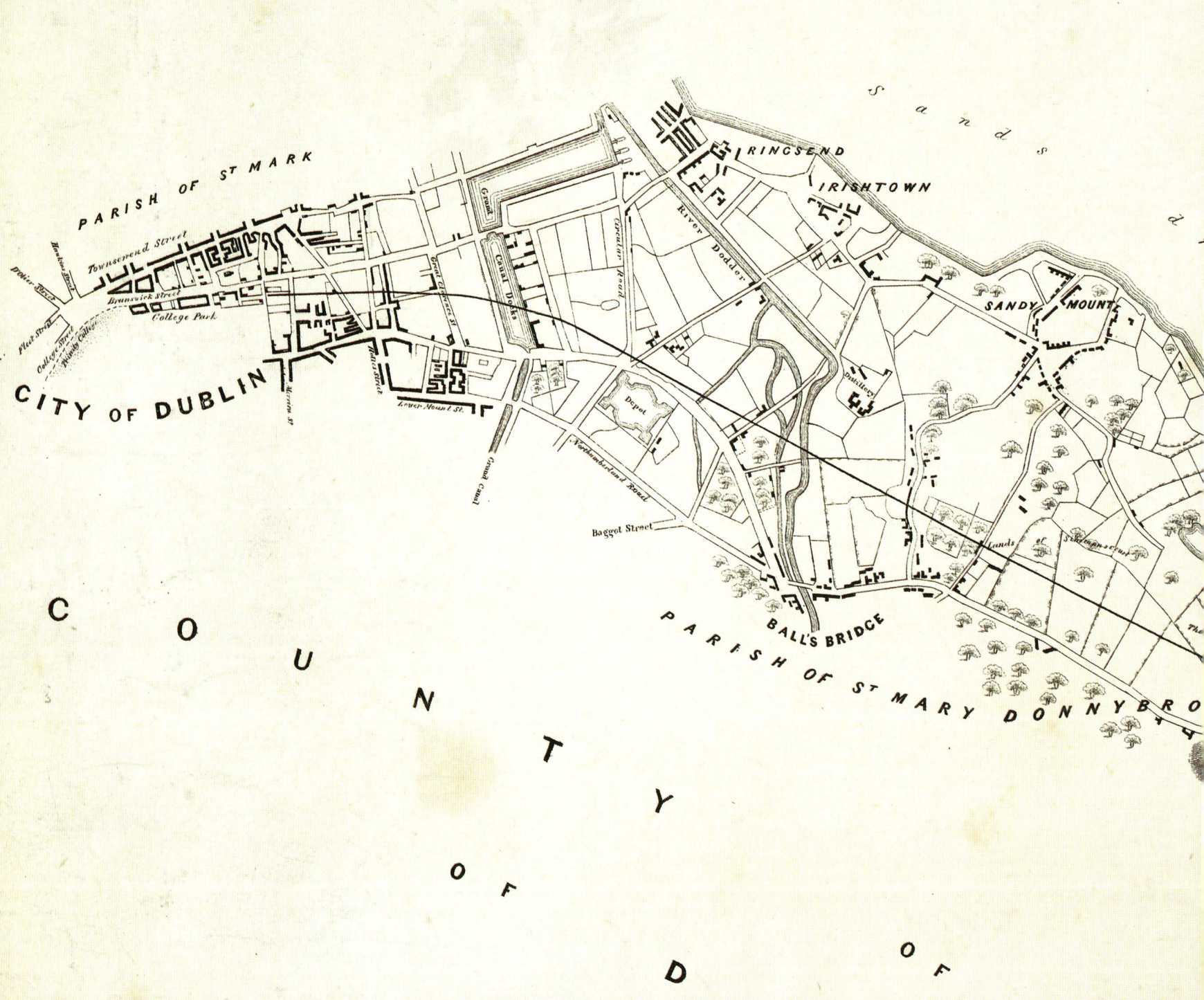

# 1830

## The first railway: Westland Row to Kingstown

A commission was appointed by the British government in October 1836 to inquire into how railways might best be developed in Ireland. The four-person commission, led by Under Secretary of State Thomas Drummond, who had been a major figure in the Ordnance Survey, submitted its first report in March 1837. This was a preliminary report only but it outlined the challenge of developing railways so as to produce maximum economic benefit to the country in the face of fierce competition from commercial interests promoting their own particular routes. However, it did not result in the first railway line in Ireland, that was already well underway by the time the commission was established.

In the early years of the nineteenth century, development of an asylum port began at Dunleary to mitigate some of the difficulties faced by ships trying to enter Dublin port. At the same time, a growing interest in suburban living resulted in a line of small settlements along the coast in favoured locations, where the wealthy built fine individual villas but also where developers, as in the case of Monkstown, set about creating housing developments. These trends suggested that there was a need for either a canal linking the city to Kingstown (named following a visit from George IV in 1821) or a railway. The latter was still a new technology, and it was only in 1825 that the first publicly subscribed railway to use steam locomotives was opened between Stockton and Darlington. Its primary purpose was to carry coal but it also carried passengers. What was proposed for Dublin would have passengers as its primary focus, though it would also transport the mail from the packet ships to the city centre.

James Pim was one of the Pim family and a cousin of the Pim brothers who were responsible for what became the iconic department store on South Great George's Street. He commissioned Alexander Nimmo to produce a plan for the line. Nimmo was a well-known and highly respected engineer and

Alexander Nimmo, *Dublin City and Ballsbridge, extract from Plan of a proposed rail road from Dublin to Kingstown* (1830) [TCD]

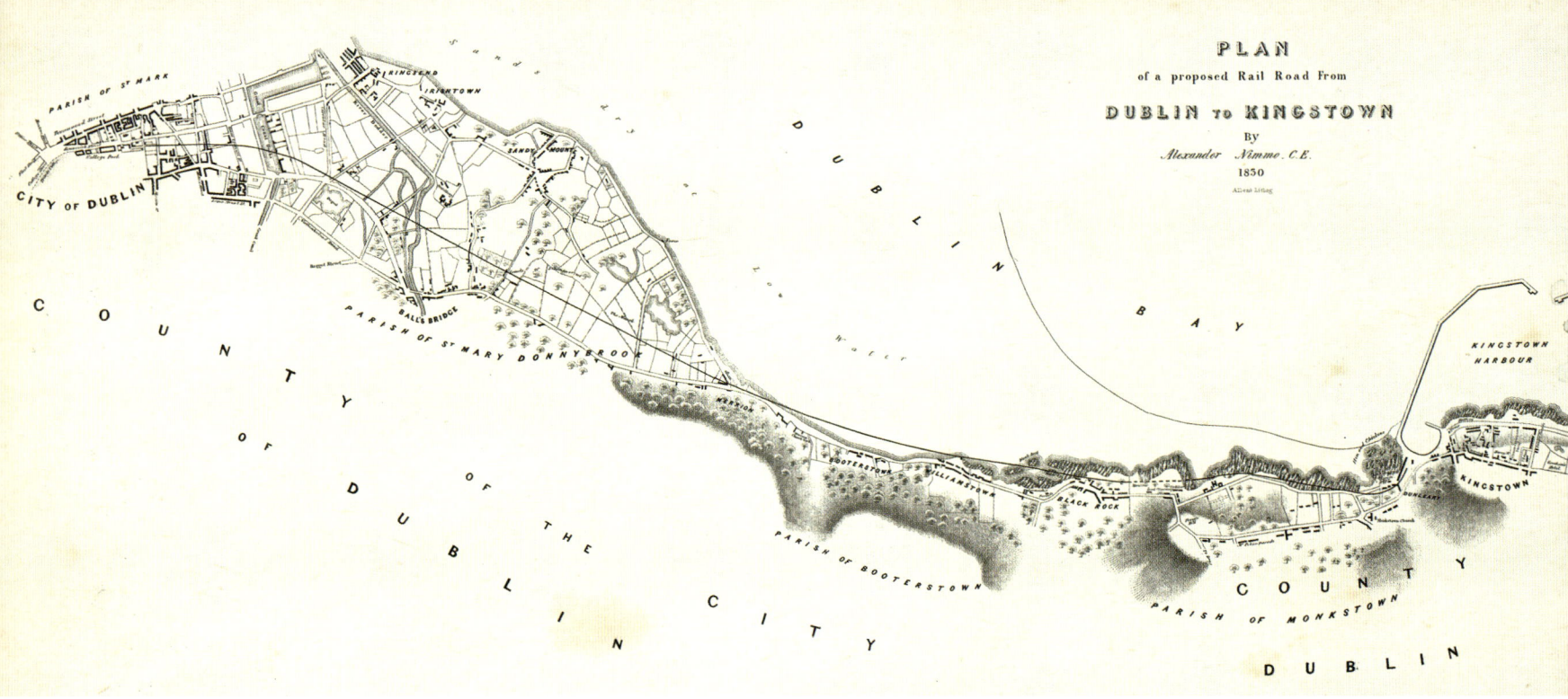

Alexander Nimmo, *Plan of a proposed rail road from Dublin to Kingstown* (1830) [TCD]

geologist who had been involved in many significant projects over the previous ten years, among which were the Commission for the Reclamation of Irish Bogs and the design of a new village, Knightstown, on Valentia Island. Choosing Nimmo as his consultant was a clever move by Pim, and it added status and credibility to the proposal.

Nimmo had the problem shared by all railway planners: how to get the line into the city in the most efficient but least costly way. The south-east quadrant of the city was where the wealthier lived, but it was also substantially built-up at this time (see the discussion on municipal governance). Since there was no perceived need for a railway service for those living close to the city centre there was no need to try and run the line though expensive and probably hostile residential areas. Instead, a terminus at Westland Row was chosen which brought passengers to the edge of the city centre. From there a looping route, which had to be elevated near the centre, kept the line close to the main residential areas but on land which was easier to obtain. Ballsbridge was well served in this way, but Sandymount less conveniently. The line continued through relatively undeveloped land until it reached Merrion, where its junction with the main roads created Merrion Gates, a bottleneck to the present day. From here to Kingstown, a long-term view would have suggested a line running along the coast but at some distance from it, leaving a hinterland on both sides. The immediate need was to serve the new houses situated along the coast and the line was built on their seaward side, involving an embankment along the tidal flats and giving passengers the experience of travelling through the sea. From Blackrock, the line continued along the coast, and this made sense given the rising topography of the land around Monkstown, to arrive at Kingstown and the west pier. This avoided the issue of how to bring the line into the town and, in any case, the expectation was that packet ships would use the west pier. Eventually, the decision of the shipping companies to use the east pier forced a reconsideration of Nimmo's plan.

Legislative approval was obtained in 1831 and the Dublin and Kingstown Railway Company (D&KR) began work on the line. The contract was given to William Dargan in 1833 and the line was completed about 18 months later. It proved to be a difficult project and cost significantly more than expected. It might have been even more expensive had the landowners acted in the manner of Lord Cloncurry, who insisted on a cutting with an elaborate bridge to maintain his access to the strand and secure his privacy.

The first train with invited passengers ran as far as Blackrock in early October 1834. The *Freeman's Journal* reported on 6 October 1834:

> On Saturday the first trial of the steam-engine 'Vauxhall' with a small train of carriages filled with ladies and gentlemen, was made on the line of railway from Dublin to the Martello Tower at Williamstown. It was eminently successful, not only as to the rapidity of the motion, ease of conveyance and facility of stopping, but the celerity and quickness with which the train passed, by means of the 'Shunts' from one line of road to another. The distance was over two miles, which was performed in a very few minutes; and the ease with which the train was brought from the most rapid motion to a complete stop was the surprise of every one that witnessed it.

The first formal service began on 17 December 1834. Initially it was an hourly service from 9 a.m. to 4 p.m. and with fares of one shilling, eight pence and six pence for first, second and third class respectively – it was clearly aimed at the well-to-do. There was only one intermediate stop at Blackrock but 'in a short time it is intended to stop occasionally at some other places for the same purpose' (*Freeman's Journal*, 16 December 1834). Parcels were also carried, while each passenger was allowed 60lb weight free of charge. Once the contract to carry mail was obtained in 1835 the service was extended from 6.30 a.m. to 10 p.m. with trains every 30 minutes. Some trains ran without intermediate stops but all other trains stopped at Blackrock while those on the hour also stopped at Booterstown and those on the half hour at Merrion, Williamstown and Sydney Parade. The extension of the line eastwards to the east pier and the present Mallin Station took place almost immediately, and was ready by 1837. It had the effect of establishing a barrier between the housing development and the harbour, but it ensured that passengers were taken into the heart of the town and close to the port.

W.F. Wakeman, extract from *Blackrock, looking across Dublin Bay towards Williamstown & Merrion – Dublin in the distance*. Dublin. *One of Five views of the Dublin and Kingstown Railway, from drawings taken on the spot by Andrew Nichol* [AUTH]

The line was a success and there was an almost immediate proposal to extend it as far as Dalkey. This encountered significant opposition but a suggestion that an atmospheric railway be used was found to be acceptable. This was a clean form of propulsion since the carriages were sucked along the route to Dalkey by means of an induced vacuum, with gravity doing the job on the return journey. It opened in 1844 and ran for ten years until it was superseded by the conventional railway line which was extended to Bray.

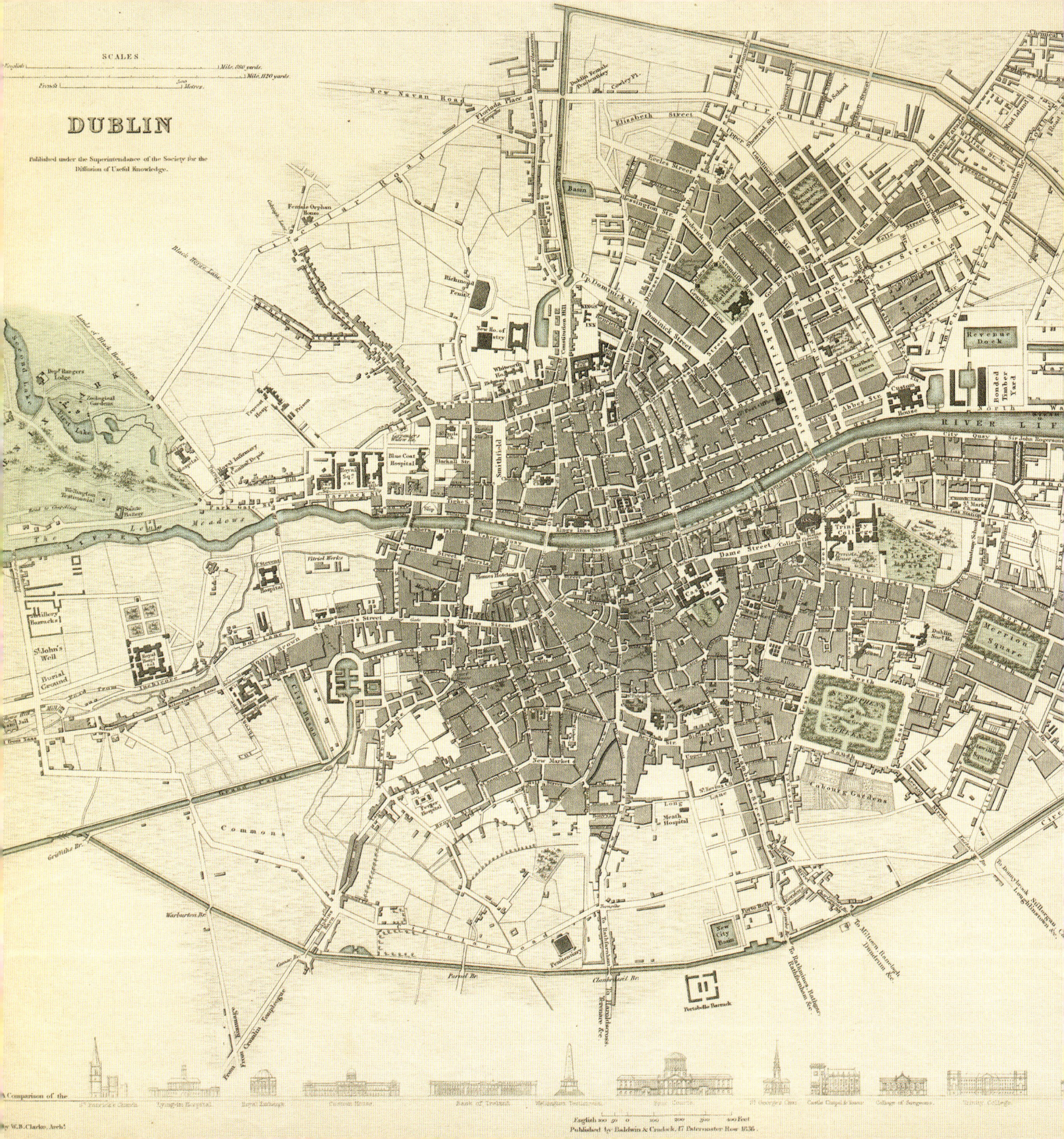

# 1836

## The Society for the Diffusion of Useful Knowledge

The Industrial Revolution transformed many cities in the UK, but not necessarily in a good way. Huge factory complexes polluted the air and water and working conditions for the newly created industrial workers were often grim. The working day was long, often more than 12 hours, and only Sunday provided respite. The children of these workers found themselves in permanent employment certainly by the age of six and perhaps even earlier. There were calls for reform and one who put his ideas into practice was Robert Owen, a mill owner at New Lanark, some 40km south-east of Glasgow. Owen not only wrote about the reform of work practices, he wanted to reform society entirely and to many he is regarded as the first socialist.

In New Lanark, particular attention was paid to providing education for the children of the model village which supported Owen's mill. The school, which opened in 1817, was well provisioned and the students followed a wide-ranging curriculum. Education, though, was not confined to the children. The adults were encouraged to attend the Institute for the Formation of Character, of which the school was an element, where they could take classes, read and discuss. Only in this way would society be transformed. The transformative power of education was recognised by others and initiatives emerged designed to provide paths to learning for those who were outside the formal education system. One such was the Society for the Diffusion of Useful Knowledge (SDUK). The use of the adjective 'useful' was important because the Victorians had no time to waste. The society was founded in London in 1826, mainly at the instigation of Whig MP Henry Brougham. It capitalised upon the improvements in printing technology to produce high-quality publications at much reduced prices. Among its early projects was the *Penny Cyclopaedia* (which cost 9d per issue) and which amounted to 27 volumes and three supplements in the years 1833–43.

---

Society for the Diffusion of Useful Knowledge (SDUK) *Dublin* (1836) [AUTH]

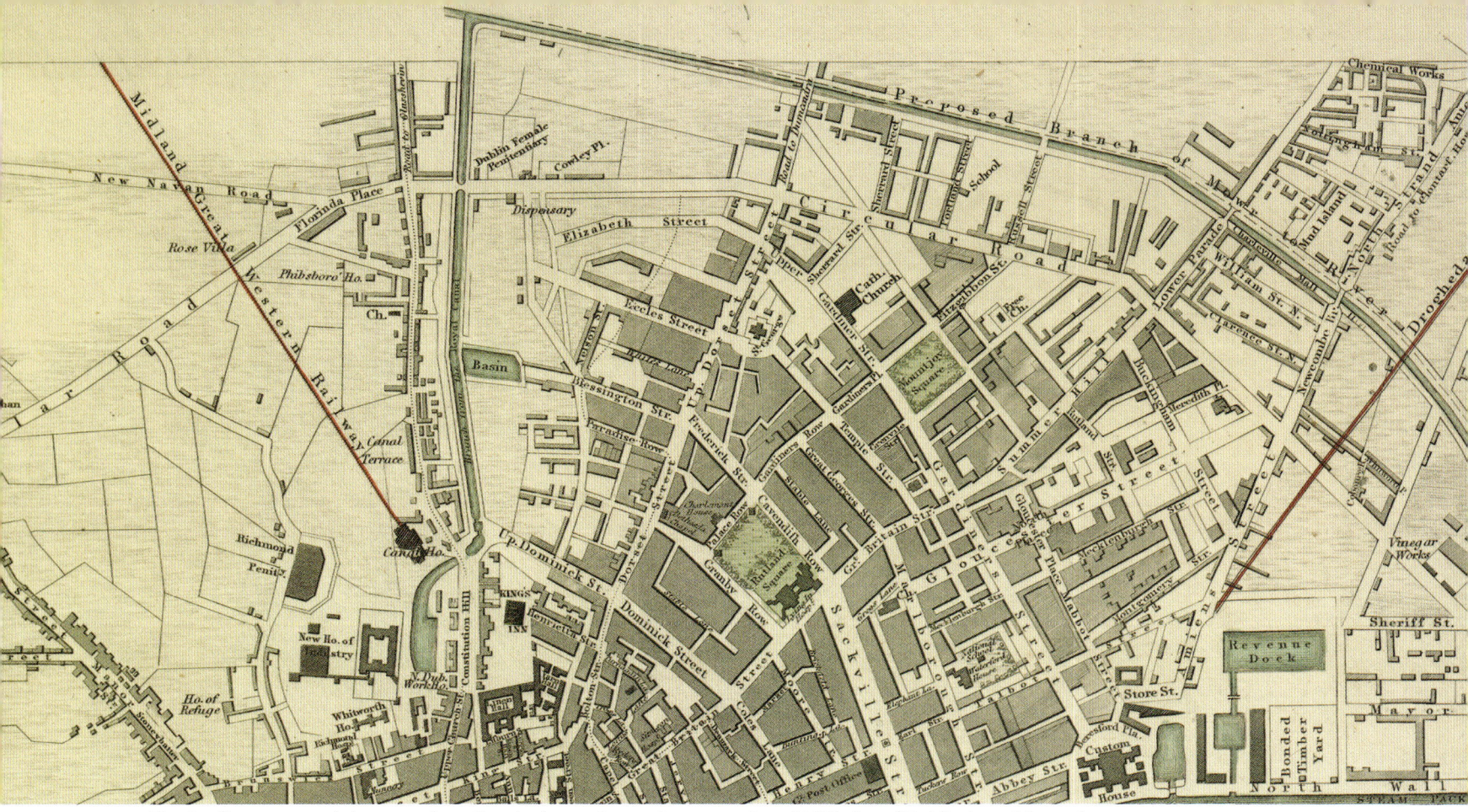
The Gardiner estate and the new railway lines in the north-east city, extract from SDUK plan of Dublin (1836) [AUTH]

While most of the SDUK's publications were standard print publications, it was recognised that maps were very important in giving people a sense of the world in which they lived. The SDUK produced, as a part work, an atlas of 200 or so high-quality maps of the world's regions. Included was a series of 51 city maps of extremely high quality. Strangely, perhaps, the addition of these maps was not enthusiastically received by the subscribers to the atlas but they have become the most sought after of all the SDUK publications today. Though the SDUK ceased business in 1846, the maps continued to be printed and, in some cases, updated, until the end of the century.

Two maps of Dublin were produced; one of the city in 1836 and the other of the city and its environs which appeared the following year. The page size was 34 × 41cm, which allowed the city map to be at a convenient scale of 1:15,000. The quality of printing was very good, better than that found in the directories. The built landscape was shown as a series of individual blocks which were intersected by the street pattern, the more important of which were named. The significant buildings of the city were indicated and named as were the various institutions. The map showed that there was still a considerable bank of undeveloped land within the boundary of the city; essentially the canals. This was particularly marked in the north-west, which was ignored in the movement of the city eastwards following the restoration of the monarchy in 1660. The presence of so many institutions was a contributory factor in diminishing its attractiveness. Visible were the New House of Industry (the workhouse), the Richmond Penitentiary and Whitworth Hospital, the Provost Hospital, the Royal Infirmary and the imposing presence of the Royal Barracks. The newly arrived Alliance Gas Works was on the right bank of the Liffey near the port, providing both light and heat to an increasing number of houses. To the north of the Liffey, the regular form of the new docklands was obvious as was the proposed route of the railway to Drogheda. This was no more than an indicative line on the map since it would take the traveller very rapidly into the bay. In contrast the route to Kingstown was built and its coastal route, which bypassed existing development, was accurately drawn. Other features of note included the city basins, providing a reliable source of water at pressure, and a number of turnpikes. A feature of many SDUK maps was the provision of engravings of important buildings at their relative scales. These appeared at the lower margin and featured the Four Courts, the Custom House, the Royal Exchange, the New Post Office, the Wellington Testimonial (showing its proposed equestrian statue) and Nelson Pillar, among others. Colour on the maps was limited to green for parks and open spaces and blue for water, though more coloured versions exist. A revised edition was published in 1853 which followed the same format but which now had the routes of the main railway companies, shown in red, as well as the proposed extension for the Midland Great Western Railway to the River Liffey, where the proposed depot was shown. The expansion of the docklands was noted and the developing suburbs south of the Grand Canal in the independent townships could also be seen. A new plate was not made for these additions and the railway lines cut crudely through the earlier text.

The *Environs of Dublin* was printed on the same page size, which meant that the image was very close to the margins – copies bound within atlases often suffered from having their lower frame trimmed. The map was also high quality with a great deal of settlement shown with their placenames, more than might be expected from a map at this scale. An attempt was made to give some indication of topography by the use of hachuring though this gives the impression of a more dramatic countryside. By this time, Kingstown was well-established as a town in its own right and the map shows its expansion along the radial routes leading to the harbour. The same is true of Blackrock but the impression is given that Booterstown is a more substantial settlement with a linear expansion inland towards the Stillorgan road. The 1853

SDUK, *Environs of Dublin* (1837) [AUTH]

edition was not greatly different and the most significant addition was the inclusion of the main railway lines. In contrast to the 1837 version, which was sold uncoloured, this version highlighted the railways in red, the main roads in yellow and the coastline in green. For reasons of economy, combinations of these colours were used to show the county boundaries.

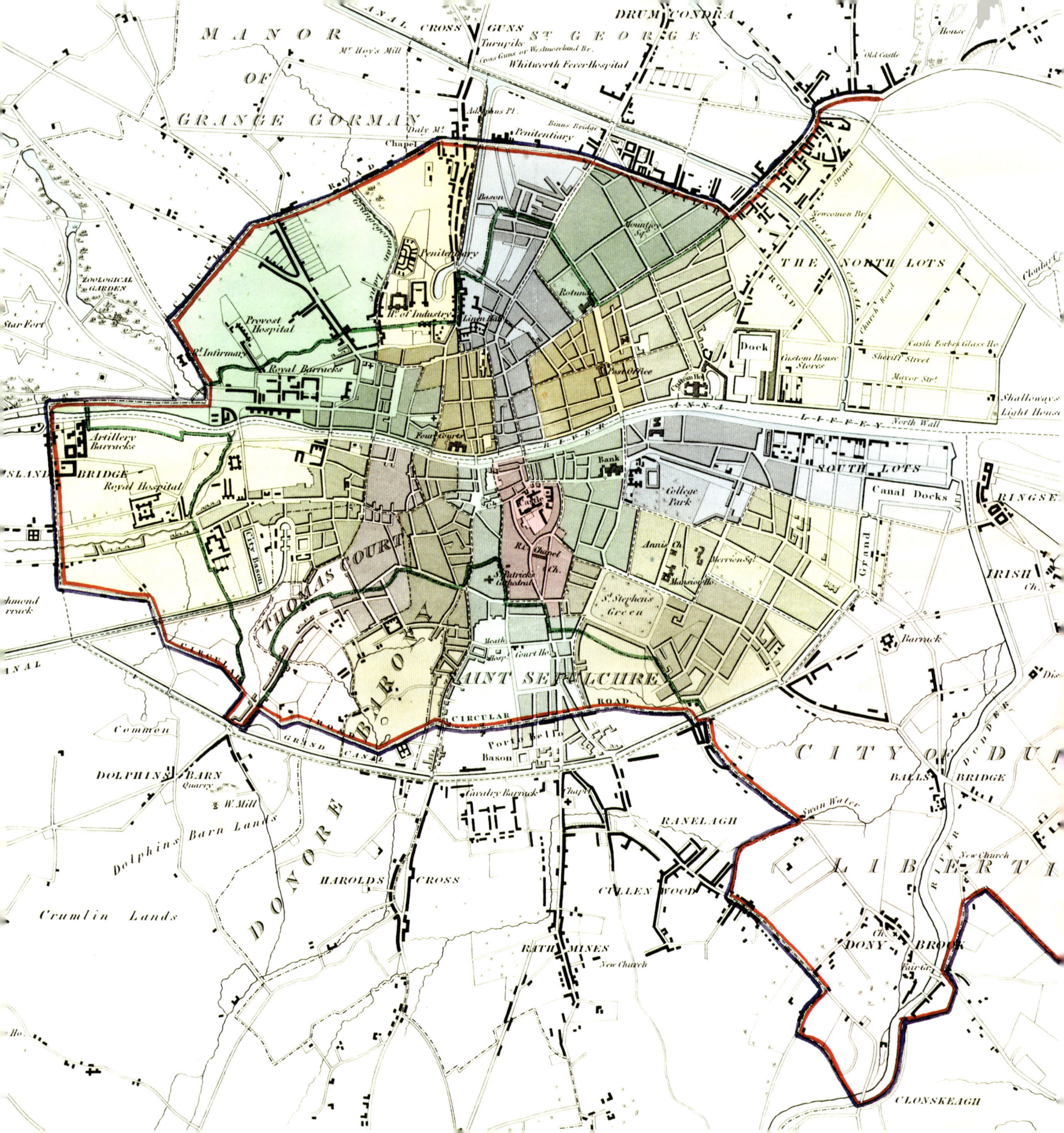

# 1837
## Reforming municipal government

In 1834, many years of complaint against the quality and lack of democracy in local government resulted in the appointment of a Royal Commission to 'inquire into the existing state of the Municipal Corporations in . . . Ireland and to collect information respecting the defects in their Constitution. . .' The report of the commissioners, together with minutes of evidence, was published in 1835 and contained an analysis of some 60 municipal corporations then surviving in Ireland. The conclusions were damning. With some exceptions, 'the governing bodies of the Irish Corporations are self-elected and irresponsible'. Dublin was the only corporation in which all the members were not elected for life but even here many held their office for that duration. The city was governed by a common council but it had an intricate system in which this council operated as two separate bodies. The Lord Mayor and the board of aldermen sat in one chamber while the sheriffs' peers and commons sat in another. There was a complicated process of selection to the various groups but each, in essence, appointed each other. The most representative group was the 96 members chosen by the various guilds and returned triennially to the Commons – the freemen. The sheriffs' peers and aldermen were members of the common council for life. The commissioners report noted:

> the absolute control exercised by the common council of Dublin over the admissions [of freemen], a control which, directed by the avowed sectarian feelings of this corporation, has practically rendered its governing body, not the representatives of the general community of the metropolis, but the head of a limited portion of the inhabitants, comparatively few in number, and avowedly exclusive in opinion.

Clearly this had to change, but it was to take some time before

---

Boundary Commission, *County of the City of Dublin* (1837) [AUTH] An alternative suggestion for electoral boundaries.

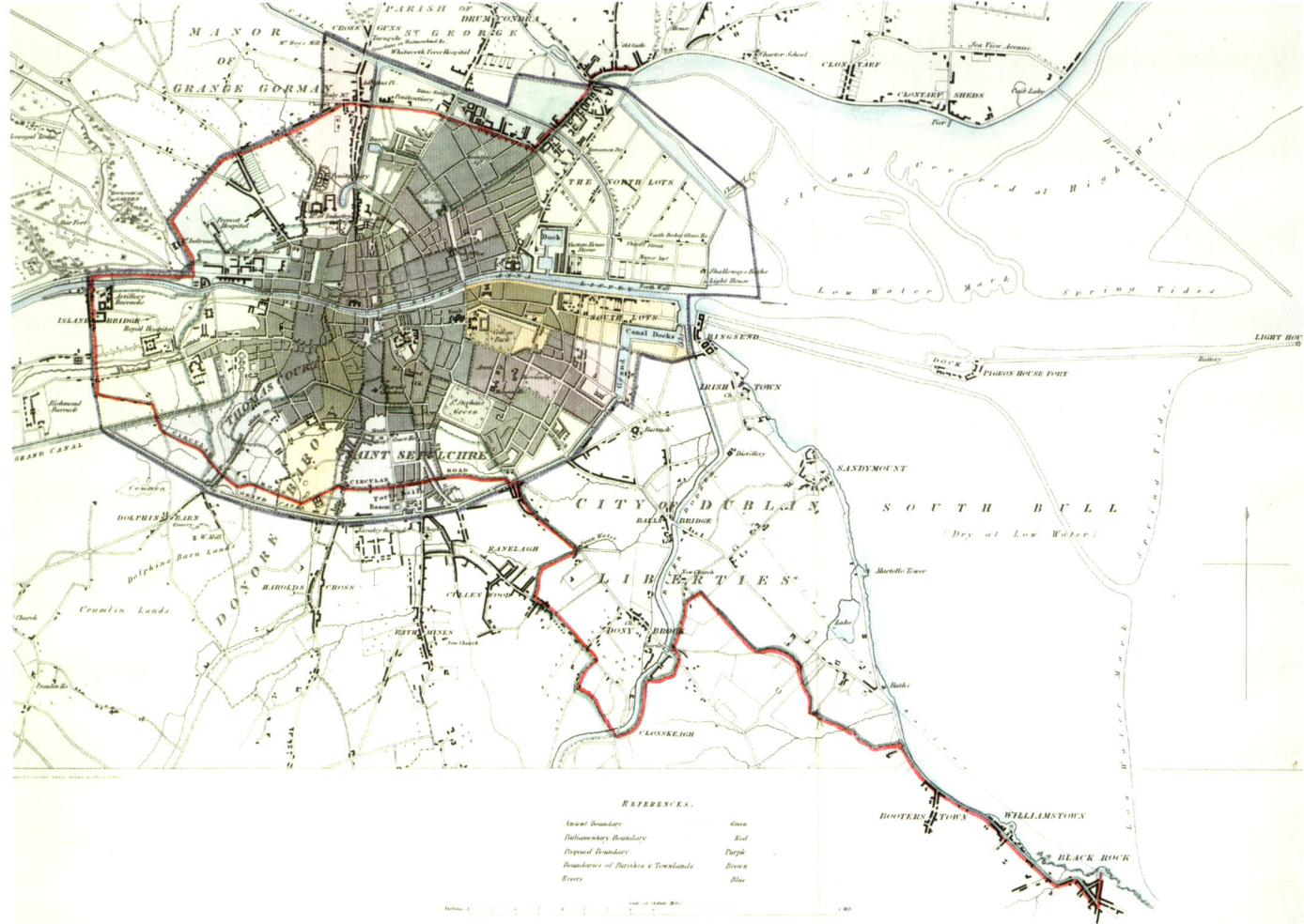

Boundary Commission, *County of the City of Dublin* (1837) [AUTH] One attempt at developing coherent boundaries for government.

acceptable solutions were found. The boundaries of the city also needed reform. As the report noted: 'the City of Dublin is a County of itself, its limits have continued without material alteration for several centuries'. Despite various attempts at clarifying the boundaries, they were still irregular and not well aligned with the built-up city. In some cases it ran through houses and gardens and did not align well with the streets. The boundary also extended along the coast as far as 'Black Rock', while to the north-east the boundary was Ballybough Bridge. The remit of the Corporation extended as far into the sea as a man on horseback could throw a dart at low tide.

This was addressed in 1837 in the instructions given by the Lord Lieutenant as to how to implement the necessary reforms to both the Corporation and the boundaries. The maps shown here, each approximately 39 × 23cm reveal the various boundaries and the suggestions for amendments. Produced by Thomas Larcom of the Ordnance Survey, each also provided an important view of the development of the city at the time. The maps show the 'ancient' boundary of the city in green, and it is clear that there was a significant disconnection between it and the city, especially in the west. That disconnection was carried over into the parliamentary electoral area whose boundary was shown in red and which enclosed an area much larger than the city. The most significant difference was a largely rural area entitled the 'City of Dublin Liberties' which contained the villages of Ballsbridge and Donnybrook and extended to Blackrock. One of the proposed solutions was to align the boundary of the city with the parliamentary electoral area. The circular roads would encompass most of the city except for a small extension in the north-east where the boundary would run north along North Strand to Fairview. This solution was not without issues though. The Grand Canal to the south was a natural boundary and this approach left a small but significant area between the boundary and the canal where development was bound to happen. More troublesome was the retention of what was suggested to become Donnybrook Ward, the large area that reached to Blackrock. Did the city need this area at this point in its development?

An alternative suggestion moved the boundary so that it followed the line of the Grand Canal to the south, leaving the 'Donnybrook Ward' to the county area. On the northside, the boundary followed the North Circular Road until it struck northwards to the Royal Canal at Mr Hoy's Mill and then along the line of the canal until it went north again along the line of the modern Jones' Road and along Clonliffe Road until it met the Tolka River. Within either of these two sets of boundaries, the city would be divided into 16 wards, each of which would have the same level of representation.

The map also showed the developing suburbs. Dublin, in the 1830s, was about to experience the European phenomenon of 'suburbanisation' as more and more well-to-do residents forsook the city centre for the quieter and less-polluted countryside. The map shows the radial developments along the main roads into Ranelagh, Rathmines, Harold's Cross and to a lesser degree north of the river in Fairview and Clontarf. In Clontarf, there were two main foci of settlement; one rising from the coast along Castle Avenue and another, which had a more commercial character, around Vernon

The developing suburbs south of the Grand Canal, extract from *County of the City of Dublin* (1837) [AUTH]

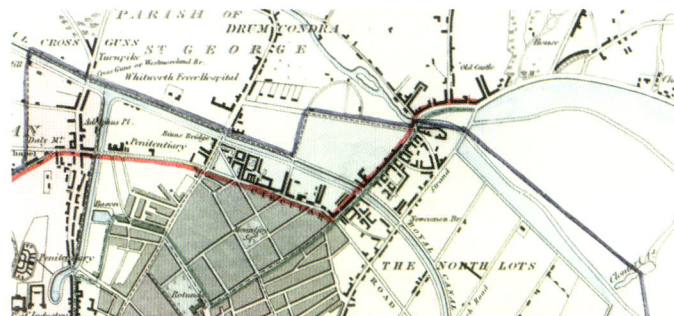

The north-east city and its suburbs, extract from *County of the City of Dublin* (1837) [AUTH]

Avenue and the 'Sheds'. Many of these suburban developments would soon be formalised as independent townships.

These proposals did not end the matter. While it was accepted that the city should be enclosed within the canals as outlined above, leaving the 'Donnybrook ward in the county', there was less agreement about the ward structure. This was articulated in the evidence given by Isaac Butt in 1847 on the occasion of the parliamentary inquiry into the proposed Dublin Improvement Bill. He pointed out that the number of wards had been reduced on the northside from seven to six, and that the size of others had been diminished without discussion. His view was that the wards had been gerrymandered 'to give one party an ascendency in the corporation greater than they should have'.

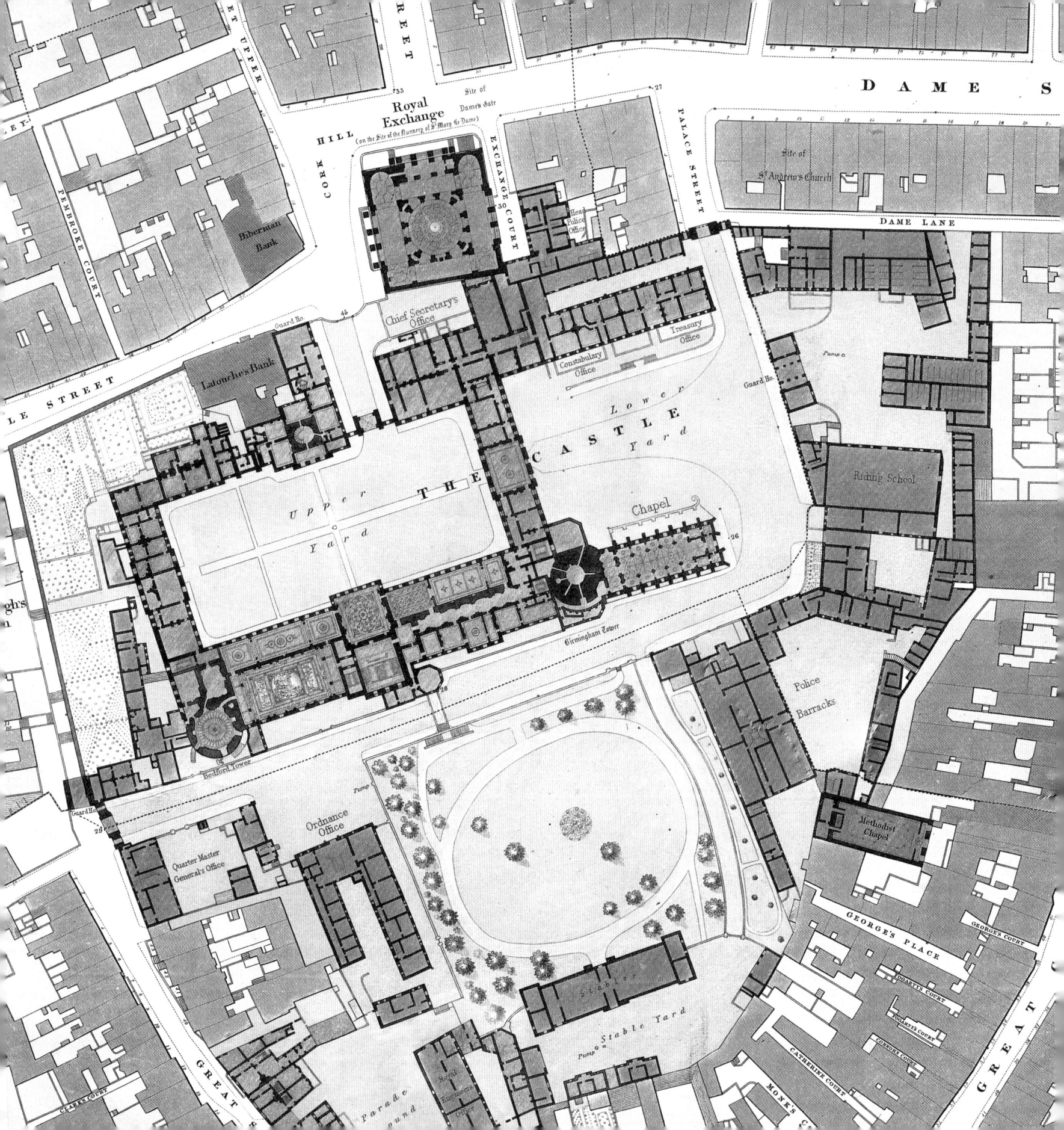

# 1843

## The Castle Sheet

While the Ordnance Survey project focused on the completion of the 6-inch series, an important precursor to taxation, it was understood that more detailed town plans were needed. Rocque had provided such a plan for Dublin, but that was in the middle of the previous century. The Ordnance Survey decided that a scale of 5 feet to 1 mile would provide all the detail that could possibly be needed while still allowing the plan to be manageable. At such a scale it was possible to show each street, house, outbuilding, pump, the layout of public buildings and gardens, and even individual trees. However, the Ordnance Survey had not originally intended to publish such maps as individual sheets. It was the Castle Sheet in Dublin that changed all that. This was one of the first such town plans produced in either Ireland or Great Britain, and it was printed at the Dublin office. The printing system used, designed by William Dalgleish, produced an image of such high quality that the OS felt it should be put on sale for reasons of professional pride. It caught the attention of the public and one commentator thought it the finest example of map engraving ever produced in the United Kingdom. Following this, there was pressure to extend the coverage to the rest of the city, and eventually another 32 sheets were produced.

This sheet was engraved in 1838 and originally published in 1840, with a revision in 1843. It covers an area to the east of the medieval city with Kildare Street on its right-hand edge. These were the streets that had developed during the eighteenth century. All the streets and lanes are named and the work of the Wide Streets Commission stands out. Parliament Street runs straight through a more organic streetscape to focus on the Royal Exchange. Dame Street, College Green and Westmoreland Street are notable for their width compared to the streets around them. In contrast, Dame Street still came

*The Castle Sheet*, Ordnance Survey 5-foot plan (1:1,056)
(1838, revised in 1843) [TCD]

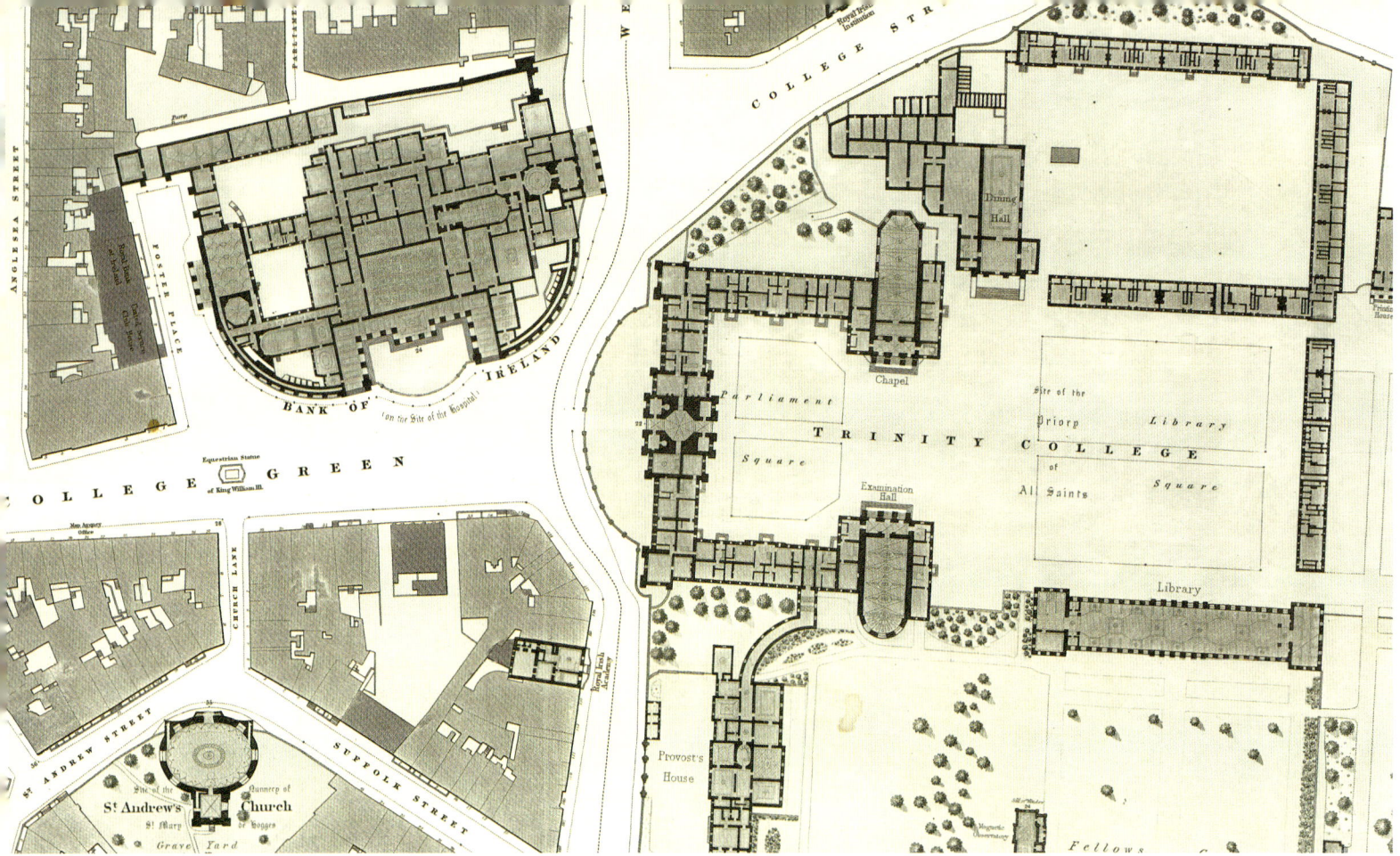

College Green and Trinity College, extract from *The Castle Sheet* (1843) [TCD]

to an abrupt end at its western edge and traffic had to take a dogleg to move up Castle Street. The problem would eventually be solved by the building of Lord Edward Street in 1886.

The medieval origins of the Castle are not obvious from the map, except perhaps for the thickness of the walls of the Bedford and Birmingham Towers, even though these are incorrectly named in a rare mistake. The layout of the Castle's buildings was a unique feature, and it was realised that printing this posed a security risk so the information was supressed in later maps. The buildings and reception rooms dated mainly from the 1750s when the Castle was extensively remodelled in the Georgian style. The apartments on the south side of the Upper Yard were used for the Viceroy's receptions and entertainments during the season; other buildings were used for the various arms of administration, such as the Heraldic Office in the Upper Yard, the Treasury and the Dublin Metropolitan Constabulary in the Lower Yard.

The internal structure of the Bank of Ireland would also have been of interest at the time. Sir Edward Lovett Pearce's parliament house had replaced Chichester House between 1729 and 1739, and the Bank was extended to the west by James Gandon, adding the curved screen wall and the Corinthian portico facing Trinity College in 1785. One of the conditions under which the Bank obtained the old parliament building was that it would alter it so it could not easily revert to its previous role. The adaptations were undertaken by Francis Johnston and he finished the façade along College Green by adding a matching curved screen wall to the east side.

The only street monument shown was the equestrian statue of William III. It was erected in 1701 and therefore pre-dated the redesign of the street, which gave it an even more impressive situation. When it was blown up in 1836, it is said that the Surgeon General was summoned to the scene by a message

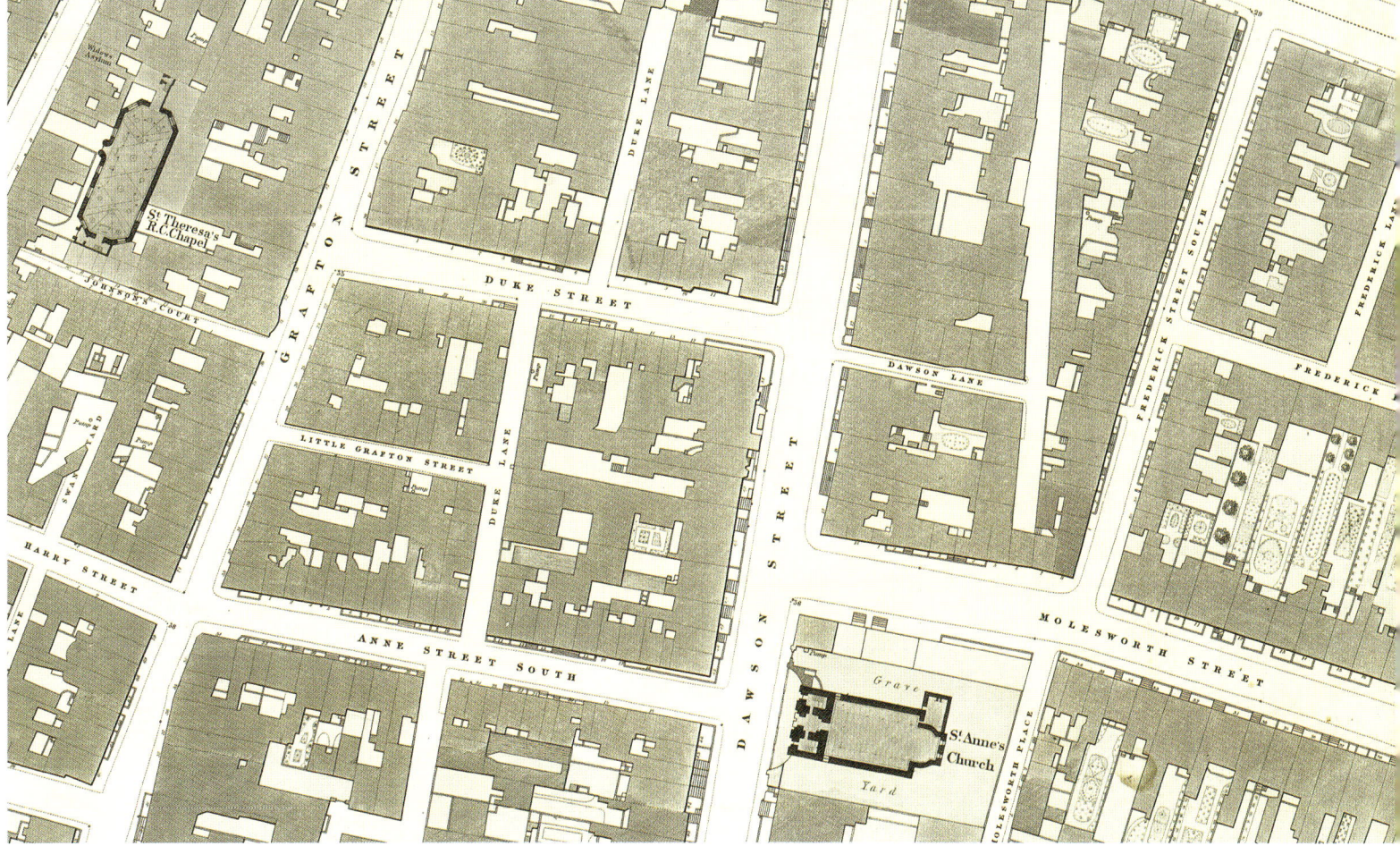

Grafton Street environs, extract from *The Castle Sheet* (1843) [TCD]

that an important personage had fallen off his horse. However, William recovered and managed to stay in College Green until 1928. Between 1752 and 1760, the west front of Trinity College was reconstructed to complement the splendour of the new parliament building. The map also offered a view of the internal layout of these buildings, even showing the roof structures of the examinations hall and the chapel. Also shown were the Rubrics, the oldest buildings in college, at the eastern side of Library Square. The northern range was replaced in 1892. St Andrew's Church, at the junction of St Andrew Street and Exchequer Street, was another imposing building – the unusual oval design (the Round Church), as completed in 1801, is shown but it was burned down in 1860 to be replaced by the present gothic design in 1866.

At this time, Grafton Street was in its transition to a primary shopping street. There had been a shopping arcade – a very unusual feature for the times – on lower Grafton Street, opposite the Provost's House in Trinity College. The Royal Arcade unfortunately burned down in 1837 and the map shows its replacement – a building for the National Bank – under construction. Despite the changes being wrought to Grafton Street, the area to its east was still residential and very much middle class. The rateable valuations for houses in Molesworth Street, where the map showed the fine formal ornamental gardens, were of the order of £85 compared to the £15 asked for the houses to the north of Dame Street, an area largely bypassed by the Wide Streets Commission.

Dame Street was also in transition. In 1843 the area around College Green was dominated by professionals, mainly bankers, stockbrokers and solicitors. Closer to the Castle were found auctioneers, valuators, surgeon dentists and merchants. South Great George's Street was dominated by wholesalers, tradesmen and craftsmen; the close proximity to the Castle having been important for business in previous generations.

# 1846

## Dublin from the air: *Illustrated London News*

It is taken for granted today that it is possible to get a bird's-eye view of the city, and it can be seen and appreciated from a variety of scales whether it be a drone, a helicopter, a commercial airline or a satellite. Brooking had offered a panorama of the city looking southwards, possibly from a viewpoint in Drumcondra. There was nothing comparable undertaken for the remainder of the century and the view of the city at the beginning of the nineteenth century, one which appeared in most texts, was a low-rise perspective of the city taken from the Phoenix Park, probably the Magazine Fort hill. It showed a city in the distance towards which the River Liffey meanders gently. The city lies on both sides of the river, but the right bank rises and we have a view of a city upon a hill. It was a tranquil scene, with surprisingly well-dressed workers in a field by the river and a well-dressed couple on slightly lower ground to provide perspective and scale.

George Petrie's engraved perspective of Dublin, for Wright's *Picture of Dublin* (1821), was taken from the north, possibly from around Grangegorman. It gave a good sense of buildings, chimneys and density through which the various domes and towers appear. It is also a low-rise image, only at roof level, and the foreground is odd, suggesting rough ground.

With only these views available, people must have wondered what the city looked like from a height. This was a problem which the *Illustrated London News* set out to solve. Herbert Ingram was a publisher (and politician) who recognised that newspaper and periodical sales rose dramatically when they included images, especially those of disasters, and so was born the *Illustrated London News*, which claimed to be the first illustrated newspaper. While it was initially successful, it found it hard to attain a critical mass of

---

*Illustrated London News*, 'The City of Dublin' (1846) [AUTH] The panorama was accompanied by a key published in the magazine together with a lengthy profile of the city.

subscribers but Ingram persevered and by 1863, The *Illustrated London News* was selling more than 300,000 copies every week. It was produced weekly from Saturday 14 May 1842 until 1971, after which it went into a slow decline. Though it was the 'London News', its stories came from all over the globe. They were richly illustrated, the nature of which changed with the technology of the times. These illustrations claimed to be accurate impressions of actual events, though they must often have been artistic conjectures.

One device used to increase subscriptions was the production of 'views of the principal capitals of Europe', and Dublin was its first subject. The panorama appeared on 6 June 1846, together with a 12-page supplement describing the history of the city and what the viewer could see in the panorama. It measured 105 × 40cm and was folded to fit into the newspaper. While written in a somewhat flowery style, it attempted a fair view of Irish history and did not shy from calling out the various examples of 'misrule' in Ireland by the British. It also offered a favourable description of the city:

> Dublin, the second city in the British empire, with respect to rank, importance, and extent, is, confessedly, the first in general beauty of appearance; for, though some of the streets of London exhibit many private mansions of superior architectural pretensions to anything of the kind to be met with in Dublin, yet it must be admitted that, taken as a whole, the latter city presents greater beauty of aspect than the main features of even 'rich, luxurious, Babylonic London' display.

It did, however, note the decline in activity, which was increasingly an outcome of the Union – there seemed to be less

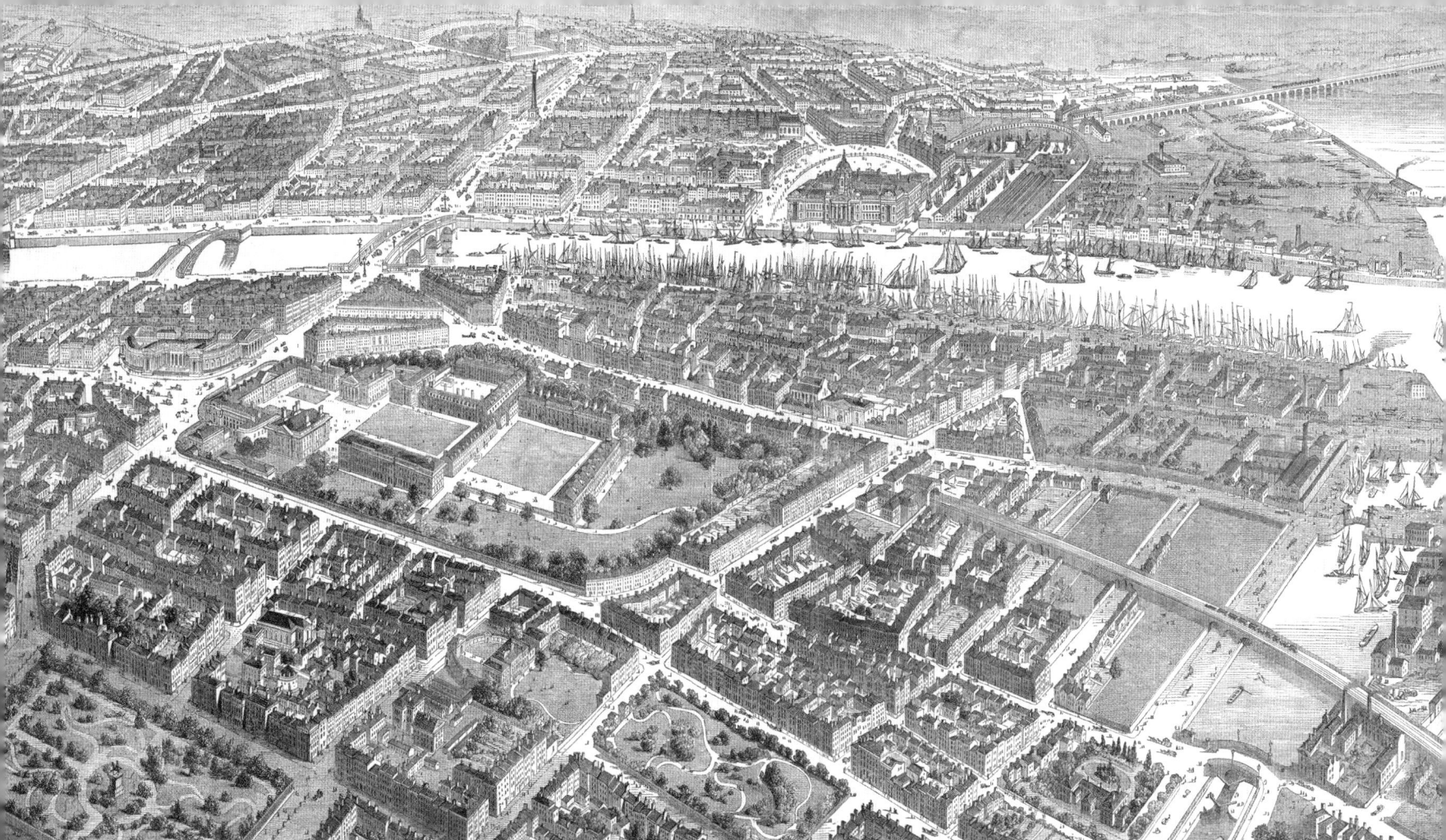

business going on and more people seemingly with little or nothing to do than in London. The ladies were well dressed and fashionable, the gentlemen less so, but it seemed that poverty was deep and the beggars were more ragged and more miserable than in London, even if it seemed that their number was not greatly different.

Dublin was viewed from the south-west, from a vantage point somewhere over Leeson Street. It was a high-level view, one impossible from any building or natural feature. It is not known how the view was obtained but it is speculated that a hot air balloon was used. It was a well-chosen vista and it allowed the focus to be on the more prosperous area of the city to the east of South Great George's Street. The viewer was shown wide, regular streets with tall, fine buildings. The streets were busy, but not crowded, and there was an air of order. Two fine squares were visible. The larger of the two, St Stephen's Green now had a more naturalistic look in its meandering paths compared to the rigid geometry of Brooking. The equestrian statue of George II on its plinth stood out. Merrion Square was shown with wandering paths and arbours. Leinster House looked impressive, but what really stood out was Trinity College, with its impressive architecture and its several squares. It looked like a city within a city.

Dublin was a port, and the quays were shown crowded with sailing ships, overseen by the Gandon Custom House, located within its half circus. The range of ships included paddle steamers and smaller vessels using the river as a means of transportation upstream. Ships berthed close to the centre and someone who paused on Carlisle Bridge could have looked upstream at the view of domes and towers, and downstream at a flotilla of craft. Looking further north, the

Thomas Bankes, 'A perspective on the city of Dublin', from *A new royal authentic and complete system of universal geography antient and modern* (1790) [AUTH] This appeared in a large number of texts over the next few decades.

George Petrie, 'Dublin from Phoenix Park', from G.N. Wright's *Ireland Illustrated* (1831) [AUTH] The view from the Phoenix Park proved popular and a number of variants exist.

detail was less but the impressive scale of Sackville Street stood out and Nelson Pillar provided a good visual orientation point. There were two railway lines, one to the north and the other to the south-west. The Grand Canal and its harbour were busy, as was Grand Canal Dock, with a steady stream of barges, beside the City Basin. The noble proportions of the Royal Hospital terminated the view to the north-west while the city melted away to the north beyond the imposing blocks of the Royal Barracks. Though the main urban area seems densely built, there was still plenty of development land within the line of the South Circular Road.

The view of Dublin was precisely rendered for the most part but there was an area to the west of the city where the detail seemed to be obscured by a low cloud. Here the landscape was hazy and while the general pattern of the streets emerged, the houses were not crisply rendered. This could be a romantic touch by the engraver, adding a bit of mystery to the city. Or it could be a veiled (literally) reference to the fact that this was the poorest part of the city, a place of decaying housing inhabited in poor conditions by a dense population. This was where many of the noxious and polluting industries were located, not on a grand scale because Dublin never developed large-scale industry, but nonetheless dangerous. Perhaps the artist was capturing a pall of pollution.

The panorama proved very popular with the *ILN*'s subscribers, having been produced in large numbers for some years afterwards, and it has also proved to be an enduring and valuable research tool. Comparisons with the street directories suggest that the representation of the city is good, and the quality of the image is such that it tolerates significant magnification very well, providing an unparalleled view of much of the city at mid-century.

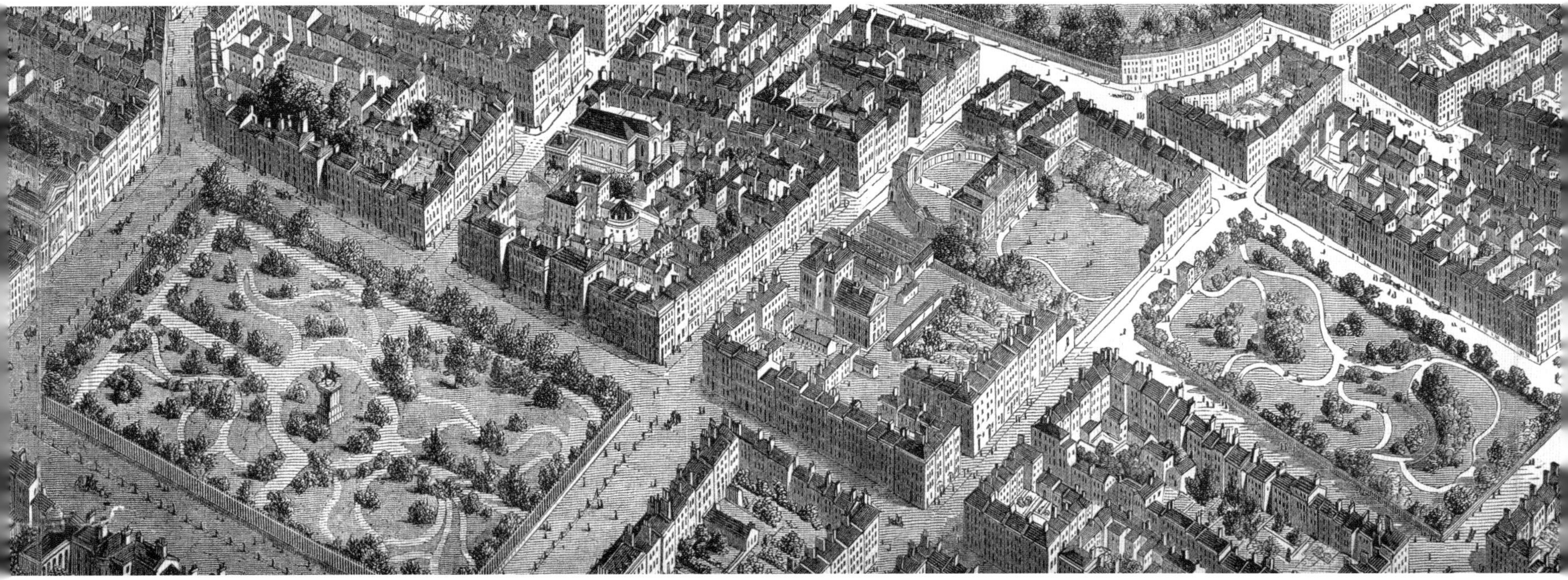

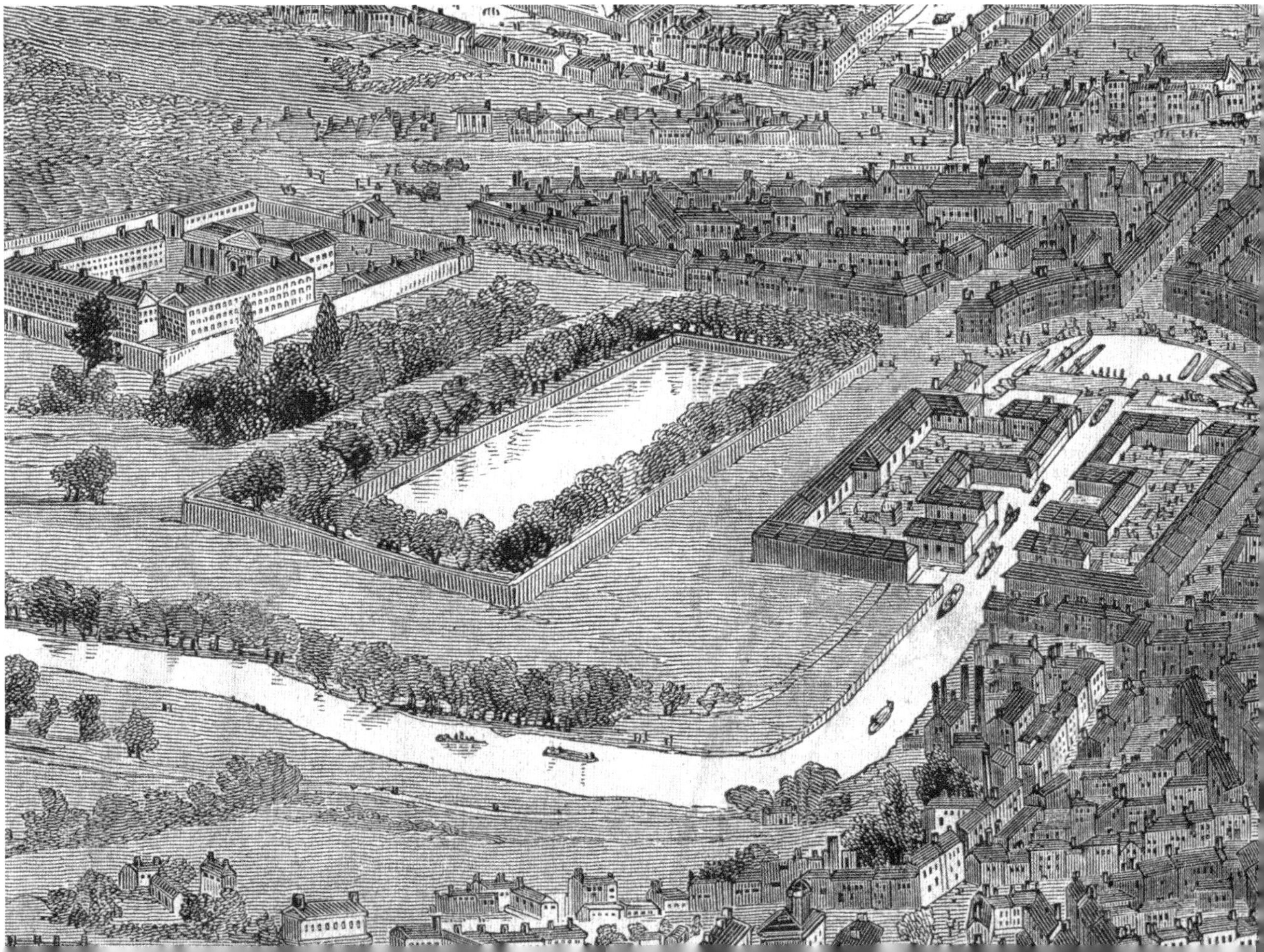

1846

*Above.* ILN, extract showing St Stephen's Green and Merrion Square (1846) [AUTH]

*Right.* ILN, extract showing City Basin and Grand Canal Harbour (1846) [AUTH]

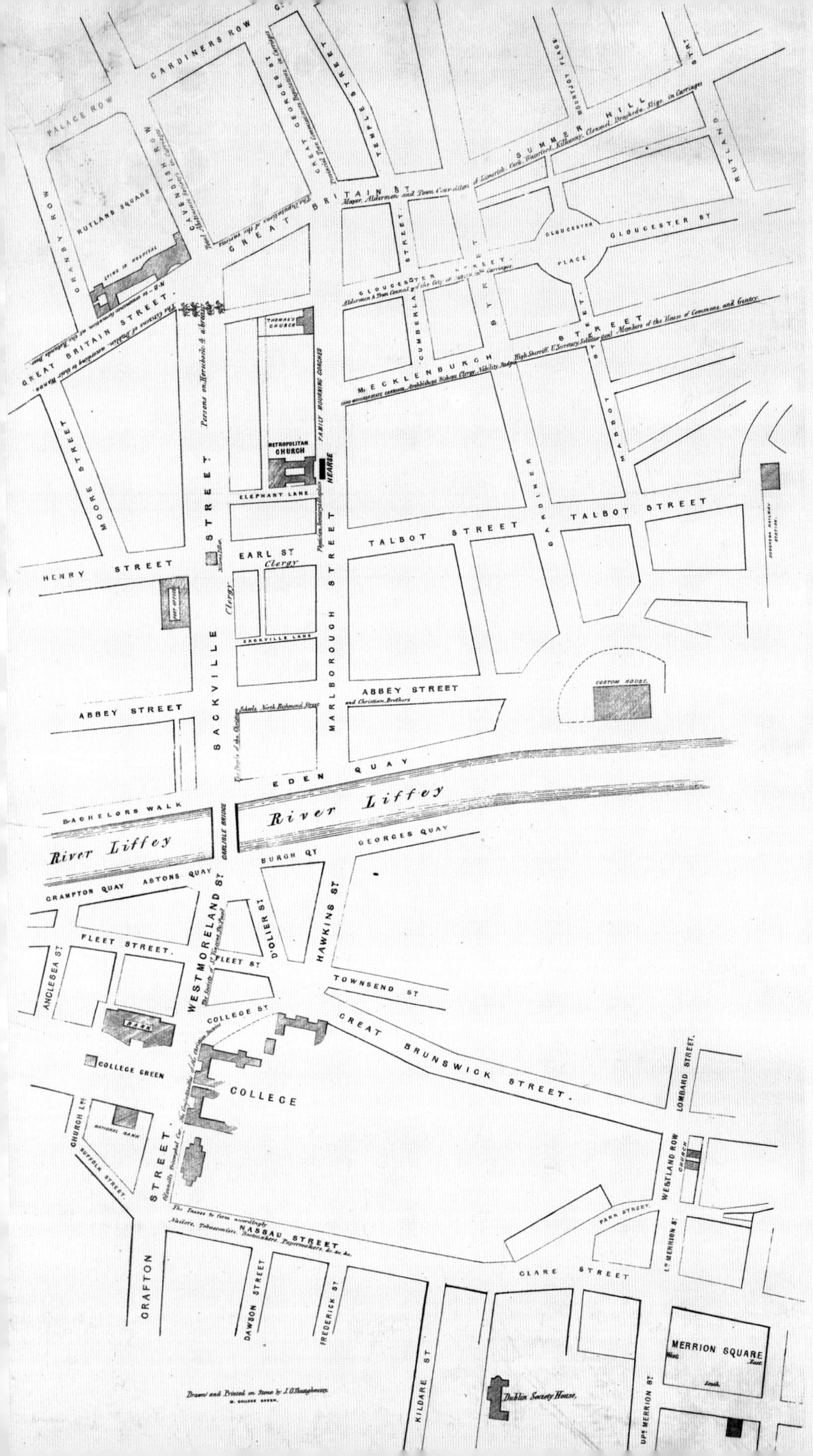

*Top.* Momento of Daniel O'Connell, *Illustrated London News*, 14 August 1847, 104 [AUTH]

*Middle.* Laying in state of Daniel O'Connell, Metropolitan church, *Illustrated London News*, 14 August 1847, 104 [AUTH]

*Above.* The Triumphal Car outside 58 Merrion Square, *Illustrated London News*, 14 August 1847, 108 [AUTH]

# 1847

## The Liberator

During the year known as 'Black '47', the Great Famine was at its peak and the newspapers reported daily on stories of death and starvation. For a while, however, one additional death captured the attention of the nation and became the focus of national grief. Daniel O'Connell died in Genoa on 15 May 1847 en route to Rome. He had been hailed in his time as 'the Liberator' for his success in achieving Catholic Emancipation in 1829 but he had not succeeded by the time of his death in another huge project, the repeal of the Act of Union.

It was very quickly decided that his funeral would be a massive state occasion, a time to celebrate the man and his achievements, though not in an overtly political way. Central to this was a ceremonial public funeral to be held in Dublin which would be organised by the Dublin Cemeteries Committee, an organisation in which O'Connell had been a vital figure. The map shown here was produced for the event.

This ephemeral map – a separate publication, printed on poor-quality paper, for just one occasion – says it was 'drawn and printed on stone by J. O'Shaughnessy, 12 College Green'. Presumably it was commissioned by the organising committee. John Joseph O'Shaughnessy was a well-established engraver and printer who had lithographed a portrait of O'Connell in 1844. Lithography was faster and cheaper than copper-plate or steel engraving, and was ideal for mass produced illustrations and images. O'Shaughnessy's plain but innovative map had a small role in the organisation of O'Connell's last public event. It was not designed for longevity, however, and though many would have been kept as souvenirs, it seems that next to none have survived.

It took time to organise the funeral, and it was only at the end of July that O'Connell's body was transported to Dublin. The event was covered in detail by the *Illustrated London News* and because of that there are good graphical represen-

---

John Joseph O'Shaughnessy, *O'Connell Funeral Procession* (1847) [TCD]

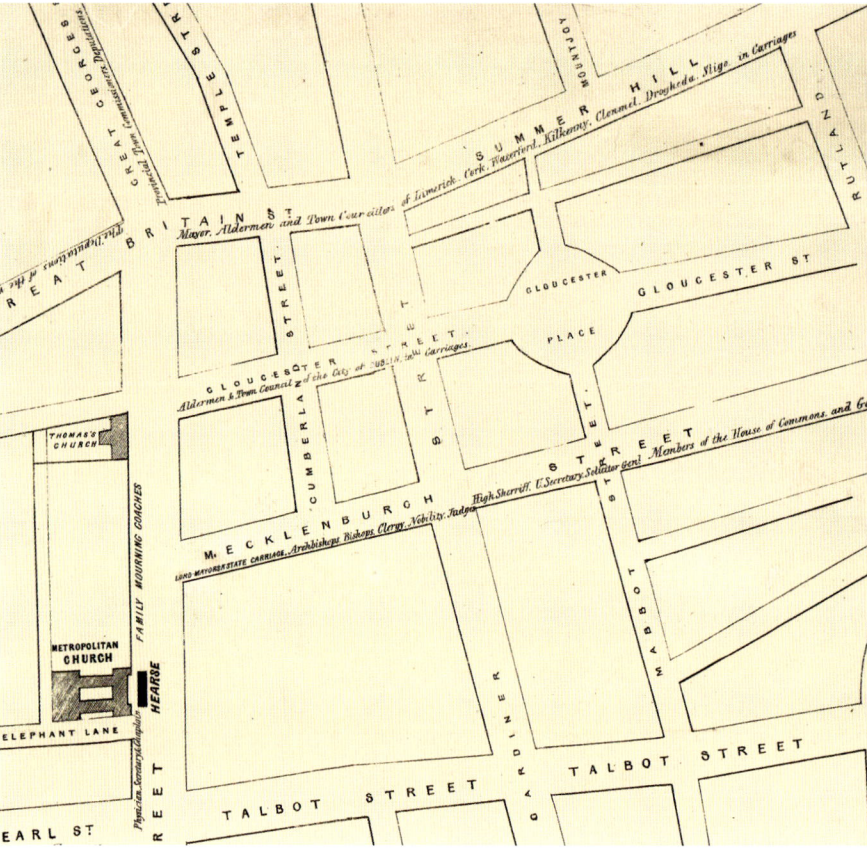

*Above.* The hearse carrying Daniel O'Connell, *Illustrated London News*, 14 August 1847 [AUTH]

*Right.* The arrangements around the Metropolitan Church, extract from *O'Connell Funeral Procession* (1847) [AUTH]

*Opposite.* A section of the crowd in College Green, the Triumphal Car can be seen outside Trinity College, *Illustrated London News*, 14 August 1847 [AUTH]

tations of what transpired. The coffin travelled from Genoa, via Paris, and arrived at Le Havre on 22 July 1847 where it received a steady stream of visitors until O'Connell was taken to Southampton on 24 July and arrived by train at Chester on the evening of the 26th. This, it seems, was ahead of schedule and it was only on 2 August that O'Connell was taken on board the *Duchess of Kent*, the City of Dublin Company's steamer. If the event had been managed in a relatively low-key manner to that point, all changed once the steamer came within sight of Kingstown. It was greeted by a flotilla of craft of all kinds, with flags at half mast and a gun salute. However, the arrival off Kingstown was just to afford the opportunity for family and dignitaries to board and, that being done, the ship travelled up the bay and into the port of Dublin. This facilitated the event to segue from the private to the public because a huge crowd awaited the ship when it docked at George's Quay. A large number of clergy went aboard to pay their respects and then the public funeral began.

An open bier without canopy, drawn by six black horses with mutes and wand bearers, was followed by members of the Associated Trades through the streets to the Metropolitan Chapel (St Mary's Pro-Cathedral). The coffin was laid on a specially built catafalque which was just inside the main entrance at the opposite end of the nave to the high altar. This allowed the throngs of people easy access and it seems that they came in large numbers until the church closed at 11 p.m. The following day promised to be a long one.

Mass was celebrated on the Wednesday morning with admission to the Pro-Cathedral by ticket only, the clergy occupying the prime locations in the nave. The *Freeman's*

# 1847

*Journal* listed some 1,500 of them; this was an occasion which could not be missed. The church was still relatively new, having been completed only in 1825. The city had not yet settled on its name, especially on whether it was 'church', 'pro-cathedral' or 'cathedral'. O'Connell had an important symbolic relationship with the church in that following his election as Lord Mayor in 1841, the first Catholic Lord Mayor of Dublin in centuries, he formally celebrated his election by travelling in state to the church for High Mass.

The map became important for the events on the following day. It was intended that there would be a funeral procession to Glasnevin cemetery, beginning at 11 a.m., one which would take a circuitous route through the city centre and in which everyone who wished could participate. A strict order of precedence was established and published in the newspapers. The various groups were allocated to the streets around the city centre and these would join the procession in sequence. The map showed precisely where each of these groups was to locate. It also showed good vantage points and some building owners sought to capitalise on this by renting out viewing spots.

The head of the procession was close to Merrion Square, where it was led by the City Marshall and followed by the members of the Associated Trades. They had already decided the order for the various trades and they lined up in their allocated spots along Nassau Street. These were followed by the Triumphal Car, which waited outside Trinity College. In 1844, O'Connell, with others, had been found guilty of conspiracy against the government and sentenced to 12 months' imprisonment, of which he served three. This verdict

The vault in Glasnevin cemetery, *Illustrated London News*, 14 August 1847 [AUTH]

was overturned by the House of Lords at the beginning of September. His release was marked by a celebratory parade through the city with O'Connell seated on this Triumphal Car.

This was followed by the Confraternities of the Christian Doctrine, the Carmelite Confraternity, the Society of Vincent de Paul, Christian Brother pupils and Christian Brothers. At this point the core of the procession was reached. This had assembled at the Pro-Cathedral and comprised the family and various dignitaries, who would travel by coach. The Lord Mayor's state carriage led bishops, clergy, nobility, judges, members of the Bar, the Under Secretary, members of the House of Commons and gentry. Notably the Lord Lieutenant and the Chief Secretary were absent. These were followed by aldermen and town councillors of Dublin and other cities and towns. These had been lined up on Mecklenberg Street, Gloucester Street and Great Britain Street. At the end of this group lined up at the top of Sackville Street Upper were 'the persons on horseback, four abreast' and they were an effective

# 1847

border between the carriages and the ordinary citizens who were organised along the Rotunda side of Great Britain Street, supposedly according to the ward in which they lived.

The procession moved off at 11 a.m. but it would have been a long time before the people outside the church were called upon to move. It passed O'Connell's house on Merrion Square and then went on a tour of the city, heading along St Stephen's Green, Redmond's Hill, Aungier Street and into Dame Street via South Great George's Street. From there it made its way to Kingsbridge and then began its journey back into the city via the north quays, Capel Street and Bolton Street. The reports suggest that large crowds came to view at all points of the journey. At last the cortege reached the North Circular Road, where the carriages left and headed back towards town.

Access to the cemetery was equally managed with the crowds gradually filtered until only a select few had access to the vault into which O'Connell was placed. Burial in Glasnevin (Prospect) cemetery was available to all, believer or not, but there was an effort to attract prestigious burials by the committee. One such initiative was the sunken circle containing vaults in which O'Connell was now laid to rest. This had been created to the design of Patrick Bryne, the Glasnevin cemetery architect about 1840. Each vault measured 8 × 8 feet and there were strict regulations about the type of monument or tomb that could be permitted on the surface of the circle above the vaults. While there was speculation in 1847 that O'Connell would eventually be moved 'home' to Derrynane in County Kerry, plans were already in preparation for a suitable monument in the cemetery. It took between 1855 and 1869 to complete the project, which took the form of another sunken circle of vaults on top of which stands the 180 foot round tower. The O'Connell vault was particularly impressive and designed for display, and it was to here that the Liberator was moved in 1869.

The memorial tower over O'Connell's vault, late-nineteenth-century postcard [AUTH]

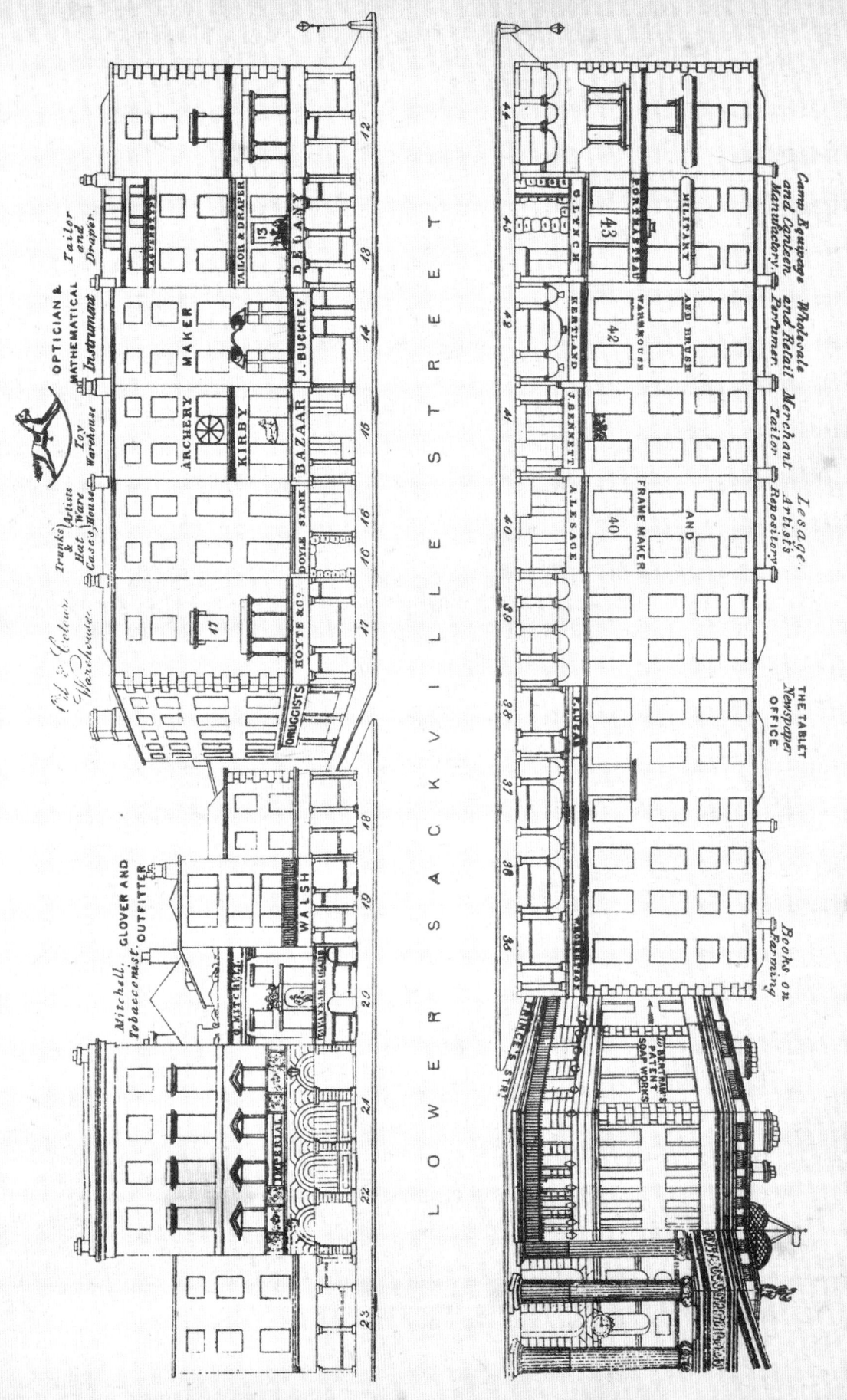

# 1850

## New City Pictorial Directory

Ultimately street (and other) directories became synonymous with the edition produced by Alex Thom and Company, and a 'Thoms' has become a generic noun for such publications. The *Post Office Dublin Directory and Calendar* had been printed by the 'Letter Carriers of the General Post Office' since 1833 and, confusingly, published by Alex Thom into the late 1850s in parallel and in competition with Thom's own publication. It was a bold decision for anyone to try to establish another such publication, yet, that is precisely what Henry Shaw did and in 1850 he published the *New City Pictorial Directory*. It does not appear to have been a successful venture, and no further edition is recorded. It might have been destined to be no more than a footnote were it not for one distinctive element that differentiated it from the other directories – it was pictorial.

Shaw was already the publisher of the *Commercial Journal and Family Herald*, which circulated between 1848 and 1872. According to Shaw the aim was to present a newspaper that would be for all classes, parties, creeds and professions, and would appeal equally to the politician, the farmer, the merchant and trader, as well as those interested in national literature. This was to be accomplished within a two-page publication – just four sides of newspaper – and there were two publication options. A single sheet was available for 1d, containing the literary material on one side and advertisements on the other (Shaw claimed a circulation of 7,500 families within Dublin); the full publication, though, added the commercial and political news, for which he had a further 1,500 subscribers, mainly in the provinces. An annual subscription to the single-sheet version cost 4s per annum for persons living in town, while the journal cost 15s per annum. In 1850, those taking out a full subscription would also receive a copy of the directory; not a bad investment!

The directory was introduced as being 'novel in design . . .

---

Henry Shaw, extract from *New City Pictorial Directory* showing part of Lower Sackville Street (1850) [AUTH]

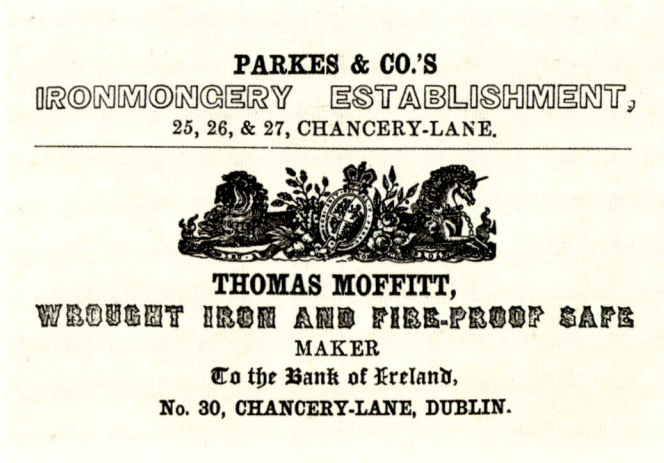

Advertisement for businesses on Chancery Lane, extract from *New City Pictorial Directory* (1850) [AUTH]

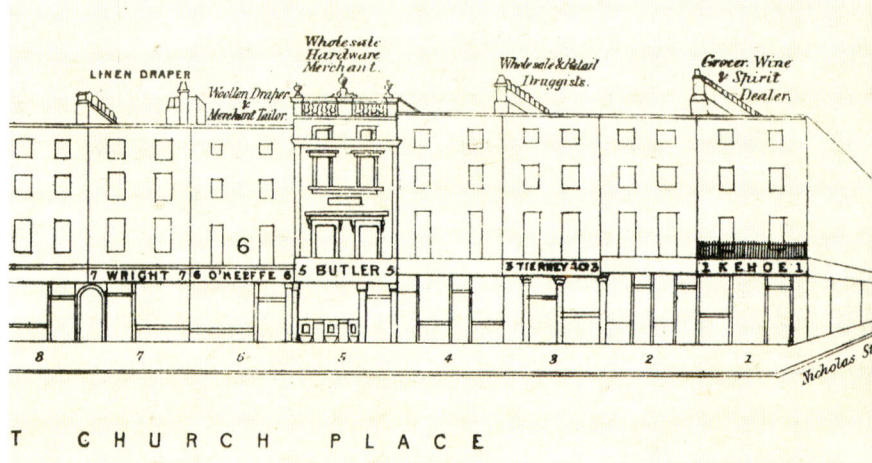

Businesses on Christ Church Place, extract from *New City Pictorial Directory* (1850) [AUTH]

equally novel in presentation'. The novel element in presentation might not seem as useful as the publication suggested: the indexes were arranged alphabetically, but were not paginated. Shaw spent a long time in the 'Explanation' reasoning why this was a good idea, but it seems like an overblown justification for what was the publisher's own convenience. In any event, the directory contained a personal review of the year 1849, a calendar of events for 1850, a listing of government departments, a banking and a law directory, followed by the street listing. The final element was an alphabetical list of nobility, gentry, merchants and traders.

What *was* novel was the inclusion of engravings of the street frontage. Shaw claimed that all the engravings had been made from drawings taken on the spot by his own artists. He also claimed that they were accurate and artistic. However, not all streets were dealt with in this way and Shaw confined the engravings mainly to business streets, to the exclusion of wholly residential areas. The detail provided also varied considerably. A business which took an advertisement in the directory could expect a detailed drawing of their premises with their business noted, otherwise the buildings were shown in outline. Notwithstanding these limitations, the directory provided an unparalleled view of the landscape of the city centre, in particular noting the alteration of the uniform and regular streets of the Wide Streets Commission to suit the more artistic notions of Victorian commerce.

The images for Parliament Street and Dame Street showed that the streetscape created as the first projects of the Wide Streets Commission was largely unchanged. The building line was mostly intact and change had been limited to embellishing the ground and first floors. Some shops had chosen to use some or all of their façade for advertisements. Dame Street towards Trinity College was developing into an insurance and banking district. Driven by the desire for insurance companies and banks to have more prestigious premises, there would be significant alterations and demolitions in the coming years. The Royal Insurance Company had an elaborate façade at No. 32, but bigger and more extravagant buildings were now on College Green, such as that of Gray and Company, bankers.

The main streets were included in the directory, but so too were quite a number of smaller, less important streets which otherwise would not have been recorded but for the presence of a business who had paid for an advertisement. The advertisements within the text itself were also fascinating. They were often bigger than would be usual and gave a great deal

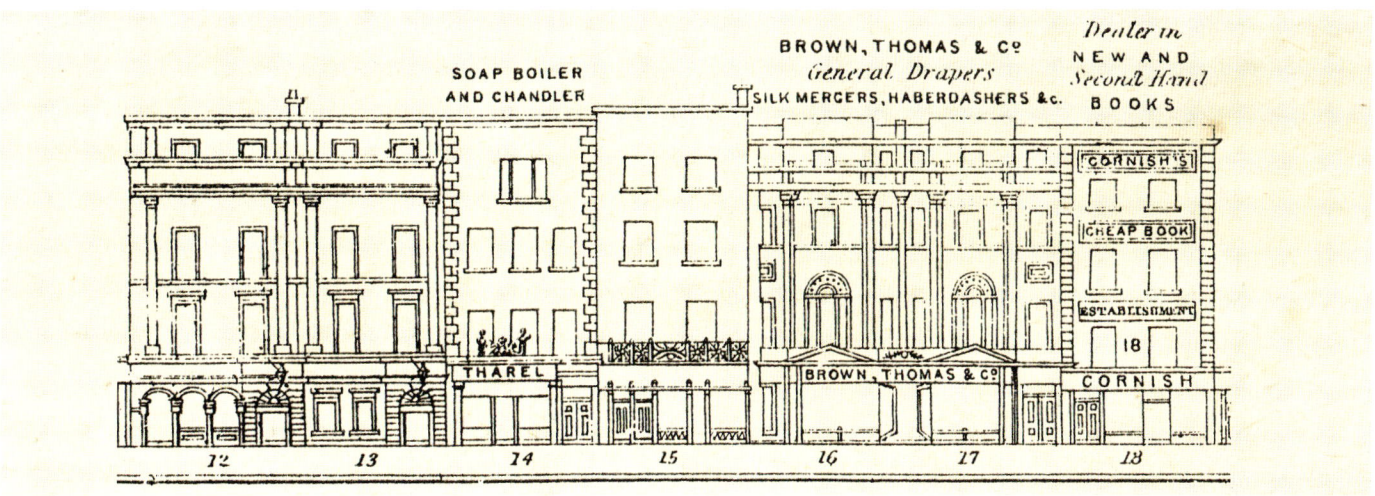

Henry Shaw, extract from *New City Pictorial Directory* showing part of Grafton Street (1850) [AUTH]

of information about the firm. Among the smaller streets was Chancery Lane, where the four-bay house of the Lord Chancellor was shown with its Dutch Billy gable and the note that it was now in tenements. The businesses of mention were Parkes' ironmongery establishment and William Curtis, brass founder and gas fitter. The latter made 'brass furniture for railway engines and carriages' among other services. What makes the page of Christ Church Place particularly interesting, is while noting the business of J. and J. Wright ('superior fast-colored ginghams, umbrellas, japans, and oiled silks') it gives a view of Christ Church Cathedral before the romantic restoration of George Edmund Street in the 1870s.

Grafton Street was well on its way to becoming a shopping street and while most of the buildings still showed their Georgian character, with the occasional Dutch Billy survival (as at No. 92), many were now being remodelled to suit better their retail character. This was especially so at the Nassau Street end, and Shaw showed elaborate façades on Nos 12–13 (the Royal Hotel and Tavern did not pay for an advertisement and so were not named!) and on Brown Thomas at 16–17. Cornish, a dealer in new and second-hand books at No. 13, displayed that he was a 'cheap book establishment' on the building's façade. Across the street, Switzer and Co. ('woollen drapers, tailors and general clothiers') had modified only the ground floor and the upper storeys still looked like the original house.

Lower Sackville Street had been transformed from an exclusive residential street into a retail street, but the streetscape was still recognisably Georgian with most of the various shops fitted into the lower ground floors of the houses. Some changes had been made to façades to facilitate advertising. Mitchell's Tobacconist at No. 20 seemed to be particularly decorative while the Imperial Hotel, though it stood out at four storeys with four bays, would soon become a much more substantial presence on the street. Across the road, the Metropole Hotel had yet to arrive and the corner site was occupied by William Robertson, a bookseller whom we were told sold 'books on farming'. Other distinctive buildings were No. 3 which was now Bewley and Evans, apothecaries and chemists to the Queen, or No. 49 where advertising proclaimed that it was the Wholesale Lozenge Manufactury of Graham, Lemon and Company – the Confectioner's Hall. This was to become an institution in Dublin for the next century and more, but this time Shaw got the entry wrong in the text of directory.

# 1861

## Dublin in three dimensions

One of the more unusual maps of Dublin was that designed and sold by Daniel Edward Heffernan in 1861. Relatively little is known about Heffernan who was a civil engineer and surveyor, and who lived at 12 Charleville Road, Rathmines, between 1863 and 1869. He developed an interest in publishing for the developing tourist market and produced a tourist guide to Wicklow in 1860, and another to the Lakes of Killarney in 1861.

His 1861 map of Dublin was distinctive because it was, as he described it, 'an isometrical view' of the city, meaning that buildings were shown in three dimensions. Nothing like it had been produced previously. Shaw's 1850 directory of the city had produced pen pictures of the streetscape for much of the city, but of the façades only. The map was large format at 38 × 26 inches and designed more for framing than actual use in the field. He claimed that he had included all the important 'public buildings, governmental [sic], ecclesiastical, and commercial; its streets, squares, and monuments; with the Birthplaces of Burke, Sheridan, Wellington, and other distinguished persons to whom Dublin has had the honour of giving birth'. He surrounded his map with vignettes of many of the buildings depicted on the maps. There were 26 of these, clearly drawn with an eye for architectural detail.

Heffernan needed sales for his map, so as well as advertisements in the newspapers he prepared a prospectus. This had a number of statements of praise from publications which had seen proof copies and a list of subscribers. Publishing a list of subscribers was a long-used encouragement to others to join the company of those leaders of society who had the good sense to buy the map. The list was a good cross section of the upper echelons of Irish society, and included the Lord Lieutenant, the Duke of Devonshire, the Marquess of Kildare,

Daniel Heffernan, Trinity College, Merrion Square and
St Stephen's Green, extract from *Dublin in 1861* (1861) [ABL]

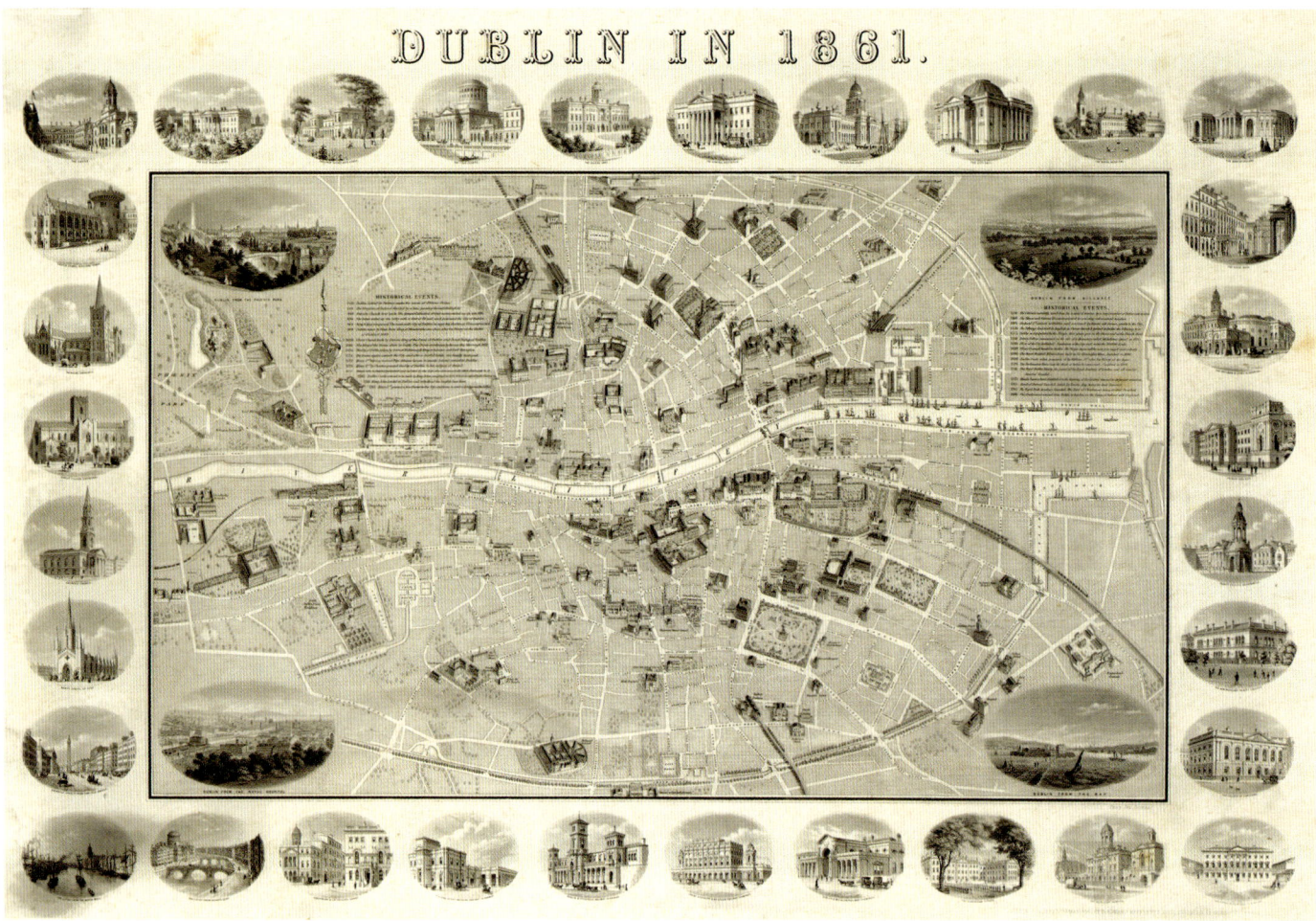

Daniel Edward Heffernan, *Dublin in 1861* (1861) [ABL]

a large selection of the bishops and senior lawyers, as well as corporate entities such as banks and public authorities. Notable among the private individuals who bought copies was William Dargan who bought six copies, and Thomas Gresham who bought ten copies. (Heffernan had probably met Dargan when he drew a ground plan and front elevation of the Great Industrial Exhibition Building in 1853 while Gresham was the proprietor of the Gresham Hotel, which was shown prominently on Sackville Street.) The map was not cheap. An 'India proof' cost five guineas, an ordinary proof three guineas, while a standard paper copy was two guineas. A copy of the weekly *Illustrated London News* cost 5d at the time.

Within the frame of the map there were four perspectives, one in each corner and each showing the city from a different angle. The lower right offered a view towards the city from the South Wall, showing the substantial nature of the fort there. There were two views from the west, one from the Royal Hospital in the south-west, the other from the Phoenix Park in the north-west. The view from the Royal Hospital showed the many church spires in the city, and the imposing bulk of the Royal Barracks stands out. The view from the Phoenix Park updated the classic view of Dublin which had appeared in so many texts since the end of the eighteenth century, the main change being the dominating influence of the Wellington Testimonial.

The map showed and named the main and secondary streets but without as much detail as would have been possible on a sheet this size. This placed all the emphasis on the three-dimensional rendering of the buildings. The railway stations were a relatively new addition and Amiens Street was shown with its two halls and its towers. The neoclassical façade of Broadstone was drawn but the viewpoint of the map required that Kingsbridge be shown from the side. This missed the Venetian palatial façade but it showed the impressive size and architectural design of the building. A similar problem prevented any of the detail of Westland Row Station being shown, but the confined nature of the site emerged as it does with Harcourt Street. The map provided a rare view of the formal geometry of the hospitals, workhouses and prisons. The Richmond Penitentiary was organised as the spokes of a wheel in contrast to the separate squares and rigidly delimited areas of the adjacent North Dublin Union Workhouse and Lunatic Asylum. The South Dublin Union Workhouse had a similar rigidity of design but the internal plan was different and the Richmond Bridewell needed no label; its shape clearly proclaimed its purpose. In contrast, the Mountjoy Convict Prison on the North Circular Road was just beyond the limits of the map and appeared only in outline.

Public buildings such as the Four Courts, Custom House and Royal Hospital were drawn in fine detail and the viewer was given a depiction that would be hard to get even with a personal visit to the location. The port was not as busy as in the bird's-eye view for the *Illustrated London News* but the camber on Carlisle Bridge was clear. The drawing of the various elements of Dublin Castle showed just how complex it was, largely unseen by most Dubliners. Dublin's role as a garrison city emerged from the sheer size of the barracks. The Royal Barracks on Parkgate Street was huge – Heffernan named the two main squares – and equally impressive was Beggarsbush with its star shaped bastions, while Islandbridge Barracks was also substantial. Heffernan included most of the churches,

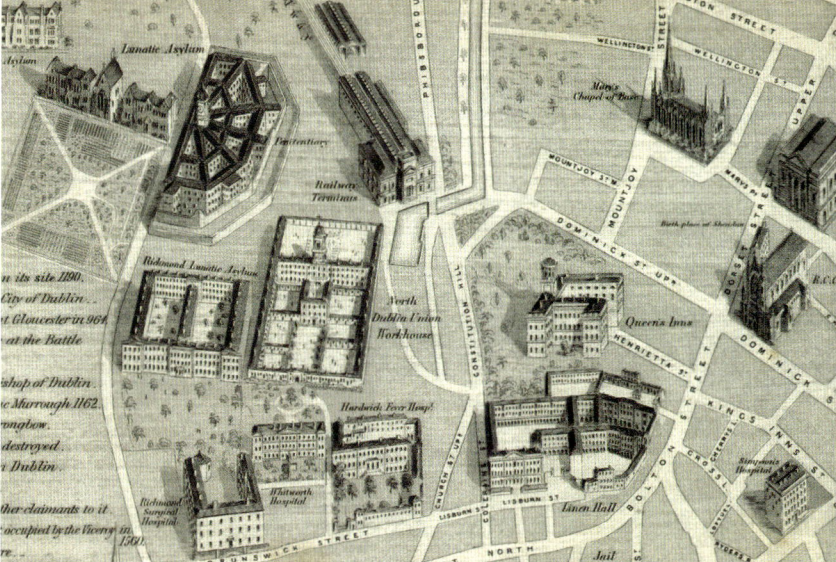

Daniel Heffernan, the North Dublin Union Workhouse and Broadstone Station, extract from *Dublin in 1861* (1861) [ABL]

indicating their denomination, as well as the hospitals, though he chose not to identify those with a specialist function.

The campus of Trinity College would be recognisable to today's visitor but Stephen's Green was yet to get its extensive remodelling. The statue of George II in the centre of the park was well drawn, especially for a map this size.

Heffernan also chose to include some of the distinctive features of the city, such as the Tenter House on Cork Street. This had been a central element of the linen industry but by 1861 the industry was in long-term decline and that building had been bought as a refuge for the homeless. There was a Magdelen Asylum on Leeson Street and, in complete contrast, the SG and US clubs on St Stephen's Green. Mr Gresham's hotel was shown but the Shelbourne Hotel, which was on the edge of St Stephen's Green, was not. Perhaps having a subscription was important.

Despite all of these drawings, Heffernan still had some space left and he chose to fill it with panels outlining key dates in Dublin's history. These were a personal view of events and some of them were quite speculative.

# 1865

## Suburban living: Rathmines

Thomas Larcom's map of the *County of the City of Dublin* some 30 years previously (*see* 1837) showed how housing was expanding radially along the main routes out of Dublin but that it was not particularly extensive and did not depart much from the main roads. This, however, was about to change, and before the end of the nineteenth century extensive suburbs would be found around, but mainly to the south, of the city. A number of factors drove this process. Across Europe, as early as the latter decades of the eighteenth century, the upper and middle classes began to abandon the centre of industrial towns in favour of healthier suburbs. Industrial towns were becoming noisy and polluted, and congested with increasing numbers of factory workers whose conditions were often poor. Moreover, central property could be more valuable as industrial and business sites than as residences. To this was added the fashionable view, promoted by the Romantic poets, that urban life was to be avoided and that only in the countryside could life be lived as it should be. This process was given greater impetus by the arrival of the railways, omnibuses and the trams, which made commuting possible for those whose business required them to access the city.

These fashions influenced Dubliners despite the fact that though the city never developed an industrial working class, though it did become the destination for large numbers of poor migrants. These poor found housing in the fine, single-family Georgian houses in many parts of the city which were increasingly converted to multi-family tenements. This was an additional push factor for the better-off, for whom suburban living with its better quality housing, was an attractive and fashionable option.

A further element copper-fastened this trend – governance. The suburbs were only a stone's throw from the city, and it was but a short walk from the centre of town to, say, Rathmines. However, it was a walk which involved crossing

Central Rathmines, extract from Ordnance Survey plan, 1:2,500, sheet XVIII-15 (1865) [TCD]

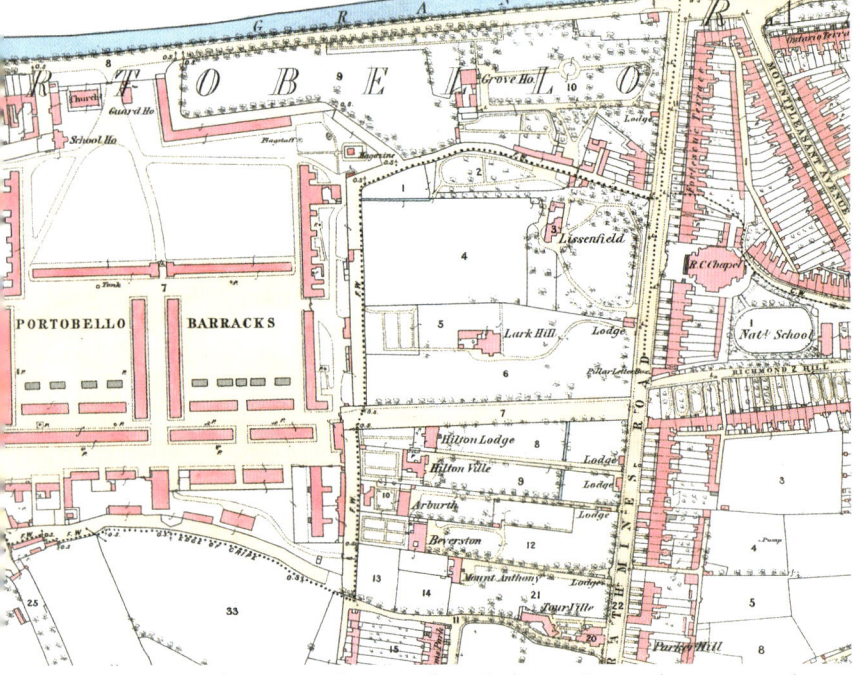

Rathmines Road, extract from Ordnance Survey plan, 1:2,500, sheet XVIII-15 (1865) [TCD]

a legal boundary, for the suburbs came to be recognised as independent townships, urban districts independent of Dublin Corporation. This allowed the suburbs to set their own levels of taxation, pass their own by-laws and, most importantly, ignore the increasing problems of poverty and poor living conditions in the city.

Development in Rathmines was evident from early in the nineteenth century and the intensity increased as the decades moved on. In 1847, a local Act, passed in the UK Parliament, created the township of Rathmines under the management of a board of commissioners, largely speculative builders. They had limited control over planning (and there is no indication that they wanted it) but they could manage the social and economic environment in which development would take place. The township was extended by the Rathmines and Rathgar Improvement Act of 1862 and the St Catherine's Improvement Act of 1866 so by that time the population totalled just under 13,000. This might not seem to be a partic-

ularly large population, but it represented a significant proportion of the city's middle-class.

The plan discussed here is sheet XVIII-15 of County Dublin, and dates from 1865. This was a first edition of the Ordnance Survey 1:2,500 plans or 25.344 inches to the mile. This scale was intended for use in urban areas where the greater detail provided was both necessary and useful. This early edition is particularly interesting for the quality of the engraving and the fact that it was hand coloured. This proved to be an expense which could not be maintained and later editions were both in black and white and less artistically engraved.

The map covers St Peter's parish, which was bounded by the Grand Canal from Clanbrassil Bridge in the west to Leeson Bridge in the east. It shows just how close this suburb was to the city and how artificial was the distinction between city and suburb. Rathmines Road was the main artery, a long straight street which replaced an older winding lane in the second decade of the nineteenth century. The great bulk of Portobello Barracks was a dominant landscape feature. It was completed between 1810 and 1815 as a cavalry barracks and must have added some colour to the locality, though barracks tended to be associated with less-desirable activities. When gas lighting came to the township after 1852, special provision was made for lanes with a reputation for immoral purposes. The development of Rathmines Road was nearing completion by 1865, especially on the east side. Terraces were the preferred form of housing, with individual names a reminder that building happened in a piecemeal, rather than a sequential, manner. Rathmines Terrace, opposite the original village centre, was one of the first and dated from the beginning of the nineteenth century. Newington Terrace, also one of the earliest developments, was out of line with later building; a reasonably regular streetline with modest front gardens. However, the map does not capture the degree of variability in height or roof lines. In contrast, the west side of the street was much lower density with villas, also early developments, on extensive grounds. The Swan River was the main reason for this. It flowed northwards parallel to the main road until it turned sharply east above Richmond Hill. For a small river, it was prone to extensive flooding and continued to be a barrier to development until it was largely culverted by 1858, two tiny slivers of blue still visible on this plan. This was just as well since by then it had become a sewer.

Rathmines is notable for the number of residential squares, more than in any other part of Dublin. These were designed as private spaces for the enjoyment of the surrounding residents, but they also show the limitations of a development process driven by speculators. None of the 'squares' evident on this map are square. Their rather odd shape results from the boundary of the plot that the developer was working with. Nor are they complete, emphasising the piecemeal nature of development. Belgrave Square was far from complete on its west side with a rather awkward boundary to be dealt with. The experience of living on the square was somewhat diminished on its north side by the fact that Castlewood Avenue was a main road. A similar comment might be made about Grosvenor Square, which, though laid out, had housing only on part of its northern edge while Leinster Square was not a square at all. This ensured that Rathmines had a more charming organic street pattern rather than the greater regularity that might be expected in a managed suburb. Roads joined at odd angles, minor roads were wider than more significant routes and connections that would be useful remained unmade.

Rathmines was a solidly middle and lower-upper-class area – the top end of the upper class was more likely to be found in nearby Pembroke. There was no attempt to build speculative houses on any kind of scale for the working classes because, in any event, most of them could not afford either the rents or the transport costs to work (if they had any). Indeed, in 1867, the commissioners ordered that all the remaining thatched cabins and wooden sheds be removed. As the century progressed, some working-class housing was provided for those who needed to work in Rathmines and small terraces appeared in quieter areas – Castlewood Terrace for example appeared behind Cornish Terrace on Castlewood Avenue. A visitor walking around Rathmines in 1865 might

Grosvenor Square, extract from Ordnance Survey plan, 1:2,500, sheet XVIII-15 (1865) [TCD]

have formed the impression that it was quite built-up but the map shows that developers had focused on the main routes only, leaving large tracts of undeveloped land behind. Providing good access to this for future housing was going to prove a challenge.

The city was easily accessible, especially after 1850 when there were a number of competing omnibus companies. Residents could go to business and/or enjoy the facilities that the city offered, but leave its poverty behind in the evening. However, though Rathmines was a middle-class suburb, having developers as commissioners meant that there was a penny-pinching attitude to major infrastructural projects. They refused to buy a supply from the new Vartry water scheme and embarked on their own water supply project, which was not complete until the late 1880s and proved to be hugely expensive. Likewise with sewerage and rubbish removal, but enough money was available to build a statement town hall, complete with clock tower, in 1897.

# 1890

## Impressions of Dublin

The *Illustrated London News* had a long-term competitor in *The Graphic*. It too was a weekly illustrated newspaper and, like the *ILN*, it provided readers with a mixture of local, national and international news. It saw itself as somewhat superior to the *ILN* and when launched on 4 December 1869 cost 6d per copy, compared to 5d for the *ILN*. It employed noted artists to produce its engravings and, not only were they included to illustrate the news, they were intended to be framed and displayed. To that end, the magazine was printed on paper which held the ink better and allowed finer lines to be reproduced with crispness. It appeared weekly until 23 April 1932, after which it went into a gradual decline, appearing in a variety of forms until it finally disappeared.

In 1878, *The Graphic* felt that it was time to provide a feature on Dublin. There was no particular reason for this, in March of that year it had produced a similar supplement on Plymouth. The Dublin supplement took up 13 pages in a run of 16 pages and appeared on 17 August 1878. It accurately described itself as 'Dublin Illustrated' because the volume of illustrations matched that of the text. As was to be expected, the text began with an account of the historical development of the city from the earliest times and then the supplement continued as a tourist guide outlining the important buildings and discussing some people of note. The engravings were particularly good, not only in quality but also in animation and giving a sense of life to the city. There was an excellent view along Sackville Street from Carlisle Bridge, which featured a range of the people of the city, including a few lost academics from Trinity College. There was a busy scene outside the Custom House which was shown to advantage, complemented by a fine long view of the Four Courts. College Green was depicted thronged with people against the backdrop of the Bank of Ireland, while Trinity College was given an equally imposing treatment. The view of Christ

---

'A Bird's Eye View of Dublin', extract showing Francis Street towards St Patrick's Cathedral, *The Graphic*, 20 December 1890 [AUTH]

Church and the Synod Hall captured the difference between this older housing and the Georgian city. The was also a perspective over the south city from the top of Nelson Pillar, though it is a pity that is was impressionistic rather than detailed.

Some 12 or so years later, *The Graphic* decided that it was time to revisit Dublin but this time to offer a bird's-eye view. This appeared on 27 December 1890, and was slightly larger than the *ILN* panorama at 38 × 123cm, though it was not directly comparable to that of the *ILN*. This time the view was towards the east, looking out to the broad expanse of Dublin Bay. This gave a majestic view of the port, the harbour and both Howth and Kingstown. It is atmospheric rather than detailed. The impression was given of Dublin being held between two arms of higher ground which tended to close off the bay in the distance. The harbour was wide, defined by the two great walls. However, while the overall impression is artistic, the value of the engraving as an historical document is much more limited. The detail that is present in the *ILN* map was simply not there. Most of the city was merely suggested by shading of different density and the occasional puff of smoke from chimneys in the distance. In that distance there was a glimpse of Ireland's Eye, behind Howth and the northern railway line running across the estuary with water on both sides; Fairview Park had yet to be constructed. It was a cloudy day. This was a great pity because the image was created by Henry William Brewer, who was held to be one of the most outstanding architectural draughtsmen of his day and who possessed a detailed knowledge of architecture. It was not a want of skill or knowledge that caused him to produce this more impressionistic view of the city.

The research value of the engraving was saved by the

foreground. The view was taken from somewhere in the vicinity of High Street – the spire in the immediate foreground is that of the Church of St Augustine and St John – but since that is the tallest structure in the locality there is no clear evidence of what was used to obtain the view. The scene was of the medieval city, which had slipped into terminal decline by the 1890s. This was the poorer area of Dublin, though poverty was by then widespread across most of the city. The houses were from an earlier time and were mostly in tenements. Such were the conditions here that Dublin Corporation was about to embark on a programme of clearance and rebuilding, and the engraving details the area just prior to that moment. To the right is Francis Street with its three-storey and mainly two-bay houses and tall chimneys. There is a good image of the Church of St Nicholas of Myra (Dubliners tend to replace the 'of' with 'and' since they had no notion of what the name signified) built in 1829. The imposing building to the far right is St Patrick's Cathedral but the relative position is misleading and distance is foreshortened, so perhaps a telescope was being used. The view of High Street is good, with the two churches of St Audoen and Christ Church and its Synod Hall well rendered. The warren of densely packed streets and lanes between the seventeenth-century tower of St Audoen's and the river is captured well. This area, with its houses of different sizes and ages, was ruinous and badly in need of clearance, and it was no surprise that it was one of the earlier redevelopment projects of Dublin Corporation. The area was in marked contrast to the fine proportions of the Four Courts, across the river. One other feature of note: standing oddly in the middle-left distance, which Brewer felt worthy of inclusion, was the multi-storey block of the Jervis Street Hospital.

# 1893

## Public health: typhoid fever epidemics

By the third quarter of the nineteenth century, Dublin had the unenviable reputation of being one of the unhealthiest cities in the empire. The 1885 UK-wide Housing Inquiry had confirmed the picture of a poor city with poor people, and the century had seen the decline of many of the once single-family Georgian houses into teeming tenements. Dublin had its middle and upper classes, and there was a thriving business and commercial sector, but though these people used the city by day, they retreated by evening to their independent suburbs. That ensured that poverty was concentrated in the area under the control of Dublin Corporation, and it struggled for many decades to address it.

In line with Victorian thinking elsewhere, the issue in Dublin began to be seen as one of public health. Disease of all kinds was understood to be associated with poverty and poor housing but it was acknowledged that the spread of disease did not recognise social boundaries and that it was in the interests of all that public-health problems should be tackled. In 1864, an important step was taken in the appointment of a Medical Officer of Health for the city, Dr Edward Mapother. Mapother's research quickly identified the most unhealthy areas in the city and he oversaw the beginnings of Dublin Corporation's role as a housing provider. However, it was his successor, Sir Charles Cameron, who took over in 1874, who was the dominant figure in public-health policy development and monitoring until the second decade of the twentieth century. Cameron held many different roles, often simultaneously, but among them he acted as Medical Officer for Health and Public Analyst. This gave him a key role in the creation of housing policy and the choice of locations for public-housing schemes.

The period during which Mapother and Cameron operated

---

Charles A. Cameron, 'Instance of Typhoid Fever in Dublin 1893',
*Report upon the state of Public Health in the City of Dublin* (1893) [AUTH]

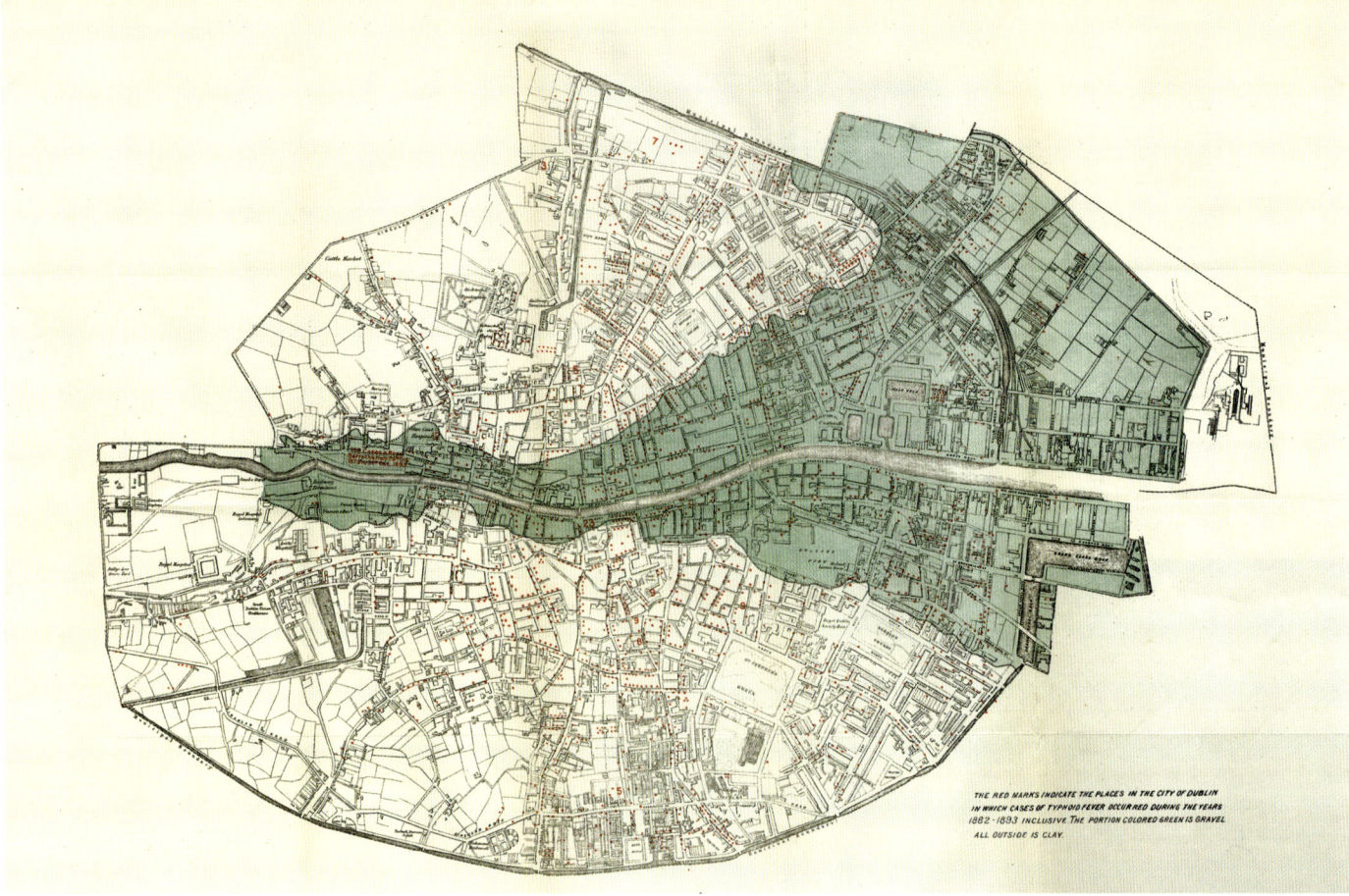

was one in which there was a transition from a belief that many ailments were caused by airborne animal poisons (miasma) to one in which the problem was the spread of bacteria. Their work was reported regularly to the city council via its No. 2 Committee (Sanitary), later its Public Health Committee. With a range of topics that increased over the years, the reports discussed problems with various industries and workrooms, and the remedies applied. Slaughter houses, of which there were many, were a major concern because of their potential to promote airborne infection by these poisons. The increasing numbers of tenements demanded considerable attention, and so, under Cameron's direction, there was a regular inspection system for tenements and lodging houses, judged against published standards with fines for breaches. Each year a report on the state of public health summarised all that had been done and analysed the various data. Tables of meteorological data were included because it was believed (and supported by the data) that the weather played a significant role in the progress of particular health issues.

By the 1890s the role of bacteria in disease was now better understood. However, there was still a general acceptance that the methods and approaches used against 'fever poisons' could also be used against bacteria. This meant clean water, clean buildings and clean air. The annual death rate in Dublin was high – higher than most industrial cities in the UK – and Dubliners had to be concerned with a variety of diseases. For example, Cameron's *Report upon the state of Public Health* for 1893 showed that 7,139 persons had died in the city in the previous year, of which 830 were due to zymotic disease (acute infectious diseases). In the early 1890s, there was a particular concern with typhoid fever, and this caused Cameron to produce the map shown here.

The map provided with the annual report showed that the disease was widespread across the city and not just in the poorer areas, especially to the west of Christ Church Cathedral. It is a dot map, showing individual cases. While this is useful in identifying spatial concentrations, it says nothing about the rate of infection because population densities were much higher in the poorer areas. Cameron expressed the view that fatalities would be higher amongst the lower classes because they did not have the physical strength to fight off the effects. Typhoid fever is a bacteriological infection from the bacterium *Salmonella typhi* and is spread by eating or drinking food or water contaminated with the faeces of an infected person. Cameron's report contained a summary of an analysis undertaken by the Dublin Sanitary Association. He agreed with most of what they said and it was accepted that poor sanitation and poor hygiene were significant risk factors. The soil and ground water were implicated because the bacterium thrived in diluted sewage but Cameron's ideas on how this might influence disease were somewhat different to those of the Association.

The map shows the distribution of gravel and clay soils because the concept of miasma or poor air was still felt to be relevant. The bacterium was endemic in the soil of the city, having been spread by centuries of cess pits and middens. Paradoxically, recent improvement in the sanitation of the city following the introduction of the Vartry water system had actually made conditions better for the bacterium in that it did not thrive in concentrated sewage. As the level of sewage in the soil decreased, so the ability of the bacterium to spread improved. It was everywhere in the city, but since the flow of sewage was easier in the gravels than in the clays, the bacterium did better in those soils. These less dense soils also allowed the bacterium to get into the air and enhanced the spread of the disease, which explained why the occurrence was worse in 1893 than it was in 1891. As Cameron put it: 'The number of deaths caused by typhoid fever, recorded in Dublin in 1893, exceeded the number registered in any previous year. The rainfall in 1893 was extremely small, and the summer was unusually warm and prolonged meteorological conditions which, probably, were the causes of the excess of typhoid fever.' A little further in his report he commented:

> In my present Report I give a map showing the distribution of 3,461 cases. The difference between the gravel and the clays, though shown very decidedly, is not as marked as in the map of 1891. This is probably due to the exceptionally hot and dry year of 1893, in which the superficial layers of the clay dried to a greater extent than usual, and permitted of the escape of the typhoid organism.

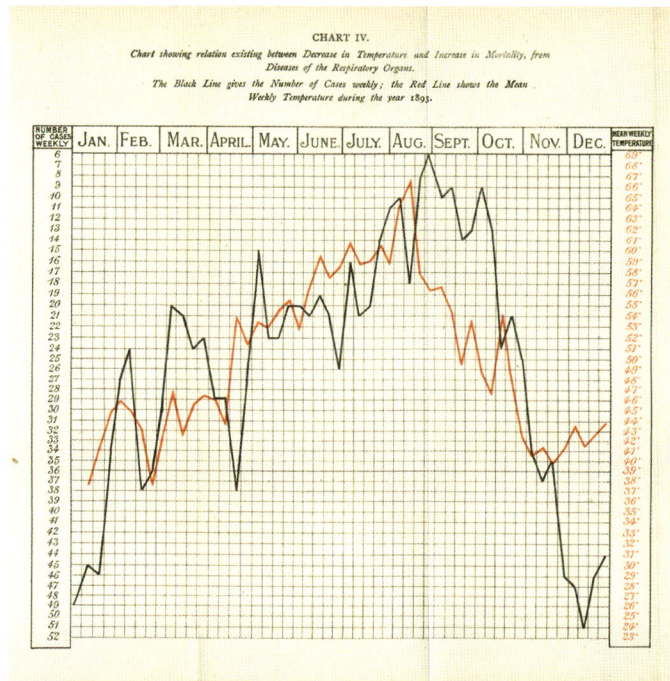

Charles A. Cameron, deaths from 'diseases of the respiratory organs' in 1893, *Report upon the state of Public Health in the City of Dublin* (1893) [AUTH]

This analysis explains his focus on good air, clean water, rubbish removal and on the covering of yards and closes with either concrete or tarmacadam. In the latter case, it was the simple expedient of preventing the organism from getting into the air. Typhoid fever was only one of myriad causes of premature deaths, especially among the working classes, and the death rate remained stubbornly high into the first decade of the twentieth century. This caused Dublin to become increasingly out of line with other cities in the UK and it prompted two major inquiries, one in 1900 and a follow-up in 1906.

# 1900

## The tram service

If the railways were somewhat peripheral to the city centre, the tram system provided excellent connectivity. A horse-drawn tram service began in 1872 from Rathgar to College Green and quite quickly a number of companies emerged, each with a particular spatial focus. Unlike the rail companies which remained independent until 1945, the individual tram companies were quickly amalgamated into the Dublin United Tramways Company in 1881, in which William Martin Murphy was a major influence for many years. A guide for visitors to the Gresham Hotel in 1886 drew attention to routes to places as far flung as Dollymount, Rathfarnham or Blackrock. There were 14 different routes, mostly southside, but it was possible to travel to Drumcondra every 20 minutes from 9 a.m. to 11 p.m. for a fare of 3d. Electrification of the lines began in 1899 and the system, as of 1900, is shown on the map here.

The map was an element in a pocket guide for visitors and was designed to be ephemeral, with new editions produced regularly; it would not have withstood lengthy use. It was folded to 12 × 9.5cm but opened to 48 × 38cm – somewhat unwieldly in outdoor use. On one side was the map showing the various routes, while the other side was devoted to advertising the various attractions of the city. Included also were advertisements for various railway tours offered by rival companies. The map shows that the network was quite dense within the city but that density varied considerably. The coverage was particularly good in the south-east of the city and into the Pembroke and Rathmines township: the better-off parts of the city. It was less dense in the north-eastern part but a tram was in easy reach of most of the residential areas there. In contrast, trams were strikingly absent to the west of the centre. This can be explained in terms of the more difficult topography in the north-west and the relative absence of population. In the south-west there were people but these were

---

Alex Thom and Company, map of Dublin with tramway lines, extract showing the central areas (1900) [AUTH]

The main terminus for trams at Nelson Pillar, extract from an early twentieth-century postcard [AUTH]

the poorest in the city and the tram companies probably took the view that they were less likely to have the necessary fares or the need to travel.

By means of a tram journey, it was possible to enjoy the amenities of Howth, Dollymount, Blackrock and Kingstown, and those reaching Howth could even reach the summit of Howth Head by a tram from Sutton (run by a rival company but not really in competition). Likewise, the traveller could enjoy a different experience and travel to Lucan via the Electric Railway Company, where, after a pleasant 40-minute journey, they could enjoy the amenities of the sulphur spa.

By 1915, there were 21 different routes, travelling on the major roads and many of these routes continue to the present-day, utilising the same numbers for the services. It was essentially a radial system, which suited the city at the time, with its main focus on the Nelson Pillar. This was the terminus for most routes; a role it continued to play until the Pillar was unfortunately destroyed in 1966.

For most people it seems that the day started somewhat later than today, if the times of the trams are anything to go by. The standard timetable for most services from the city began after 8.00 a.m. and Howth could only be reached after 9.00 a.m. Only four trams travelled on the Rathmines line prior to 8.30 a.m. Early services were available from selected locations, reflecting particular business needs. Cars ran from Clontarf at 5.30 a.m., 6.00 a.m. and 6.30 a.m. to Nelson Pillar, returning at 6.00 a.m., 7.00 a.m. and 7.30 a.m. It was possible to get a tram from Donnybrook to Nelson Pillar at 5.30 a.m. every morning except Thursdays when the service was 30 minutes earlier because it ran on to the cattle market on the North Circular Road. There were special cars from Dalkey (4.45 a.m.), Terenure (5.00 a.m.) and Sandymount (5.30 a.m.). In addition, the early cars from Howth to the Nelson Pillar were 'express' cars with only limited stops on the way and no stop until the tram reached Dollymount.

During the normal day, trams were frequent. The service from Rathmines to Terenure ran every three minutes and this was probably the best service. Intervals of between five and ten minutes seemed to be the norm. Cars ran to Clonskeagh every seven or eight minutes and to Palmerston Park or Dalkey every five minutes, but those travelling between Drumcondra and Rathfarnham had to wait 12 minutes between trams. This journey could still be completed for 4d in 1915 whereas it cost 5d to go to Howth and a return ticket to Dalkey could be had for 8d. The shorter lines such as those to Clonskeagh, Palmerston Park or Drumcondra could be travelled for 2d.

It was not only people that travelled by tram, there was an impressive parcel system too. While many businesses in the centre had their own delivery service, it was also possible to have goods sent by tram, and parcels were collected and delivered in Dublin and the suburbs as far as Greystones at a cost of 2d for a parcel of less than 7lbs, while a massive 56lbs (25.4kg) could be transported for 6d. The service was fast and an item ordered in the morning could be expected by the afternoon.

Daily service ended at times very similar to today. The last trams left the Pillar between 11.00 p.m. and 11.40 p.m. with the last services in the opposite direction leaving their terminus at 10.30 p.m. (Dalkey) and 11.13 p.m. (Clontarf). Sunday

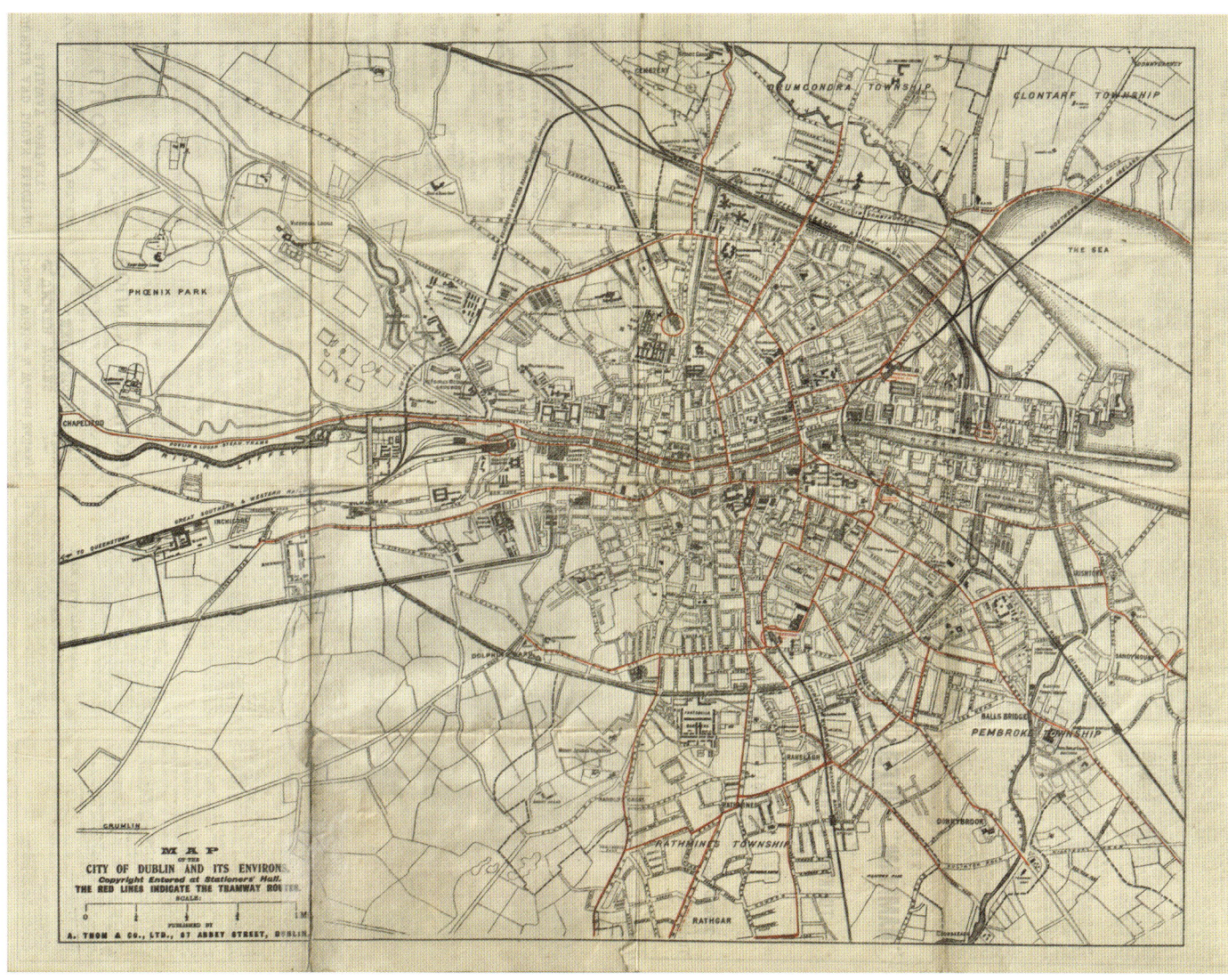

was certainly a quieter day. It being a day of rest there was no service into town until 9.25 a.m. and many did not run until 10.00 a.m. This was the start time for most services from Nelson Pillar, with the Howth service beginning at 10.35 a.m. However, it was recognised that people did not necessarily end their day earlier on a Sunday and most services ended at times similar to those of the working week.

Over time, though, the tram system came to be seen as somewhat of a nuisance in the centre and a hindrance to traffic. Buses gradually replaced the trams because of the need to develop new routes and also because of the flexibility which they offered. By the 1940s only a handful of routes were served by trams, with the greatest number on the routes towards Blackrock and Dún Laoghaire. The final city tram was the No. 8 to Dalkey which left Nelson Pillar at 11.30 p.m. on 10 July 1949. It was stripped bare by souvenir hunters as it trundled along the route. The Hill of Howth tram service survived until 1959.

# 1906

## The high death rate

Unfortunately, Dublin Corporation's early building programme and the public-health policies of Charles Cameron did not reduce the death rate as much as was needed. This prompted two studies of the problem, one in 1900 and one in 1906, the latter mostly an assessment of what had happened to the 1900 report. A committee appointed by the Local Government Board began its work in February 1900 and took evidence at public hearings on 17 occasions over the next two months. They then undertook visits to Manchester and Liverpool to see what had been done there. Part of their focus was an examination of the degree to which the causes of the death rate in Dublin differed from those in other cities. They found that Dublin was better in some regards than others but that diseases of poverty were much worse in Dublin than elsewhere. The death rate from tuberculosis was particularly high, followed by that for enteric fevers. Insanitary conditions, poor housing and intemperance all contributed to the incidence and the spread of disease.

Their recommendations related to ensuring better water and sanitation in tenements, together with the erection of wash houses and swimming baths, the cleaning of streets and lanes on a regular basis, and the removal of rubbish. This was to be done in conjunction with the building of many houses for the labouring and poorer classes. These conclusions cannot have come as any surprise and it is not immediately apparent what the 'added-value' of the inquiry was except, perhaps, to focus attention on the issue.

The 1906 inquiry was undertaken by Surgeon-Colonel D. Edgar Flinn, Medical Inspector of the Local Government Board, with a focus on the sanitary arrangements of the city and their administration. The report is useful for the updated and summary statistics on a range of areas including the work

---

The worst areas of poverty, extract from map to accompany *Report of the Sanitary Circumstances and Administration of the City of Dublin* (1906) [AUTH]

The living and dietary of labouring classes, extract from table for North Anne Street, *Report of the Sanitary Circumstances and Administration of the City of Dublin* (1906) [AUTH]

| Occupation of Head of Family. | No. in Family² | | Average Weekly Income of Family (including Children's Earnings). | Average weekly expenditure on food. |
|---|---|---|---|---|
| | Adults. | Children. | | s. d. |
| Labourer, | 2 | 1 | 12s., . . . . | 6 0 |
| Slater, | 2 | 1 | When working, £1. Very debilitated. | 8 0 |
| Labourer, | 2 | 3 | £1, . . . . | 14 0 |
| Do., | 2 | 5 | 18s., . . . . | 12 0 |
| Do., | 2 | 5 | £1 when working, | 10 0 |
| Do., | 2 | 3 | 17s., . . . . | 9 0 |
| Do., | 2 | 2 | When working, 10s. Son, 5s. | 6 0 |
| Builders' Labourer. | 2 | 2 | £1, . . . . | 12 0 |
| Labourer, | 2 | 4 | 18s., . . . . | 10 0 |
| Do., | 2 | 3 | Husband 12s., wife 8s. (charing), total £1. | 12 0 |
| Handy Labourer. | 2 | 2 | Has not worked for 3 months; 15s. when working. Wife earns 12s. | 8 0 |

of the Dublin Artisans' Dwelling Company. Flinn also provided a damning analysis of why the death rate had remained high despite the 1900 inquiry: 'It must, however, strike anyone who reads the recommendations and the record of what has been done to carry them into effect, that comparatively little has been accomplished in the six and a half years since the report was made.'

His analysis, therefore, did not introduce anything new, but the clarity with which it was presented was welcome. The principal causes of the high death rate were identified as 1) poverty, with its attendant evils; 2) tuberculous disease; 3) intemperance; 4) insanitary conditions under which the poorer classes live; 5) overcrowding; and 6) want of knowledge in the feeding and care of infants. An intensive housing programme, together with improvements to public sanitation (essentially the recommendations of the 1900 report) would do much to address the problem, though the report did not suggest how such might now be accomplished.

There was far less emphasis on the role of weather and the issue around clays versus gravels, though a version of Geological Sheet 112 (1:63,360) was included. The concern was not entirely absent, but Flinn was noncommittal. He noted that:

the accumulation of large quantities of well-water soaking into the soil for so many years, as well as the leakage from defective house drains, must cause a continuous moisture, and insanitary conditions consequent on this soakage must inevitably ensue. How far these conditions may have affected the prevalence of Phthisis, Enteric Fever and Rheumatic affections is a question requiring earnest consideration.

Even more intractable was the problem of poverty. Dublin had many poor people because there was no demand for unskilled labour in what was a service, and not manufacturing, economy. It was estimated at the time that each adult needed to spend three to four shillings per week on food as a bare minimum, while about two shillings was needed per child. As Flinn put it,

owing to the comparatively low rate of wages paid for unskilled labour, the amount of weekly income earned by the ordinary labouring man is, as mentioned before, insufficient to maintain a family in a fit and healthy condition, and young children of tender years have in consequence to bear privation, and even if they are healthy when born the lack of sufficient food soon tells its tale, and such children grow up weak and become an easy prey to disease.

Flinn presented the results of a survey of the income and expenditure of people in the poorer districts to emphasise his point. In one house in North Anne Street, near the Four Courts he found 53 people living there and food expenditure was at or below the minimum. One labourer with a wife and two children reported a weekly wage of 15s, when working, from which he spent 6s on food. The suggested minimum would have been 14s.

All added up to a map of the city showing the area of concentrated poverty. It must be emphasised that this was a map of relative poverty – the worst areas in a city noted for the poverty of so many of its inhabitants. The map was drawn

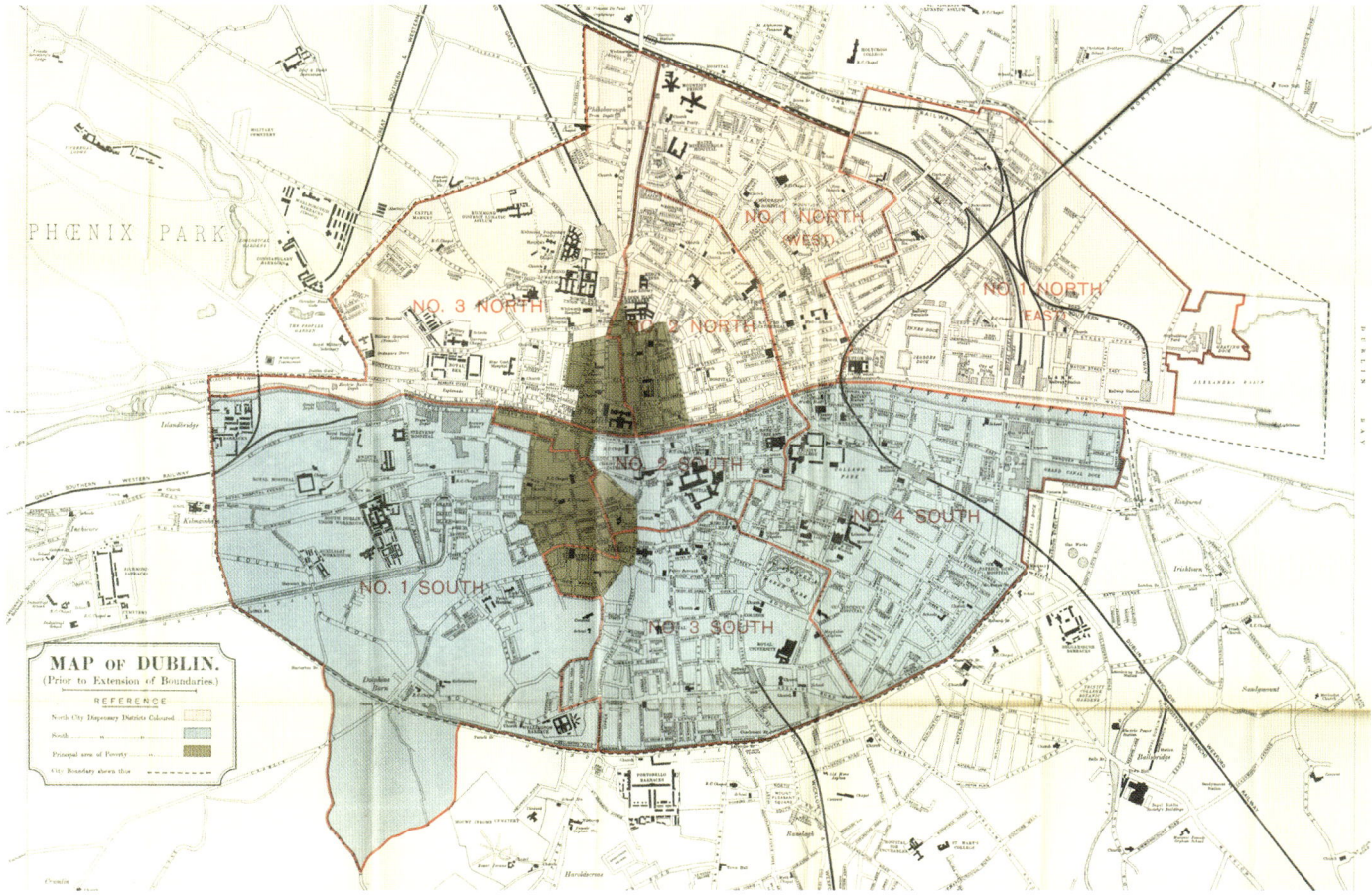

on a Thom's directory map and showed the various dispensary districts in the city. Nobody could have been surprised by the pattern. The worst areas of poverty existed on both sides of the river. On the northside, this was in the environs of the Four Courts and bounded by Smithfield, North King Street and Capel Street. It was a commercial as well as a residential district and included the fish and vegetable market as well as the old Ormond Market. It was busy and bustling but even a casual observer would have noticed how decayed were the houses and how the business of the markets added to the insanitary nature of the many narrow streets. South of the river there was an area of similar size that stretched from the river down beyond the Coombe to Mill Street and which was bounded by Bridgefoot Street and Patrick Street. This was the area captured very well by Brewer in his engraving for *The Graphic* in 1890. There was a dense population inhabiting what had once been merchant's houses but which had been in decay since the early years of the nineteenth century.

Both areas were the focus of efforts by Dublin Corporation. A significant housing scheme was already being planned for the Church Street and Ormond Market areas though it would take another ten years before they were completed. The area south of the river had been identified as needing attention by Dr Mapother some 50 years previously. The Dublin Artisans' Dwelling Company, with the help of Dublin Corporation, had built excellent housing in the previous 25 years but this had done nothing for the plight of the very poor. However, as Flinn was completing his work, two schemes along Patrick Street, focused on those with lower incomes, were nearing completion. The Iveagh Trust had built a flat complex, together with a hostel for single men, a wash house and swimming pool, and an institute for education, while Dublin Corporation had built a similar flat scheme on the adjacent site.

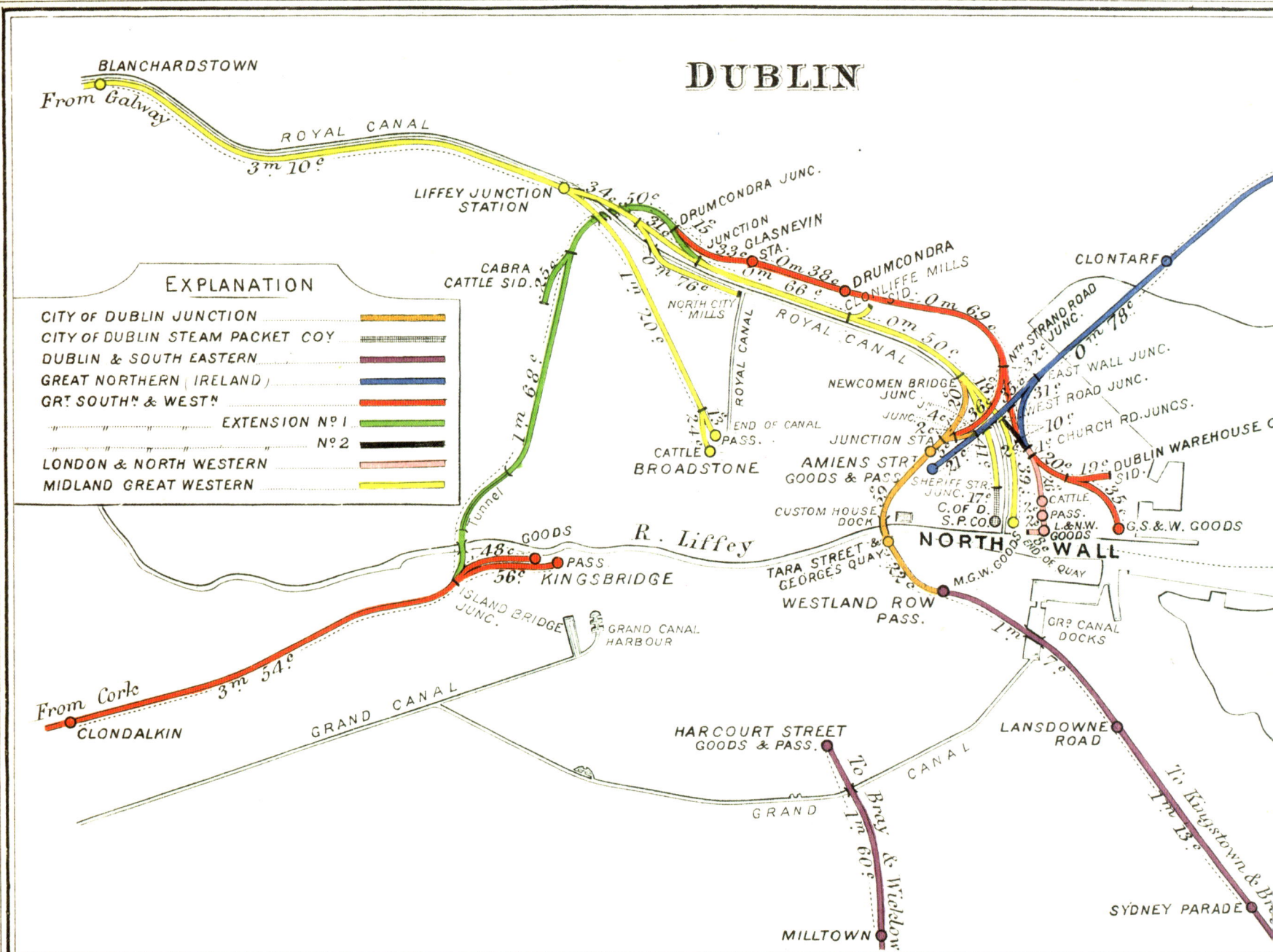

# 1907

## Dublin's railway system

The railway first arrived in Dublin in 1834 with the construction of the Dublin and Kingstown Railway between Westland Row and Kingstown (Dún Laoghaire). By the time the Railway Commission undertook its analysis in 1837–8, interest had grown dramatically in railway building and a large number of joint-stock companies were vying to get permission to lay track. The commission was concerned that this would overheat the market very quickly and that unhelpful competition between companies would be to the detriment of all. They were mindful that the existence of two canal companies serving Dublin had ensured that neither was particularly profitable. The commission therefore recommended two main railway lines, one linking the south and the other going north. They did not object to other linkages but were not convinced by the business case. The line to the south would have its terminus at Barrack Bridge (Kingsbridge) in Dublin. This location was chosen for its relative proximity to the city centre and it was suggested that it would be possible to link it with the Kingstown railway at some point in the future. The railway would link Dublin, Cork, Limerick, Waterford and Kilkenny. The latter linkages involved such an indirect route that it must be wondered how the commissioners felt it would work at all. The northern line would travel via Navan, Castleblaney and Armagh to Belfast. A branch line would link Kells to Cavan and Enniskillen. They reckoned, however, that a line connecting Drogheda to Dublin might be extended to Dundalk and on to Belfast via Newry.

The generality of the commissioner's report was accepted but the railway companies were undeterred and proceeded with their own plans. The Dublin to Drogheda line was completed in 1844 and had its terminus at Amiens Street. The potential routes for this line had been indicated for sometime on directory maps. A coastal route was the least troublesome, with a terminus at Amiens Street, though this meant that its

*Dublin*, Railway Clearing Company (1907) [AUTH]

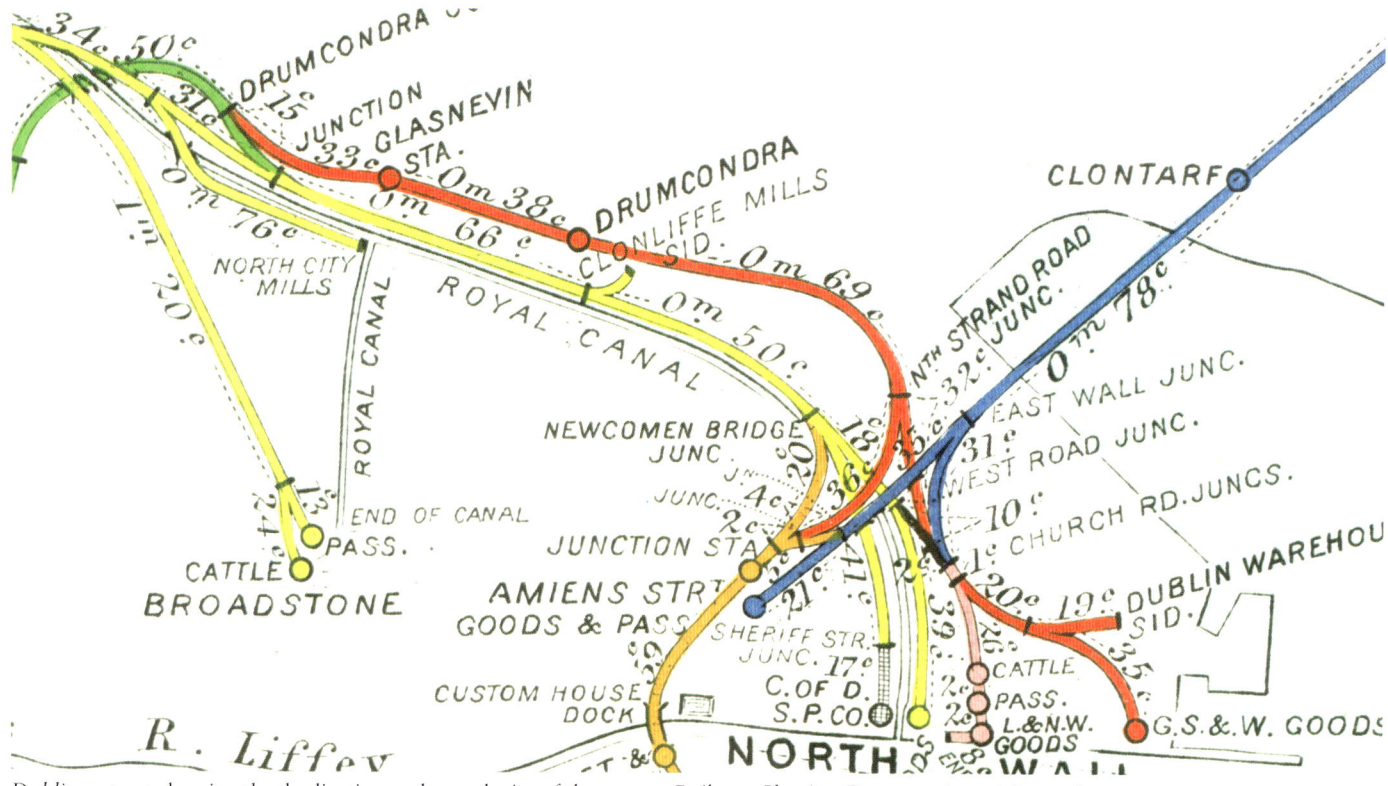

*Dublin*, extract showing the duplication and complexity of the system, Railway Clearing Company (1907) [AUTH]

first passenger station was in Clontarf, on the edge of the township. The terminus was close enough to Sackville Street and the main business district, though not in an entirely salubrious location.

Railways coming from the west or south had difficult choices too. The city began at Kingsbridge with the Guinness brewery on the southside and the great mass of the Royal Barracks on the northside. It would have been difficult, if not impossible, to run the railway any further into the city and it was no surprise that when work began on the Great Southern and Western Railway Company's line to Cork in 1844 that its terminus was Kingsbridge. This network grew by both direct building and acquisitions to become a major transport hub despite being some 2.25km from Sackville (O'Connell) Street. Its great competitor was the Midland Great Western Railway which began its Dublin to Mullingar and Longford line in 1846. The absence of development in the north-western sector of the city and the presence of the North Dublin Union allowed the company to swing its railway line deep into the city. Its terminus at Broadstone was still a considerable walk from the city centre but it allowed for a connection with the Royal Canal, one of the assets of the company. The Harcourt Street line developed into an important suburban railway, though its ultimate destination was Wexford. This was operated by the Dublin Wicklow & Wexford Railway Company which headed inland at Bray. Not only did it manage to get its customers to the edge of the main shopping and business district, it also provided excellent connections for the well-developed and better-off suburbs along its route.

As a result, by the middle years of the nineteenth century, Dublin had five significant railway terminals but they were mostly peripheral to the city and the lines were not connected.

Only two could be said to offer commuter services, both serving the middle-class southern city. If a company wanted to extend its operations though, it had two choices: it could either build its own line, or it could use the network of another company. It often made sense for them to co-operate since this was the cheaper and more convenient approach, but both solutions were used in Dublin and this is where the map becomes a very important document.

The map here is a schematic of the railway system as it was in 1907. These maps were produced regularly by the Railway Clearing Company, a firm designed to facilitate the use of infrastructure by different companies by collecting tolls on behalf of the user organisations. They appeared from time to time at A4 size (approximately) in bound volumes and showed all of the major networks in the UK and Ireland. The ownership of the various lines is shown on the map here, together with the length of the various segments.

Both the Midland Great Western Railway Company and the Great Southern and Western company needed to get rail access to the docklands. This was relatively easy for the MGWR since it could follow the route of the Royal Canal. It was much more complicated, however, for the GSW. They could not extend through the city centre so were forced to build an elaborate loop (involving a tunnel under the Phoenix Park), which from 1877 allowed them to join with the Midland Great Western Railway at Glasnevin, whose line it then used until they diverged at the docks. However, business did not always involve co-operation. By 1901 the Great Southern and Western had decided to run its own line from Glasnevin into the docklands and were able to use their significant political influence to run a line parallel to that of the Midland Great Western Railway. The two lines are shown very clearly on the map and on the ground they are barely 200 metres apart. The GSW decided to open two passenger stations at Glasnevin and Drumcondra and even managed to persuade the Drumcondra township commissioners to give up their new town hall for the station. However, passenger demand on the route was poor and both stations were closed by 1912.

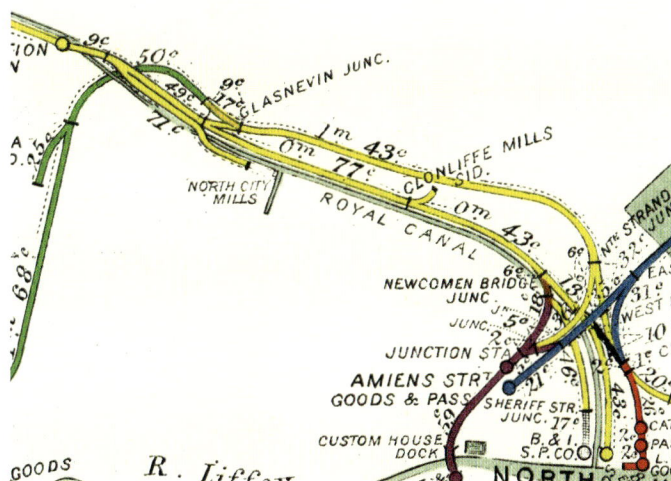
*Dublin*, extract showing the contraction of the system, Railway Clearing Company (1939) [AUTH]

Mail was transported from Kingstown to Westland Row where it had to be unloaded onto trucks and brought to Amiens Street for onwards distribution. The companies decided that this dislocation had to end and this resulted in the City of Dublin Junction railway (the Loop Line) which opened in 1891. There was exploration of an underground connection but that proved not to be a feasible option at the time. The visually striking Loop Line bridge resulted which achieved the desired connectivity at the price of rendering the new swing bridge (Butt Bridge) useless and obscuring the view of the Custom House. That the railway companies were able to achieve this against the wishes of Dublin Corporation was testament to the power which they wielded in the Westminster Parliament.

The 1907 map shows the railway system at its maximum extent. Further editions of the Railway Clearing Company were produced in 1912, 1928 and 1939, but the Railways Act 1924 resulted in the amalgamation of all the railways operating wholly within the Free State as the Great Southern Railways Companies. The 1939 map shows a much simplified system. The Drumcondra and Glasnevin stations were gone, and so too was the Broadstone link and the Royal Canal spur.

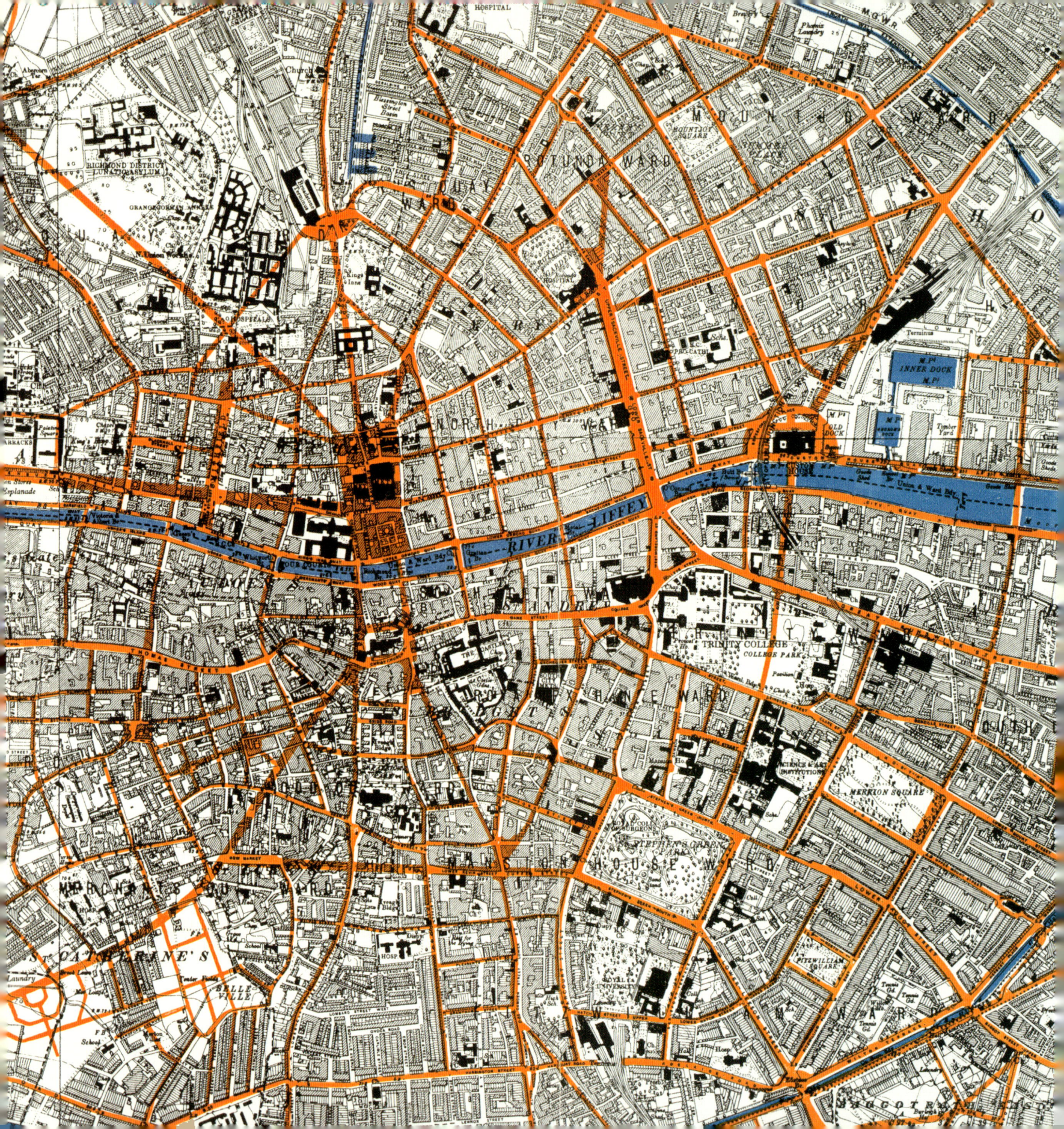

# 1914a

## Dublin of the future: civic projects

In February 1914 the report of the Local Government Board inquiry into the housing conditions of the working classes in Dublin was published. It set out in very stark terms the challenge that city faced to improve conditions for the poor. At the same time in Dublin there was a sense that it was possible for the city to have a much brighter future and this could be achieved by adopting a town-planning approach to its development. This meant taking an holistic view of the city and not simply seeing projects as individual and disconnected entities. The concept of town planning was gaining energy in the English-speaking world. In Britain, Ebenezer Howard's followers were promoting enthusiastically the ideas of garden cities and garden suburbs, while the 'City Beautiful' movement in the United States was advocating the remodelling of central areas to bring beauty and grandeur to the city. Drawing on the Beaux Arts movement in France, it argued that this would not just create a more beautiful urban environment, it would be to the benefit of both citizens and economy.

Dublin was well connected to these currents and in 1914 it was decided to hold a great Civic Exhibition which would demonstrate what could be achieved in many areas of urban life. In conjunction with this event, there would be an international competition for a town plan. This carried a substantial prize of £500 provided by the Lord Lieutenant, Lord Aberdeen, and a panel of influential experts was appointed to judge the entries. The organisation of the event led to the establishment of the Civics Institute of Ireland Ltd, which became a significant voice for coherent planning in the decades that followed. The entrants were asked to consider Dublin under three headings – Housing, Communications and Metropolitan Improvements – and eight plans were received by the close of applications. It took much longer than expected for the adjudicators to come to a final view because of the communication difficulties caused by the First World War – one of the judges lived in the United

---

Patrick Abercrombie, Dublin new and existing streets, extract from *Dublin of the Future* (1922) [AUTH]

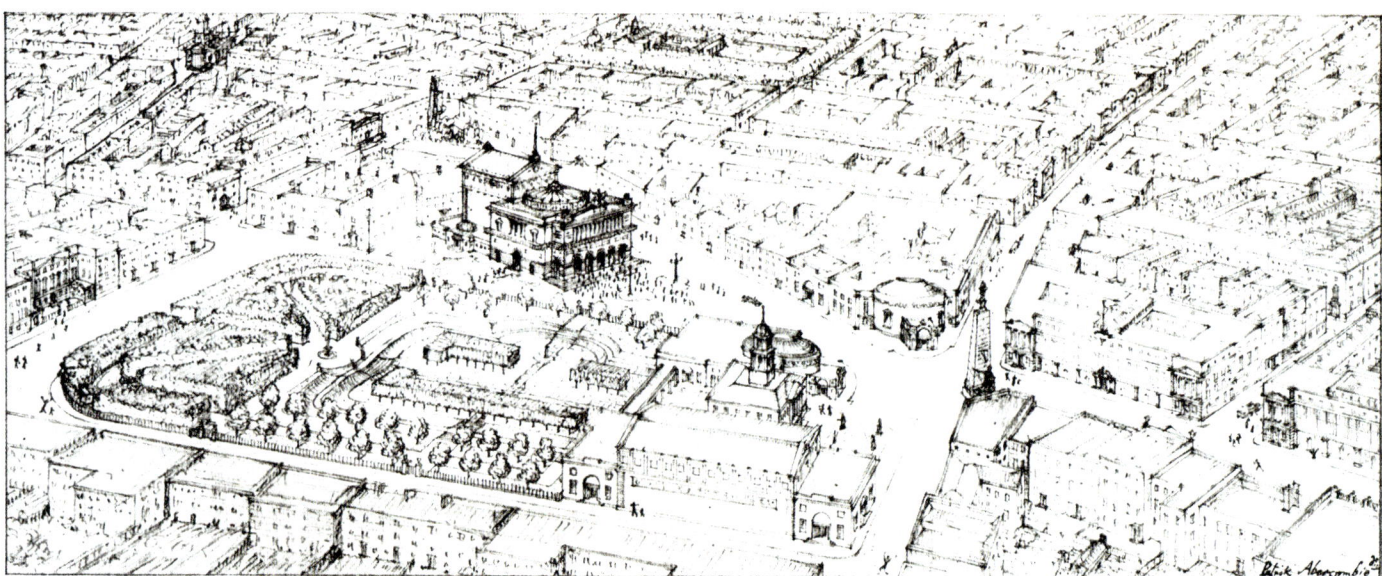

The National Theatre at the top of O'Connell Street, extract from *Dublin of the Future* (1922) [AUTH]

States of America – but they announced in September 1916 that the winner was Patrick Abercrombie of the Department of Civic Design in the University of Liverpool. This began a lifelong association between the city and Abercrombie, and his influence was immense.

It took until 1922 before the Civics Institute was in a position to publish *Dublin of the Future*. It was never suggested however that it be used as a finished town plan. It was a competition entry and was always going to be more flamboyant than pragmatic. Rather, it was hoped that it would stimulate debate and imagination as to what could be done, when circumstances permitted.

There were many ideas in the plan but a key element was that Abercrombie sought to broaden the spatial extent of the city centre by creating an alternative focus in the west of the city centre. Dublin had moved away from its historic core and Abercrombie saw significant potential in reanimating the older area, now greatly in decline and ruinous in places. His plan envisaged splitting the city centre into two distinct focuses. The medieval quarter would be devoted to traffic, and its removal would allow the area around Sackville Street to take on a more civic/cultural role. Even then Dublin had significant traffic congestion and Abercrombie's idea was to improve radial access into the city, not only by widening these roads but also by drawing them all into a massive traffic centre. Some 15 roads would focus on this area around Christ Church and the Four Courts and this would allow traffic to flow around the centre and exit at an appropriate point. Thus, the movement of traffic through the centre would be greatly facilitated. This would be no boring traffic intersection. The schematic provided by Abercrombie envisaged significant civic building and visual focuses, all in the best Renaissance style. The fact that there would probably be carnage in traffic terms was not considered. Nearby would be the great Catholic Cathedral at the Bolton Street end of Capel Street. The higher ground would permit a fine sweep down Capel Street to the river. The cathedral itself would be in classic Renaissance style but with the singular addition of a round tower – after all, this was Ireland.

This would leave a much more serene Sackville Street and the possibility of completing it as the Gardiner family might have wished. It was good practice in Renaissance urban

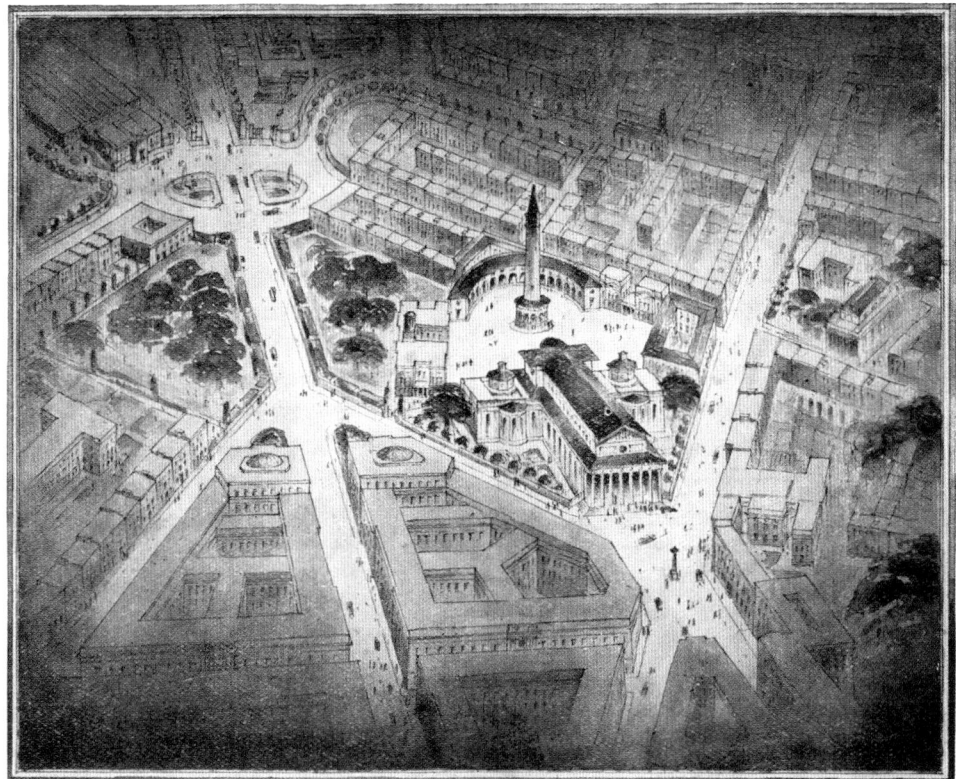

The Roman Catholic Cathedral at the top of Capel Street, extract from *Dublin of the Future* (1922) [AUTH]

planning to terminate a vista with a focus such as a monument or a grand building, depending on the scale. The Rotunda Hospital would have served this purpose but it was offset. Abercrombie proposed that a new national theatre would be constructed on Rutland Square, which would not only be a fine building in itself but would have the width of Sackville Street as an approach. The Custom House was another fine building which would receive special treatment. The Loop Line bridge would be removed, as would Butt Bridge. Instead, a new bridge would focus on the main entrance of the Custom House. Dublin did not have a monumental avenue such as the Avenue des Champs-Élysées but, faithful to the idea of Grand Manner planning, Abercrombie felt that one could be created from the Wellington Monument in the Phoenix Park using an improved Abbey Street to join up with the completed Beresford Place, thus linking the various elements in his plan.

The more prosaic elements in his plan included ideas for housing, and for suburban and regional roads. The housing plan favoured suburbanisation as the solution to Dublin's housing crisis but would have made use of Dublin Bay by filling it in from Sandymount to Blackrock. His transport solutions were comprehensive, and the very fine maps show the impact on the city. While he made use of existing roadways and routes, the impact of his plans on the city centre would have been dramatic. Significant demolitions and realignments would have been necessary but this was justified by the general state of decay.

The plan was never destined to be implemented in an holistic manner but the destruction of much of the city centre during 1916 and 1922 provided greater opportunities to realise some of these ideas than Abercrombie would have envisaged. Unfortunately, this did not occur, but other elements of his plan exercised an important influence on thinking for the next two decades.

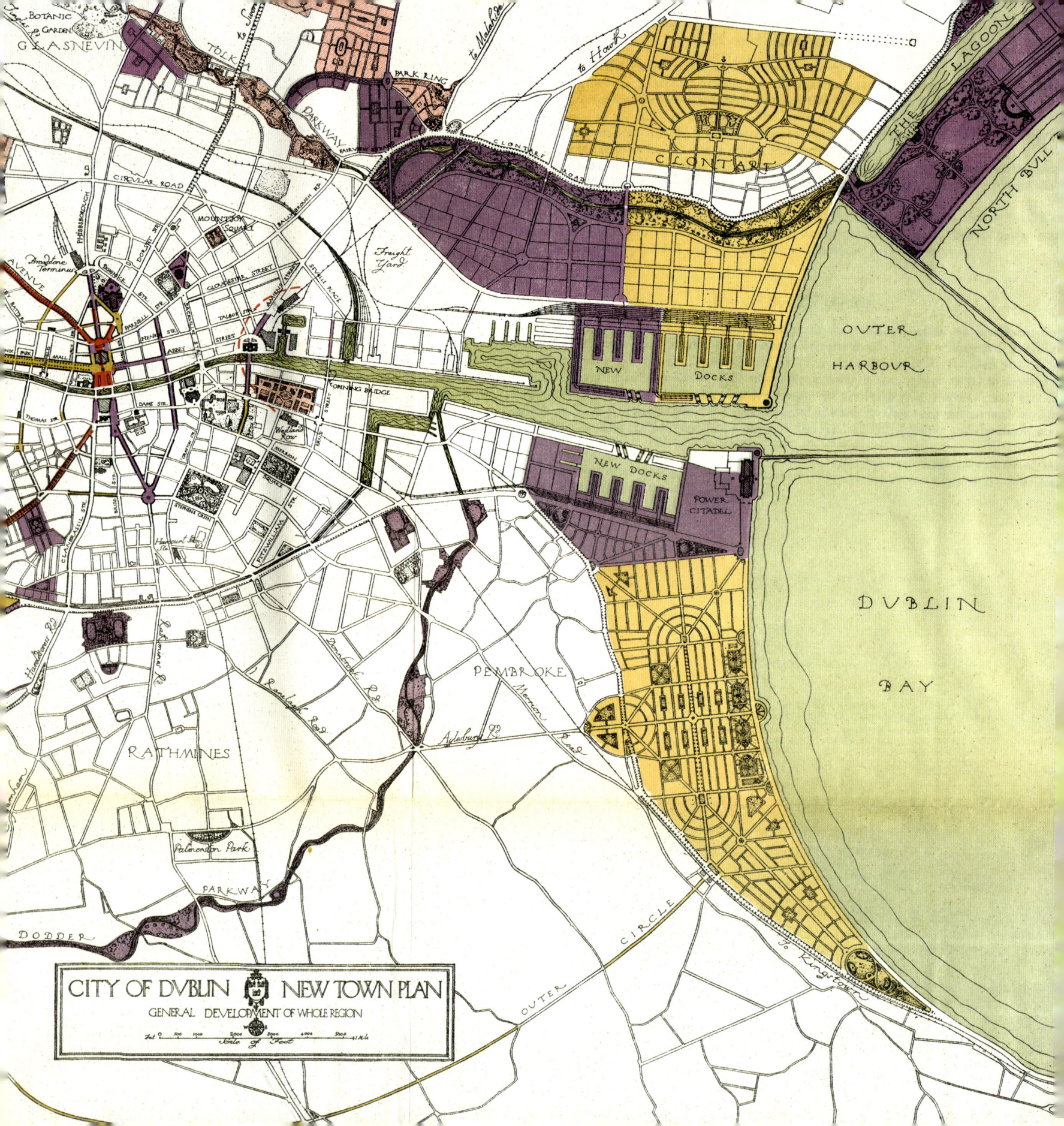

# 1914b

## Dublin of the future: the city view

There were many photographs and diagrams in the printed version of Abercrombie's plan which appeared in 1922. Most were in black and white but five large coloured maps were included. These were 55 × 42cm and were works of art in their own right. Unfortunately, they had to be folded to fit into the book and the rather flimsy paper on which they were printed made it difficult to keep them in pristine condition. Plans 2 and 3 dealt with the tram and railway proposals, while Plan 5 looked at parks near public buildings. Plans 1 and 4, however, were probably the most interesting. Plan 4 attempted a land-use zonation of the urban area, a central element of any development plan today but novel in 1914. While Abercrombie always emphasised that the plan was a menu from which projects could be taken, Plan 1 departed somewhat from this and set out periods of execution for various elements, classified into one of three degrees of urgency.

The most urgent projects were coloured red and pink. Naturally, housing was a priority but so were road improvements. Abercrombie would have been aware that Dublin was in the midst of a housing crisis and the Housing Inquiry of 1913 would have been a hot topic of discussion while he was developing his entry. Though Dublin Corporation had been building social housing since the 1880s, they had only been chipping away at the problem and a massive expansion was needed. One of the lines of debate was whether these houses should be built within the built-up area or whether suburban housing was better. The inquiry had come down clearly in favour of suburban housing, at least in the beginning, on grounds of ease and cost. Abercrombie agreed. While the plan included some proposals for inner-city projects, most of the required housing would be provided in suburban locations. The developments would be low-density, at about 12 per acre,

---

Patrick Abercombie, proposals for the south bay, extract from Plan 1 periods of execution in three degrees of urgency, *Dublin of the Future* (1922) [AUTH]

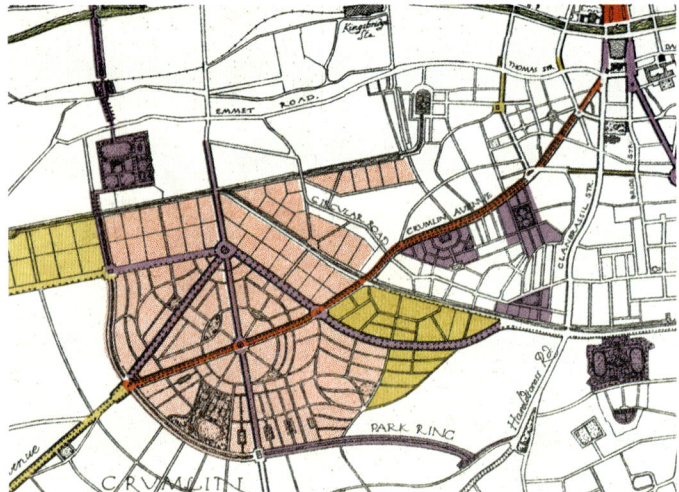

Patrick Abercombie, proposals for housing, extract from Plan 1 periods of execution in three degrees of urgency, *Dublin of the Future* (1922) [AUTH]

the developing standard at the time and one favoured by Raymond Unwin, one of the most influential architects and planners of the era. The sites identified by Abercrombie were ones which must have come up in conversation with officials from Dublin Corporation. Marino and Drumcondra were under active discussion but there had not been much public mention of Cabra or Crumlin. These would become important locations in the coming decades, though it would be the latter years of the 1930s before Crumlin was built.

The suggested layouts show Abercrobmie's love of flowing, sinuous roads within an overall well planned and symmetrical design. There was a lot of 'garden suburb' influence here and while the design might seem rigid from the air, there would have been plenty of variety and open space on the ground. The priority roads were in Crumlin and Cabra. The new housing would need them because these roads were not well developed or finished radials. Abercrombie's view was that the city needed wide radial and circumferential roads to move traffic efficiently around the city before it plunged into the traffic centre. These two radials would become 'super-normal' radials with a width of 120 feet, double that of the standard radial. Their construction would require significant demolitions as would the clearance of the area around Christ Church Cathedral as part of the traffic centre. This was also the moment to get rid of the Loop Line railway, a feature which would ruin his plans for the Custom House.

The second line of priorities were coloured purple and comprised a variety of projects. Further improvements to the road network were suggested for Crumlin and the city centre. The latter were designed to complete the inner city road system that would provide circulation around the city centre. Additions to housing were also planned for Crumlin and Glasnevin as well as the extension of the Marino scheme; the land for this had already been identified and work was afoot to obtain it. This phase would see the beginnings of the civic projects and he intended to start with the Roman Catholic cathedral at the top of Capel Street. This project had captured the imagination of parts of society and there had already been an attempt in Dublin Corporation to make a suitable site available (not the one that Abercrombie had intended). The attempt had fizzled out once the archdiocese had worked out the cost but the project was certainly one that would get public attention and there was a lot of derelict land in the area – much more after 1922.

In this phase, work would begin on the linear parks on the banks of the city's rivers. Abercrombie's idea was to reserve the banks of the river for recreation along the Tolka, Dodder and, indeed, many of the minor rivers of the city. The land was not particularly suitable for other uses, so he felt that a virtue could be made of this by creating a traffic-free continuous space for recreation. He did not appear to think much of Dublin Bay, however.

Visitors to Dublin Bay had been waxing lyrically about the beauty of the prospect offered by the broad sweep of the bay. This was especially so once the danger associated with passage into the bay had been diminished by the building of the sea walls. The shallow nature of most of the bay tempted Abercrombie to fill it in. He would leave just a narrow channel for the River Tolka to flow along the coast road to Dollymount, but would otherwise fill in all of the bay as far as Dollymount and join that area to the extended city by a new

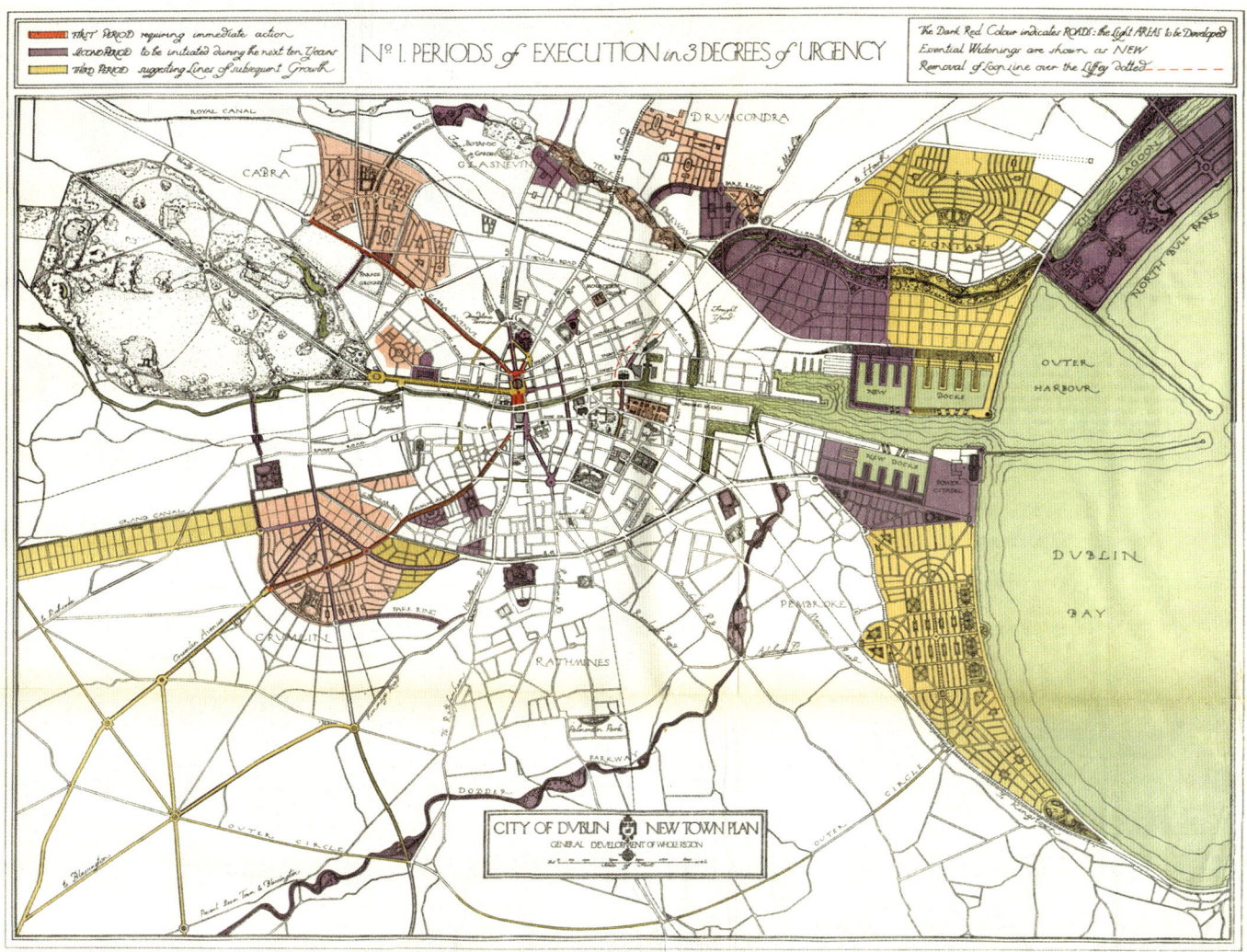

road. This reclamation would be done in phases two and three, with the land being devoted mostly to housing. He also intended taming the Bull Island. It was already an important recreational amenity for the city as a beach, but he intended a more managed parkland.

The southern part of the bay would also be filled in with a curving seawall from about halfway along the South Wall to Blackrock. In the second phase, the emphasis would be on developing the port and new docks would be created east of Grand Canal Docks. It was here that the visually striking power station would be located.

The final phase would see the completion of the infilling of the bay on both sides. Along the new area from Sandymount to Blackrock, Abercrombie envisaged a single residential area in the form of concentric elliptical roads. Green space and recreational areas would be provided between the houses and the new line of the bay. A similarly elaborate housing design was envisaged for Clontarf. The land existed for this – that later developed as St Anne's Park – but there was at this time no suggestion that it would ever fall into Dublin Corporation's hands. Abercrombie was prescient!

This would not be the last time that filling in the bay would be suggested, and many of the other ideas described in this map came to fruition, though not necessarily as he might have proposed. Looking at this map, it is easy to comprehend the influence of Abercrombie on the city of Dublin.

# 1918

## Building the suburbs: north city survey

Dublin Corporation had some significant decisions to make in 1918. Its housing programme had been severely curtailed by the reduction of credit during the First World War. While it did manage to complete some projects, the almost total freeze on state support made it impossible to contemplate expansion of its housing programme – the need for such an expansion having been made clear by the 1913 Housing Inquiry. Though the findings of the inquiry were hardly a surprise, it did have the effect of setting targets for house building, and also suggested very strongly that the only way that this could be achieved, in the short term at least, was by significant building on greenfield sites in the suburbs where land was cheaper. They did not argue against inner city renewal, that could come in time but it was too expensive to be seen as the first or primary solution.

By 1917 the easing of wartime credit restrictions meant that a renewed housing programme could be contemplated. However, there were already ominous signs of price inflation in building materials and a shortage of skilled builders. This quickly had an impact on the reconstruction process in the city centre and, for example, forced the curtailing of the redevelopment of the Earlsfort Terrace site for University College Dublin. Inflation would force a change in the nature of social housing provision but, for now, the question was where to build and how much to build.

In 1917, the Corporation undertook a land-use study of the northern half of the city. This was intended to report in detail on the geographical distribution of the various land uses and confirm likely sites for suburban development. The report was presented to the Corporation by its Housing Committee (Report 13/1918) in 1918. As well as including the maps discussed here, the report gave a brief history of the main

---

Dublin Corporation, tenements in city centre, extract from Appendix 2
of *Survey of the North Side of the City of Dublin, 1918*, Report 13/1918 (1918) AUTH]

# DUBLIN: MAPPING THE CITY

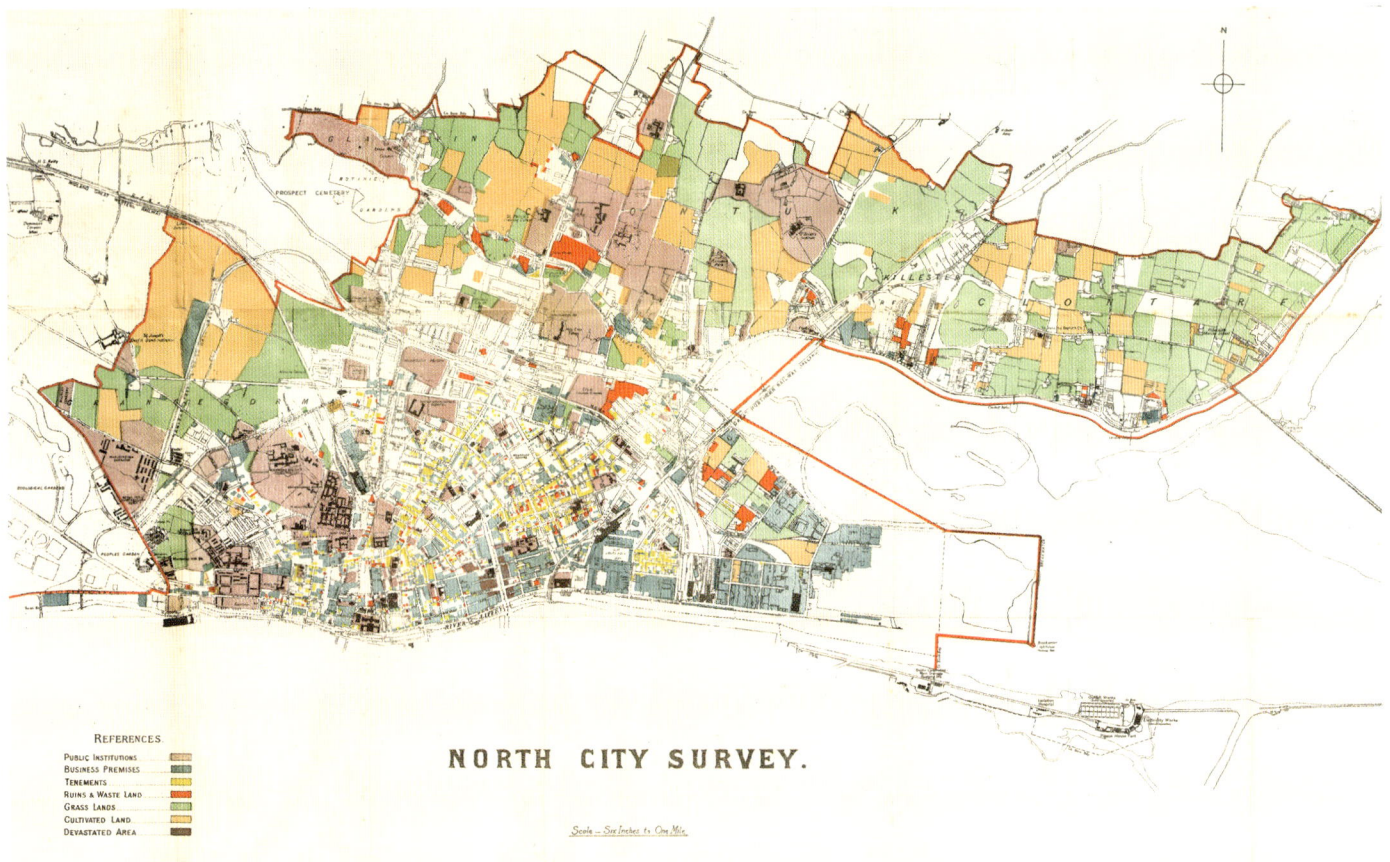

Dublin Corporation, Appendix 2 of *Survey of the North Side of the City of Dublin, 1918*, Report 13/1918 (1918) [AUTH]

streets in the city centre, setting out when and by whom they were built. This was a very useful summary for a reader seeking to get a broad outline of the development of the city.

There were two specially produced maps, both at a scale of 6 inches to 1 mile and in colour, though the printing was not of the highest quality. One map showed the nature of land use – different colours were assigned to categories and the immediate impression was one of green space and institutions with large campuses. It showed that the built-up city was still quite small and compact, and that Dublin Corporation controlled quite a lot of rural land which was under cultivation. There was certainly potential for significant suburban housing within a short distance of the city centre.

Furthermore, a closer inspection showed the widespread distribution of tenements in the city centre. Coloured in a greeny yellow, they appeared in all parts of the city within the North Circular Road. Additional information was provided in the report on individual streets, recording their decline and the rise of tenement numbers. The tenements were now both on back streets and main streets, and were found in the business districts as well as those which might be described as residential. The decline of the Gardiner estate was particularly evident with a high density of yellow to the immediate east of Sackville Street and which stretched to the circular road at Portland Row. The extent of the problem around Mountjoy Square was clear. Of Summerhill it was noted that: 'the street still retained a good class of residents up to the middle of the last century'. There were 'only 3 tenement houses out of a total number of 144 in 1850. Sixteen tenement houses appeared in 1875, 19 in 1900, and at present there are 72 such houses.' Another major cluster could be seen to the west of Rutland Square, reaching to Church Street and beyond, but there was hardly any location where they were not evident.

Together with these descriptions, there was a statistical analysis of what was needed and an outline of the possibilities. From their analysis of the housing stock, they proposed that 627 first-class tenements (those capable of being made fit for purpose) be renovated and adapted to self-contained flats. This would house 2,192 families but it would displace a further 1,599.

Adding these with the numbers in the inferior tenements, the Corporation estimated that the north city would need 10,100 dwellings. This could be further classified as:

- 4,836 families, consisting of one and two persons, requiring at least two-room dwellings.
- 3,174 families of three or four persons requiring three-room dwellings.
- 2,090 families of more than four persons requiring at least four rooms.

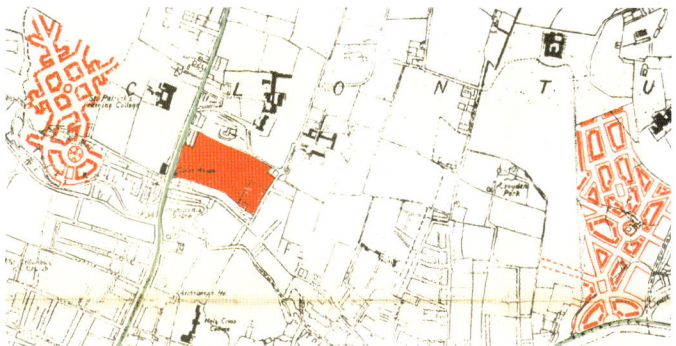

Dublin Corporation, planned suburban housing developments, Appendix 1 of *Survey of the North Side of the City of Dublin, 1918*, Report 13/1918 (1918) [AUTH]

Dublin Corporation, planned central housing developments, Appendix 1 of *Survey of the North Side of the City of Dublin, 1918*, Report 13/1918 (1918) [AUTH]

About 405 acres would be required for this number of houses and the second map showed the location of proposed schemes. A major renewal programme was contemplated for the area east of Sackville Street but the major schemes were planned for Marino, Drumcondra and Cabra, with some schemes in the docklands. These had been under consideration for some time and the Corporation had the specific advice of Raymond Unwin and Patrick Geddes about most of them. A suggested layout was shown for those schemes where some work had been done on their design. Very geometric designs were favoured with a grid being interrupted by squares and circles, somewhat different though to those suggested by Abercrombie. This might suggest that a very rigid and uniform landscape was being contemplated but they would be built in line with garden city/suburb ideas. The designs which had already been submitted to the Local Government Board for the docklands projects at North Lotts and Newfoundland indicated that the housing would be placed within a great deal of green space, and the design of the individual streets would ensure variety in the street line. The geometric patterns would make certain that large numbers of houses could be built without creating a single line of view and the impression of uniformity. The report was duly adopted and Marino, Drumcondra and Cabra were built over the following decade. The stated intention was to undertake a similar survey for the south city but there is no evidence that it was done, nor indeed any mention of it again. Perhaps the Corporation had enough to be getting on with.

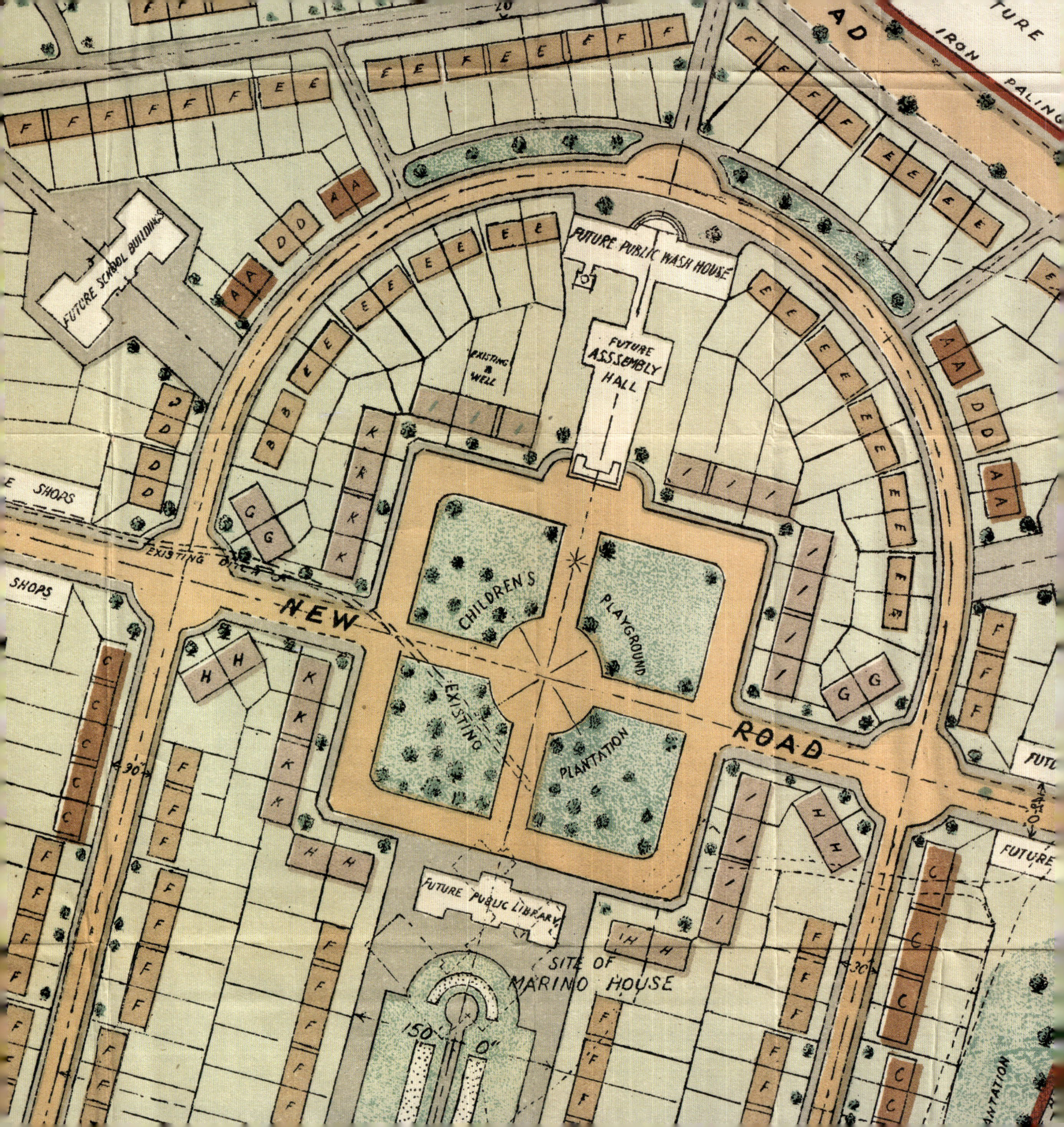

# 1919

## Marino: a model suburb

Marino was one of the sites which had long been on the agenda of Dublin Corporation as a suburban housing area. In 1910, Mr William Walker, a well-known businessman who held a lease on a 50-acre site in Marino, had a proposal considered by Dublin Corporation for a 'miniature garden city' there. Mr Walker was a partner in Sealy, Bryers and Walker, a successful printers and publishers which had premises at Upper Brown Street and 94–96 Middle Abbey Street, but Mr Walker also had a variety of other business interests. In proposing a miniature garden city, he was demonstrating his up-to-the-minute thinking. Ebenezer Howard's book *Tomorrow – A Path to Peaceful Progress*, had been published only in 1898 but the garden city movement which it engendered had been energetic in promoting the concept of garden cities and garden suburbs. Dubliners were well aware of these ideas and of other model villages.

The proposal was considered by Dublin Corporation on Monday 14 November 1910, but did not get very far. The members of the Corporation suspected that the proposal had more to do with Mr Walker seeking to make a significant sum from ground rents rather than with the provision of working-class housing. However, the advantages of the site were recognised and a process was begun to obtain the site for working-class housing but under the control of Dublin Corporation. It was always going to be a project for the middle future because of the scale suggested, but it never fell off the Corporation's agenda. When Dublin Corporation commissioned a review of their future plans in 1914 from the well-known planners, Raymond Unwin and Patrick Geddes, this site was included in their consideration and it was greatly praised. The First World War led to a restriction on credit and any further consideration of the site had to wait. Except for the urgent

---

Dublin Corporation, Marino Housing Scheme, sheet No. 1, extract showing upper square, *Report of the Housing Committee* 210/1919 (1919) [AUTH]

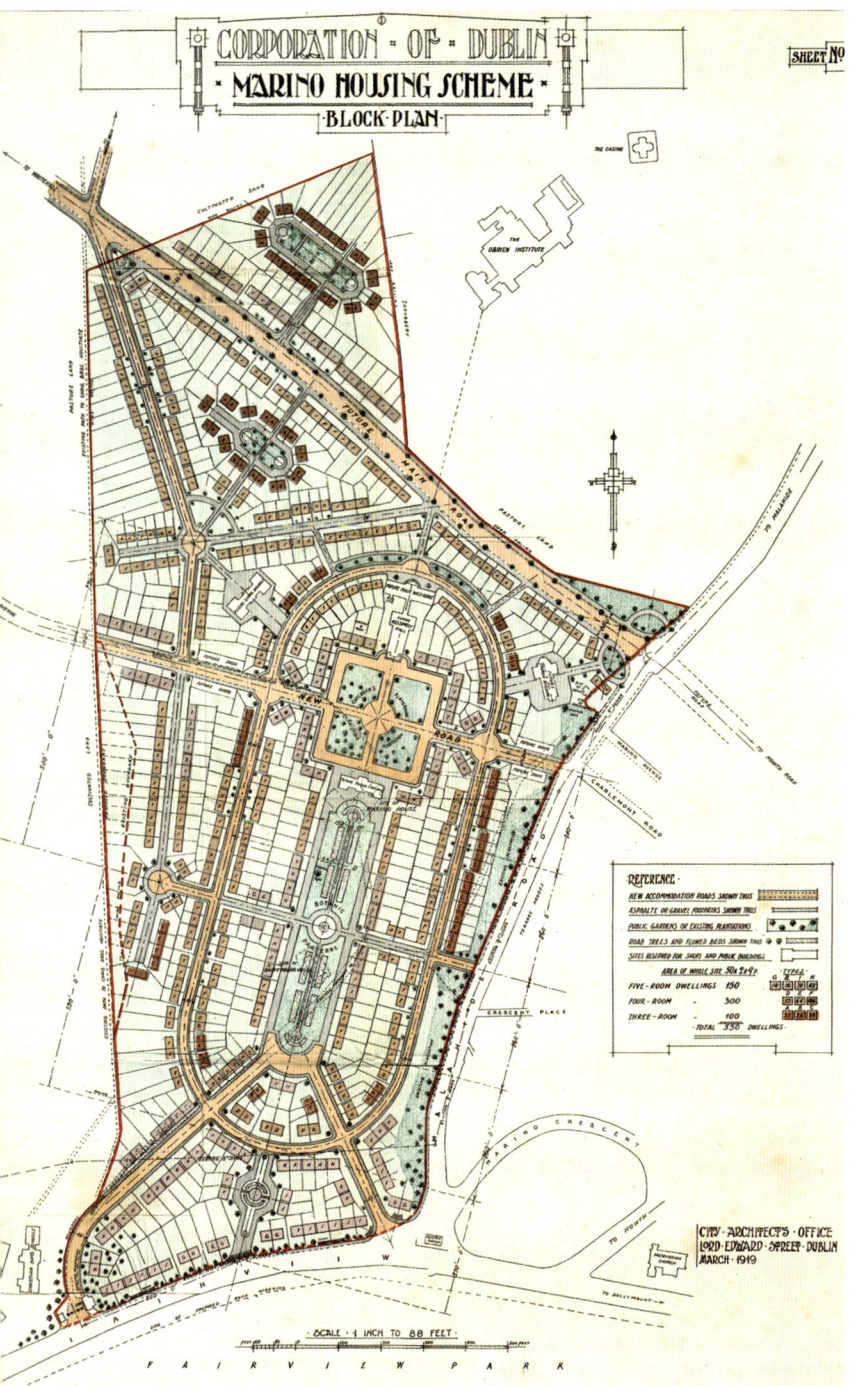

Dublin Corporation, Marino Housing Scheme, sheet No. 1, *Report of the Housing Committee* 210/1919 (1919) [AUTH]

need for housing, it was not a problem for Dublin Corporation to wait. They had come to realise that the land was ultimately theirs, part of the city estate, and it was due to revert to them within years.

The map of proposed schemes, produced in 1918 and discussed earlier, showed an outline of what Marino might look like. By then another influence had been added to the already strong garden city movement. In the UK, the Tudor Walters report, a blueprint for public housing, had been published and, though its remit did not extend to Ireland, its findings chimed with thinking in Dublin. Dublin Corporation would soon decide that the majority of their new houses would contain at least three bedrooms in addition to a good-sized living room, scullery, coal store and wc. Smaller houses would still be provided but it was now desirable to install baths with hot and cold water supplies in all houses.

These considerations influenced the plans which were placed before the Corporation in 1919 (Report 210/1919). The 50-acre site was now back in the hands of the Corporation and they saw the potential to extend the scheme to the west. For the time being, the plan was to build 550 houses, made up of 100 three-roomed, 300 four-roomed and 150 five-roomed. The houses each had a scullery, coal stores, wc, larder and bath, and there were large gardens as well as open spaces, playgrounds, trees and shrubberies. The Local Government Board, whose approval was needed for any housing scheme, reflecting Tudor Walters, now stipulated that there should be no more than 12 houses to the acre, but this plan would have had no more than 11. The scheme was bounded by two existing main roads: the one through Fairview towards Clontarf, and the Malahide Road. A new road was planned for the northern boundary which would link to Glasnevin. This became the 100-foot road, a statement avenue bordered by two lines of trees and named Griffith Avenue.

Accompanying the report was an impressive outline plan for the development, drawn at a scale of 88 feet to 1 inch (1:1,056). While the Hampstead Garden Suburb was an inspiration, this plan owed much to model towns such as Port Sunlight and Letchworth. The main feature was an extended oval, running approximately north–south, with two rings of houses. The outer ring faced onto the surrounding roads while the inner ring faced onto a monumental linear park with a formal square at its northern end. There would be a library and an assembly hall in this park. The original Marino House was now ruinous and would be replaced completely. To the north-west smaller geometric patterns of houses were suggested with intersecting roads to provide an overall focus for the development. The plan was notable for the variety of housing types, not quite as many as provided in Port Sunlight but certainly more than might be expected in a typical local authority housing scheme. There would be ten different house types, four for the five-roomed houses and three each for the four-roomed and three-roomed houses. The larger houses were concentrated along the Fairview edge, the northern edge and around the main square. An interesting effect was obtained by having a particular design for each side of the square. Elsewhere, variety was ensured by breaking up the sight lines by using setbacks and by limiting the runs of individual designs. This was true also of the smaller house types.

The large houses were semi-detached with variation achieved by differences in the roof line, window design and door placement. The smaller houses were terraced but also with similar types of variation. These houses had a covered passageway to allow access to the rear. Within all house types, there were differences in layout and presentation, and included with the plan were schematics for each of the house types showing internal layouts and façades and even a line drawing of one design by H.T. O'Rourke.

The entire scheme was estimated to cost £535,500. The Corporation's analysis suggested that the maximum rents that could be afforded were 10s, 8s and 6s for the five-, four- and three-roomed houses respectively. This came nowhere near the actual costs and it illustrated the difference between post-war incomes and post-war inflation in building costs. This meant that the scheme would require a substantial annual subsidy from the rates. The Corporation took fright at the cost implications and, while endorsing the broad design values, decided to save what it could by concentrating on three

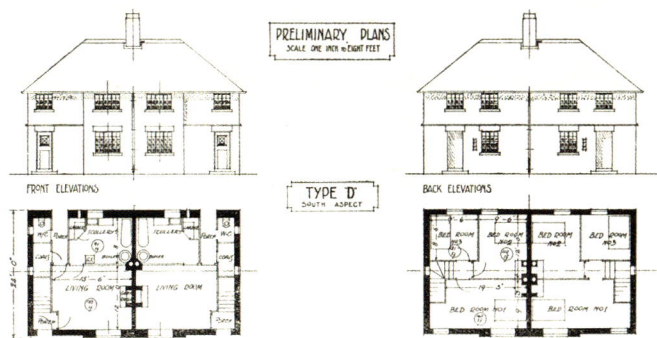

Dublin Corporation, Marino Housing Scheme, sheet No. 3, extract showing Type-D houses, *Report of the Housing Committee* 210/1919 (1919) [AUTH]

housing types only and altering the mix slightly to 111 three-roomed, 291 four-roomed and 148 five-roomed. This would have the effect of saving just over £37,000, which was significant in the context of the Corporation's budget.

It took some years before the scheme became a reality and the final design was different because it proved possible to obtain a much large site – the Croydon extension. Many of the principles of the 1919 scheme however were maintained. There was variety in house types, though all had five rooms, and in the layout of the streets. There was considerable open space, both individual and communal, and density was maintained at 11 per acre. It was considered a model suburb and was perhaps the best scheme that Dublin Corporation built for many years. The biggest change was that the houses were for sale rather than rental. By the time Marino was nearing completion, the Corporation had decided that its suburban housing areas would be for tenant purchase; a policy first implemented at Fairbrothers' Fields. People would buy their houses via an annuity scheme – essentially a mortgage with Dublin Corporation. It restrained the cost of house building as a charge on the city's rates but it restricted social houses to the better-off working classes. It was hardly ideal, and this was recognised but it was seen as the only feasible approach in the 1920s when the state and city's resources were under greater than usual pressure.

# 1924

## Léar Sgáil conntae Bhaile Átha Cliath

The map included here is signed Cathal Mac Dubhghaill and is a rare example of a map in Irish. Born Cecil Grange MacDowell he adopted his Irish name after about 1910, and earned his living as draughtsman in the Engineering Department of Dublin Corporation. He was also organist and director of the choir at St John's Church, Sandymount, from 1900 to 1908. An active Republican who wrote rebel songs, he was interned in Frangoch in Wales, where he drew landscapes, following his service at Boland's Mill during the 1916 Rising. Upon release he was the arranger of *A Soldier's Song*, written by Peadar Ó Cearnaigh to music by Pádraig Ó hAonaigh. The address on the map was given as Gort Glas, An Dumhach – a reference to Greenfield, 28 Claremont Road in Sandymount, where he once lived. There is no date on the map to suggest when it was produced but Mac Dubhghaill was using an older base map. The absence of the Loop Line bridge and the configuration of the rail network suggests that this base was produced before 1890. Mac Dubhghaill's map is similar in design to that included in atlases of the time. One which bears a close resemblance is that included in the *Atlas and Cyclopaedia of Ireland*, produced by P.W. Joyce and published in 1900. This was one of a number of American publications with a political purpose, designed to foster interest in the country and to raise its international profile. The atlas contained essays on each county, discussing its history and geography, accompanied by a map and photographs.

Mac Dubhghaill's map translates as *A map of County Dublin*, though it also provides some information on neighbouring counties, probably for design reasons. It showed the baronies in different colours, though not all of them were named. He included a quite detailed road network, but there was no differentiation and would have created surprise for anyone who used it to plot a journey. The scale was shown in

---

Cathal Mac Dubhghaill, Léar Sgáil conntae Bhaile Átha Cliath (1924) [TCD]

Cathal Mac Dubhghaill, extract showing south city from Léar Sgáil conntae Bhaile Átha Cliath (1924) [TCD]

both English and Irish miles – including the latter was a political statement despite Irish miles still being in use in much of the country, though not in Dublin, until the English measurement became standard in 1926. The rivers, canals and railways were included and named, though railway stations were marked but not named. Many placenames were included in different sized script, presumably to indicate size or importance. What were regarded as significant settlements were shown in large script with an indication of building. Apart from the naming of rivers, very little information was given on the physical geography of the county, except for spot heights. Higher locations were shown both with somewhat exaggerated hachuring and the height in feet. Thus Cip Lubhair is shown at the bottom of the map with a height of 2,483 feet. The modern Irish usage is Cipiúr, but Mac Dubhghaill often used his own version of names.

This copy of the map was heavily revised and it seems that a new edition was being prepared. While most of the revisions are in Irish, some are in English and one, in particular, gives a reasonable indication of the date. Beyond the top neatline at the margin of the map, there is a note 'include 100' road'. The 100-foot road, later Griffith Avenue, was a central element of Dublin Corporation's Marino development. It was intended to provide a suitably grand northern boundary to the scheme which also become an important west–east link from Glasnevin to Clontarf. This would date the revision to the early 1920s as the road project took shape after 1924. It could be a little earlier, if it is assumed that Mac Dubhghaill made the revisions, given where he worked.

Much of the revision comprised the addition of a considerable number of extra placenames. It does not appear to have involved the addition of a new class of placename, but rather an attempt at completeness. Only a few corrections were suggested to the previous edition, such as the change of the height of a hill in north Dublin to 586 feet instead of the 568 feet shown. Some baronies but not all – Uppercross and Rathdown – were marked for deletion, but Cuala, in large letters, would be included in South Dublin (Cuala was a territorial division in Gaelic Ireland south of the River Liffey, encompassing the Wicklow Mountains and reaching as far south as Arklow). There were changes to some existing entries, such as Carraig Dubh (Blackrock) which was marked for promotion in size and a change in location to facilitate the inclusion of Baile Liam. Other location shifts were similarly marked, such as around Greenhills, the intention being to make the text fit better.

Not all the locations, including Baile Liam, Coimead or Fannings Walls (for which there is a note to 'lookup'), would be particularly important at the time and it is unclear why places such as Cluainluibh (Clonliffe) needed to be included at all. Given that there were very few deletions and a large number of additions in the margins which were not marked on the map, it is likely that the completed map would have been quite crowded and fussy.

There was not much change suggested to the infrastructure information. The Loop Line joining Amiens Street and Westland Row stations was added in pencil and there was a suggestion that the railway stations might be indicated by 'stad', but not by name. The Tolka River, already named, was extended closer to its source, and the Pinkeen River was added while the Seantruibh River was now named. The Camac had been included but not named previously and there was a suggestion that it might be named Camóg, followed by a question mark.

There is no indication that this revised map was ever published. Mac Dubhghaill's health had never been good since his internment, where he contracted tuberculosis, and he died in 1926 following an accidental fall from his hotel room window while on a recuperative stay in Nice, France.

Cathal Mac Dubhghaill, title area from Léar Sgáil conntae Bhaile Átha Cliath (1924) [TCD]

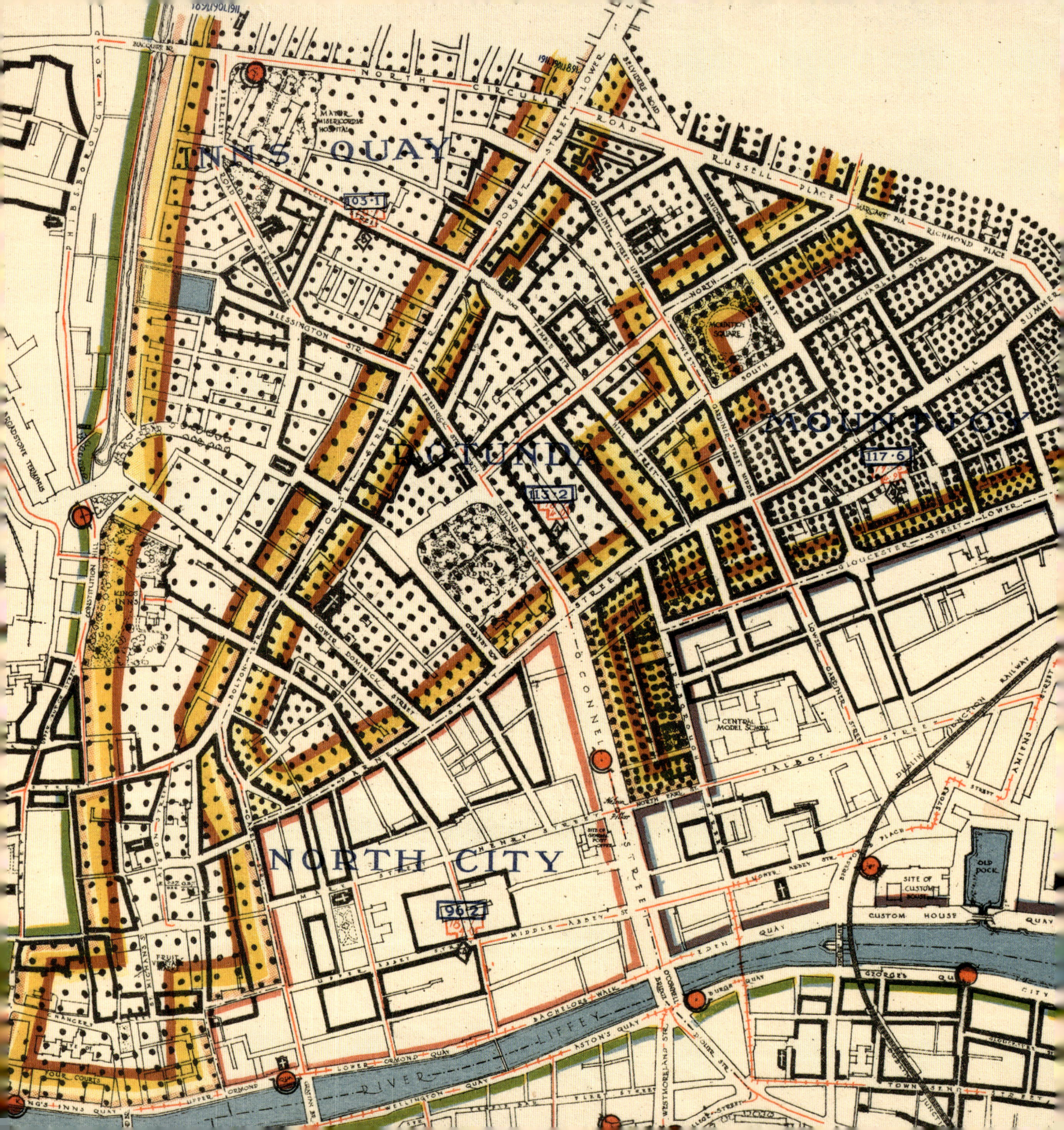

# 1925a

## Hygiene: the Civic Survey

The Civics Institute of Ireland was incorporated in March 1914 with the initial aim of providing a legal framework for the Civic Exhibition that was due to take place during the summer of that year. Its broader aim was to act as a promotor of good practice and ideas, especially in the developing area of town planning. One such activity was the international competition for a town plan which they published in 1922 as *Dublin of the Future*. As discussed earlier, this was not a fully fledged town plan, more flamboyant as a competition entry, but it set out an agenda which could be realised. Before that could happen however, a civic survey was needed to produce a baseline for development. Work on such a survey began in 1923 with the establishment of a Civic Survey Committee whose members were drawn from various professional and social bodies as well as elected representatives. It had the support of both local and national government, and the army provided very useful aerial photography. The chair of the committee was H.T. O'Rourke, the City Architect, and it is clear that he was deeply involved in the day-to-day work of the survey. The survey was published in 1925 and the imprint of the University of Liverpool showed the continuing and developing relationship with Patrick Abercrombie.

The outcome was a report which provided a snapshot of life in the city of Dublin under six headings: archaeology, recreation, education, hygiene, housing, industry & commerce, and traffic. It was a detailed analysis which also set out what the Civics Institute believed was the appropriate policy response to the city's development needs. The book was richly illustrated but of huge importance and interest was the series of maps. These maps were bound into the volume but were also available as a separate portfolio. Those which focused on the city centre were based on the Ordnance Survey 25-inch series while those that offered a more regional view were based on the 6-inch series. Although the maps were very colourful, they

---

'Hygiene', extract showing population density in the north city (1925) [AUTH]

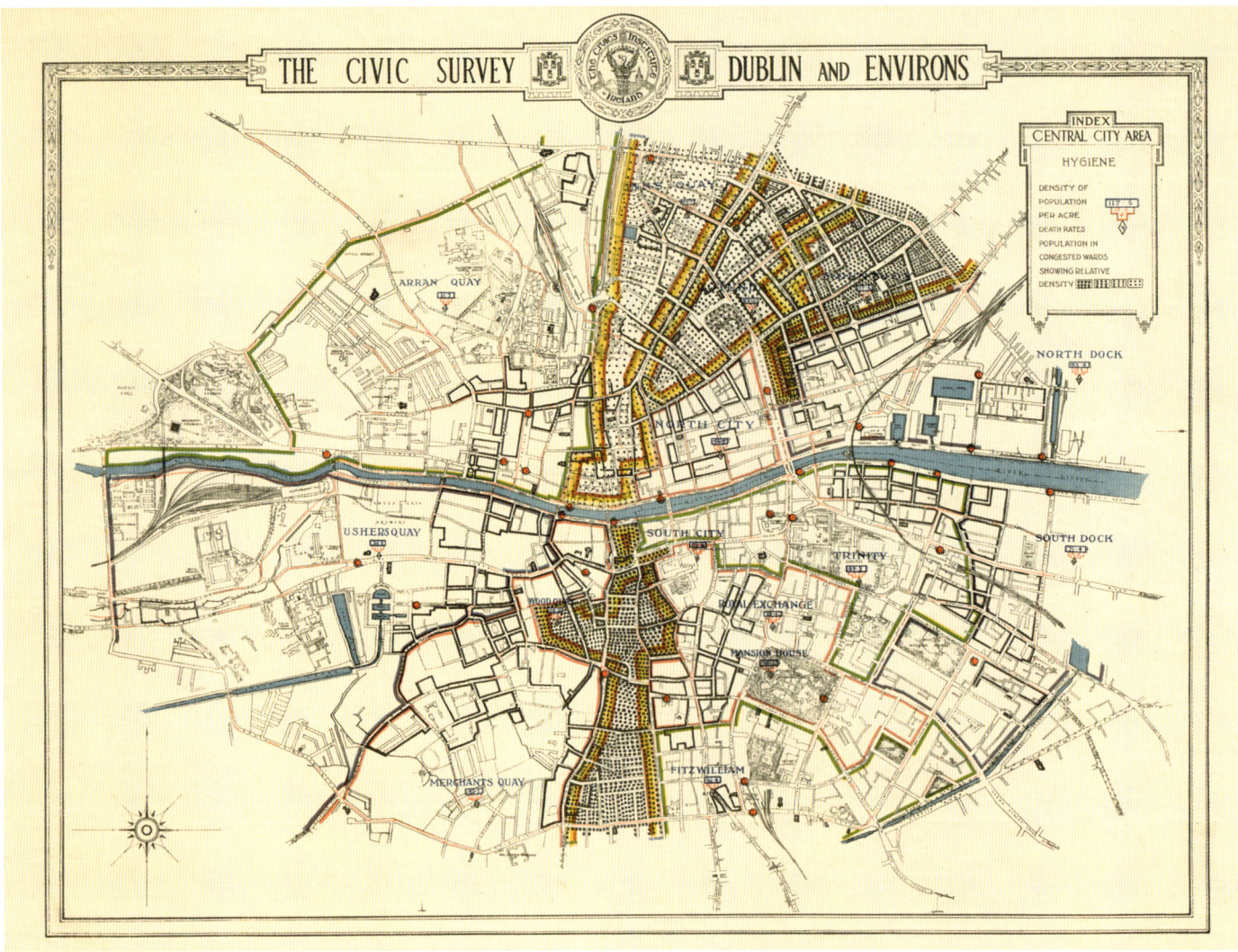

'Hygiene', map from Civic Survey, Civics Institute (1925) [AUTH]

were rather crudely produced and not to the same fine level as commercial or Ordnance Survey maps. Their importance lay in the information they presented and they gave a view of the city and the region unavailable elsewhere. Each map was 56 × 42cm – a useful size allowing a lot of detail to be shown – and the city was contained within an attractive frame with a header which highlighted the Civics Institute, and its crest of the Phoenix and the promise of 'Resurgam' ('I shall rise again'), together with the association with Dublin Corporation.

The map dealing with 'Archaeology' actually outlined the growth phases of the city and showed the area destroyed during 1916 and the Civil War, described somewhat coyly as 'political upheavals'. The 'Hygiene' map provided another perspective on the housing problem by showing population density. This was done by means of a dot shading system showing relative densities. The correlation between areas of poor housing and areas of higher densities was clear. Most of the north-east quarter of the city had such densities with higher densities east of O'Connell Street into Summerhill. South of the river, a zone extending from the river between Clanbrassil Street and Bride Street showed equally high concentrations of people. Solving the housing problem was not just going to be a question of replacing existing dwellings, many additional units were going to be needed to reduce these densities. In addition to the visual, the actual densities per acre and annual death rates were shown for each ward. In the survey's view, the maximum 'hygienic' density in a ward needed to be below 50 to the acre, but in Wood Quay it was 138.3, in Mountjoy it was 117.6 and in Rotunda 113.6. This compared unfavourably with figures in the better-off areas such as 19.8 in Pembroke and 11.9 in Dún Laoghaire. The survey noted that in 1922 the annual death rate per 1,000 population was 17.6 for the city, rising in places to 20 per thousand – compare this to a figure of 11.7 in the 'congested, smoke-laden city of Sheffield'. The yellow tones indicated population growth in the various census periods since 1891, with the darker tone indicating the most recent. While visually striking, this aspect of the map cannot be said to be particularly communicative.

These maps were full of information that was hard to come by and the hygiene map was no exception. No city can exist without an efficient water supply and an equally efficient sewerage system but these are rarely mapped and tend to be taken for granted. This map shows the network of sewers using a fine red line, so fine as to make it easy to miss. The survey was happy with the quality of provision and its view was 'there can be no question that the municipal area is now provided with one of the finest main drainage system in any city of the world'. They were referring to the Main Drainage Scheme which was completed in 1906. This involved building two large interceptor sewers, north and south of the river, into which the existing system was diverted. To ensure a good flow, these had to be buried at a considerable depth; in some cases as deep as 8 metres. The map showed these as a red line with perpendicular dashes running along both quays. The two

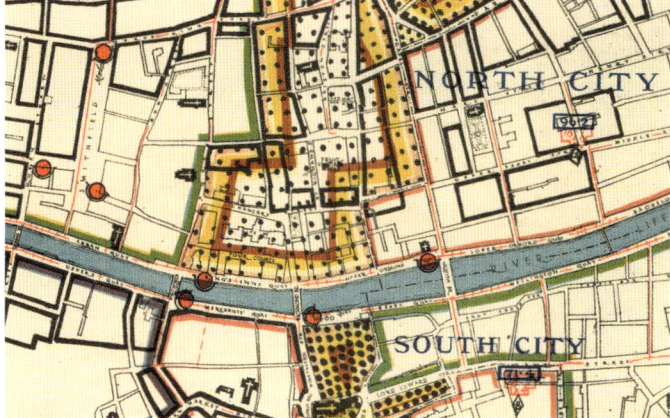

'Hygiene', extract showing main sewers and public toilets (1925) [AUTH]

sewers remained independent until the northern one reached a point near Marlborough Street, from which the map shows a branch heading off towards Clontarf. The north then joined the southern flow at a point near Hawkins Street using a complex and ingenious siphon system that was buried under the Liffey and required excavations of 20 metres in places. From there the pipe took a fairly direct line under Great Brunswick Street (Pearse Street) under Grand Canal Dock and the Dodder. It seems the engineering challenges were significant and the Civic Survey wrote of 'almost insuperable difficulties in construction'. The remainder of the network seems less dense than might be expected and for the most part, followed the line of least resistance, the roads. It seems to have been adequate for the city's needs at the time but it also meant that the housing schemes which were destined to occupy many of the locations shown on the maps needed the additional infrastructure of sewering. For more practical day-to-day needs, the map also showed the location of public toilets. A large red dot showed a network of what were mostly basic cast-iron enclosures open to the weather. The network was not particularly dense but they were regularly spaced along the quays, with some in the central area and with the two most substantial facilities located on O'Connell Street and College Green. It was a lot better than it would be in the years to come.

# 1925b

## Regional transport network: the Civic Survey

The regional maps in the Civic Survey were based on the Ordnance Survey 6-inch series and they followed the same design, with the same elaborate frame, as the city maps. The environs enclosed an area from Balgriffin in the north, Blanchardstown in the west and Cabinteely in the south. The geological map was very colourful but it emphasised the homogeneity of the city area at this broad regional scale. Most of the city was described as having 'thick Limestone boulder clay', the result of glaciation. Along the river there was a narrow band of different types of gravels which widened at the bay into alluvium overlying raised beach. This gave a good indication of the shape of the bay in earlier times and the same kind of configuration could be seen along the banks of the various rivers in the city. Gravels also dominated in the Sandymount area, with 'pale crystalline rock' outcropping at the coast southwards from Blackrock. The regional 'Housing' map further emphasised the scale of the housing problem as the zone occupied so much of the city's area. However, the map also gave hope for the future because it showed the new suburban developments.

The regional traffic map was an overlay of the various transportation modes; a complete summary of how to get around in Dublin. Though there was a traffic congestion problem in the city, and Abercrombie had suggested a radical system of new roads, nothing much had been done at this regional scale. The network was a radial system focused on the city centre, the product of centuries of organic growth. As with the other regional maps, distance from the city centre was shown as a series of concentric circles and these served to remind the viewer how compact the city still was. The greater portion of the built-up area, with the exception of the coastal suburbs, was within 1.5 miles of the city centre; easily within walking distance. The railway system, shown in dark blue, was not greatly different to that of 15 years previously

---

'Traffic', extract showing south city routes and fares (1925) [AUTH]

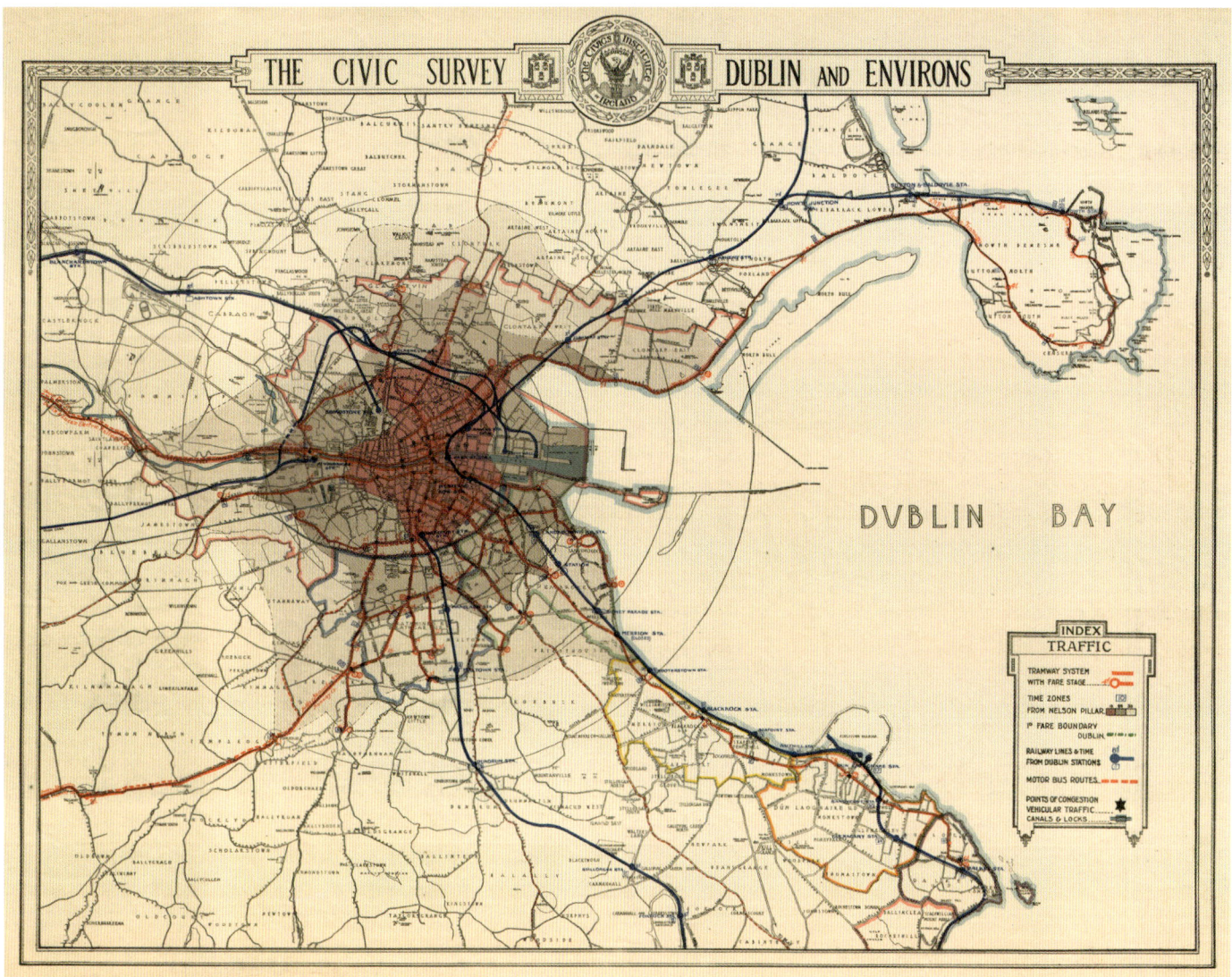

'Traffic', map from Civic Survey, Civics Institute (1925) [AUTH]

(see earlier discussion). Drumcondra Station was now closed, and so too was Merrion Station, but otherwise the same companies were providing the same range of services. The complex arrangement on the northside was due to competition between the Midland Great Western and the Great Southern and Western companies. The coastal line and the Harcourt Street to Bray line were the main commuter routes, and the map showed the cost of a standard ticket from the city centre and the length of time it would take. A sixpence would get a traveller as far as Foxrock or Dalkey; good value and very competitive compared to the tram. The Howth line was more expensive and it would cost almost double to get to Howth Harbour. Travel time was exceptionally good and the map suggested that a train could get a traveller to Howth in 21 minutes, 25 minutes to Dalkey and 18 minutes to Foxrock.

The tram network was more extensive, with a very dense

network in the city centre and lines stretching far out in the countryside. The map showed how far could be travelled in 10, 20 and 30 minutes. From today's perspective, this was pretty good – Drumcondra, Ranelagh and Inchicore could each be reached in 20 minutes from the centre via a relatively slow-moving tram. These times are possible today but only because of the priority measures given to public transport. It was particularly impressive that Templeogue and Booterstown could be reached in 30 minutes. Shown on the map were the route numbers and the fare stages associated with them. In some cases, there was a saving in travelling by tram compared to train. It seemed that Dún Laoghaire could be reached for 5d but it would take over 50 minutes. One useful feature was the 1d fare in the city centre; what in more recent times was called a 'shoppers' fare'. The area within which this was offered was shown as a green hatched line and it covered all of the commercial and shopping area of the city as far as the canals, but did not stray into the mostly residential areas to the west.

Buses had arrived as an alternative to trams with the advantage of being faster and more flexible, though not everyone liked them (see the Dún Laoghaire Civic Survey) because of their noise and pollution. Bus services were unregulated and while the Dublin United Tramways Company (DUTC) was the most important operator, there were quite a number of independent small-scale rivals. Competition between the various companies was fierce and passengers were often in danger of being run down as buses sought to get to them first. This led to new housing estates being supplied with their own bus companies, often with distinctive names rather than numbers. There was 'Adaline' which ran from Walsh Road (Drumcondra) to Eden Quay, or 'St Anthony' which travelled from Oxmantown Road to Aston Quay, though commuters there could also choose the 'Lone Ranger'. Unfortunately, the map does not capture the detail of the routes worked by these operators and shows just the major companies, including the 'Irwin Tours Company Bus'.

The unregulated nature of transport allowed some interesting services to operate. The line of the distinctive Dublin

'Traffic', extract showing services around Howth (1925) [AUTH]

and Blessington Steam Tramway (DBST) cuts across the lower left of the map. This was a steam-powered tram between Terenure in Dublin and Blessington (later Poulaphouca) in County Wicklow which ran from 1888 until 1932. This was a journey of more than 25km and took the traveller deep into the countryside. It could not compete with the Paragon Omnibus Company which began operating a direct bus service between Blessington and the city centre in 1929. The Great Northern Railway company operated an 'electric tramway' on a looped route from Sutton which took the traveller up the Hill of Howth and then down to Howth village. It was an important tourist attraction and it managed to survive until 1959. Not so long lived was the Dublin and Lucan Electric Railway Company which operated a line via Chapelizod to Lucan with a terminus close to the main gate of the Phoenix Park. However, though it is shown on this map, it closed in 1925.

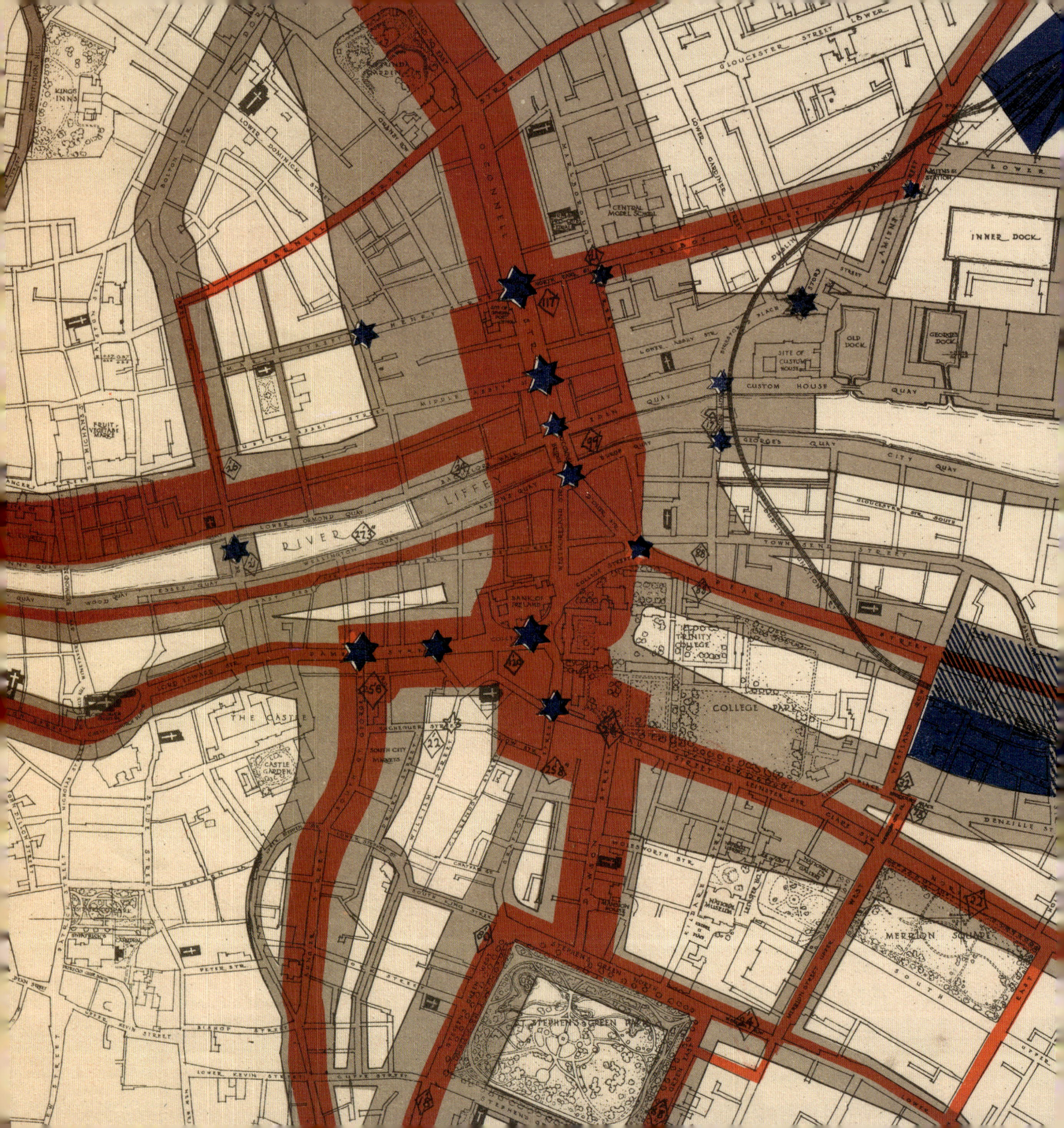

# 1925C

## Congestion and bridges

The central area traffic map in the Civic Survey formed part of a detailed analysis of transport modes and congestion issues. The Dublin Metropolitan Police had taken a traffic census in November 1923 and the Dublin United Tramways Company supplied additional data. Congestion was a major concern and the report looked at the pros and cons of the various transport modes. While there was no doubt as to the value of the trams as a mass-transit mode, the tramways were recognised as major contributors to congestion because they dominated many streets and other traffic built up behind them:

> The chief cause is the tramway system, for they are very large vehicles, and of necessity they must move slowly in narrow thoroughfares. And in addition they must drop passengers and pick them up in the middle of the road. For these reasons tramways are given the highest obstruction unit among all classes of vehicles.

The map showed traffic density in the centre of town, using a concept called 'a unit of obstruction', a measure of the impact of that kind of vehicle. Tram density was shown in red, while grey was used for ordinary traffic and blue to show traffic at railway stations. Blue stars marked the pinch points, the points of particular congestion. Trams were only one contributor to a congestion problem exacerbated by traffic tending to focus on O'Connell Bridge to cross the Liffey while ignoring the bridges to the west. The traffic density map shows that traffic travelled parallel to the river until it reached either O'Connell Street or College Green and then it converged. This street system or axis was one of the great achievements of the Wide Streets Commission and it was as well that the commissioners understood the word 'wide'. However, the axis was now congested and all signs suggested that the problems could only get worse. The problem was exacerbated by the geography of Dublin port and the location of industry in the city. The port

Central City Area Traffic, extract showing congestion points (1925) [AUTH]

Congestion on Butt Bridge in 1929, before it was rebuilt (1929) [AUTH]

operated on both sides of the river but this meant that traffic needing to cross from one side of the port to the other had to travel up and down the quays. Butt Bridge had been built as a relief to the main axis of College Green/O'Connell Street but had turned out to be wholly inadequate. Dublin was not particularly industrialised but its main centre was in the south-west and therefore traffic needed to travel along the same routes to the same pinch points.

Following the Civic Survey, action was indeed taken. Solutions had been discussed and argued about for decades but perhaps the Civic Survey helped because of the clarity with which the report stated the problem. Firstly, there was a reconfiguration of the tramlines on O'Connell Bridge. This might seem like a small thing but it moved the lines towards the centre of the bridge and thus freed up two lanes for other traffic. There were now two tracks on the west side of the bridge, separated from the rest of the traffic by a low barrier. The east side of the bridge now had a single track close to the central reservation, used as a contraflow from time to time. The configuration remained long after the era of the trams and was a feature of the bridge into the 1960s. The second initiative was the widening of Butt Bridge. It had been built as a swivel bridge in order to maintain access to the quays up to O'Connell Bridge. However, the paint was barely dry before the swivel was made redundant by the building of the Loop Line railway joining Amiens Street and Westland Row stations, and it proved inadequate almost immediately because it had only two lanes. There was soon agitation for a replacement, and it was this that produced a proposal for a bridge further east which, if it had come to pass, would have given Dublin a very distinctive piece of engineering.

Doing anything in Dublin with bridges was complicated. Dublin Corporation did not have the sole authority in the matter because bridges were seen as part of the port infrastructure and the port was under the control of the Port and Docks Board. The relationship between Dublin Corporation and the Port and Docks Board (and the other councils in Dublin) had been set out in the legislation which permitted the remodelling of Carlisle Bridge and the building of the swivel bridge. Unfortunately, that relationship had proved fractious, resulting in litigation and inaction. So it was with some trepidation that the Port and Docks Board sought the approval of the Oireachtas (legislation was necessary) to redevelop Butt Bridge. As expected, there was opposition, but nobody anticipated that the President of the Executive Council, W.T. Cosgrave, would become so personally involved. The debate was protracted but in the end the Port and Docks Board found themselves able to proceed with the redesign of Butt Bridge but only if they built another bridge further downstream. This might not have seemed such an issue except that it had to be a transporter bridge. Dublin Corporation would have to provide the necessary access routes, but it had plans of its own.

It remains unclear where the idea for a transporter bridge came from or why W.T. Cosgrave was such an enthusiastic supporter. They were not common either in the UK or the European mainland with no more than two dozen, and their era was over by the time Dublin came to contemplate them. Rather than a fixed structure at ground level, a gantry is built at a significant height above the ground. From this is hung a platform or gondola and it is the movement of the gondola

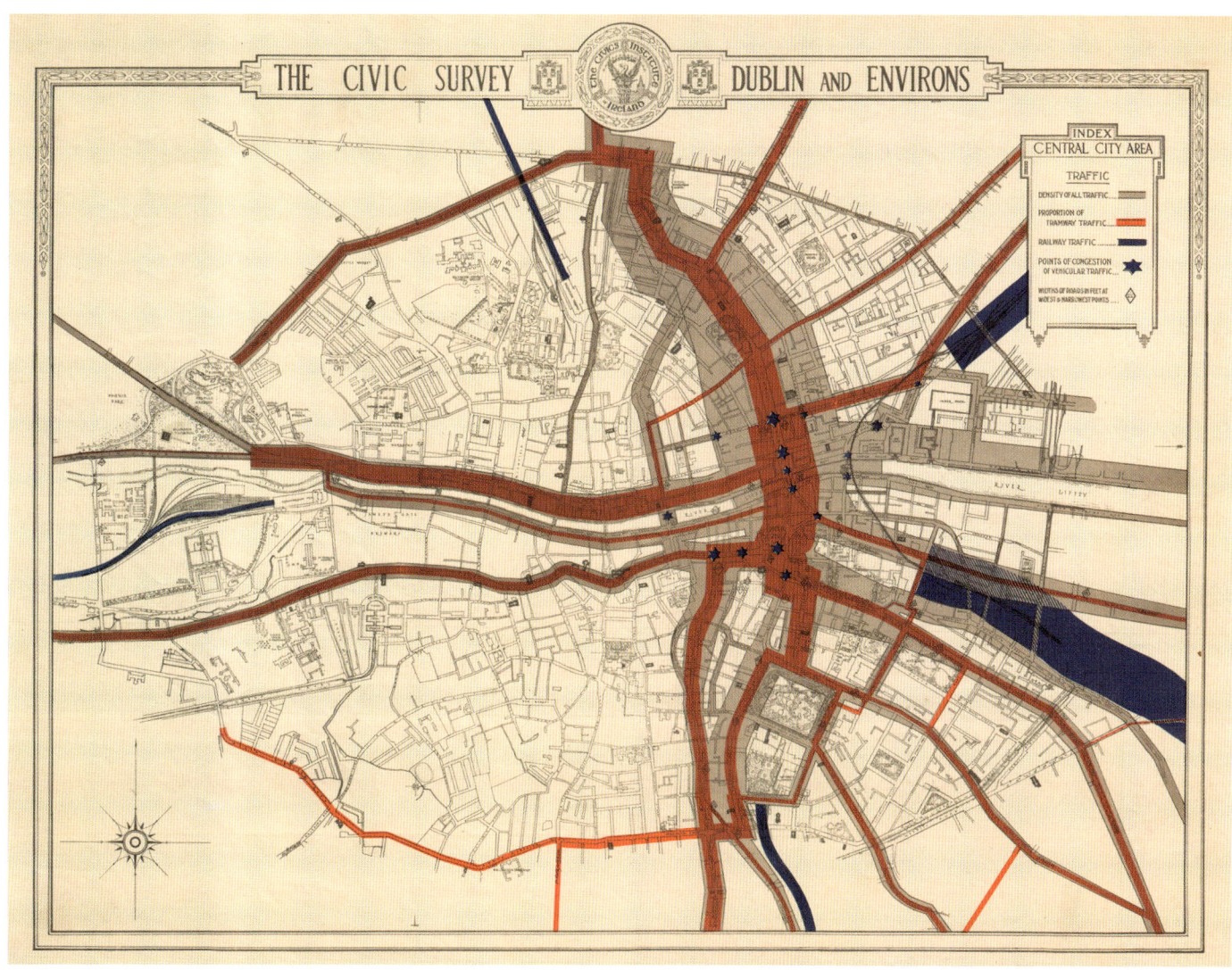

Central City Area Traffic, map from Civic Survey, Civics Institute (1925) [AUTH]

from one side of the river to the other that transports the vehicles. The advantage is that there is no disruption to the river traffic except when the gondola is in motion. The largest surviving is the Newport Transporter Bridge in Wales. Built in 1906, it has a span of 196m (645ft) and a shipping clearance of 57m (187ft).

Neither Dublin Corporation nor the Port and Docks Board wanted to proceed with this project, though all agreed that an additional bridge was needed since it did not have the capacity to move any significant volume of traffic, even if a double gondola was used. Prevarication lasted until the 1950s when the bridge project was absorbed into the wider issue of transport planning. By the time it was finally decided to proceed with the bridge, all notions of a transporter bridge (or its successor idea, a lift bridge) had disappeared and the Talbot Memorial Bridge, completed in 1978, was conventional.

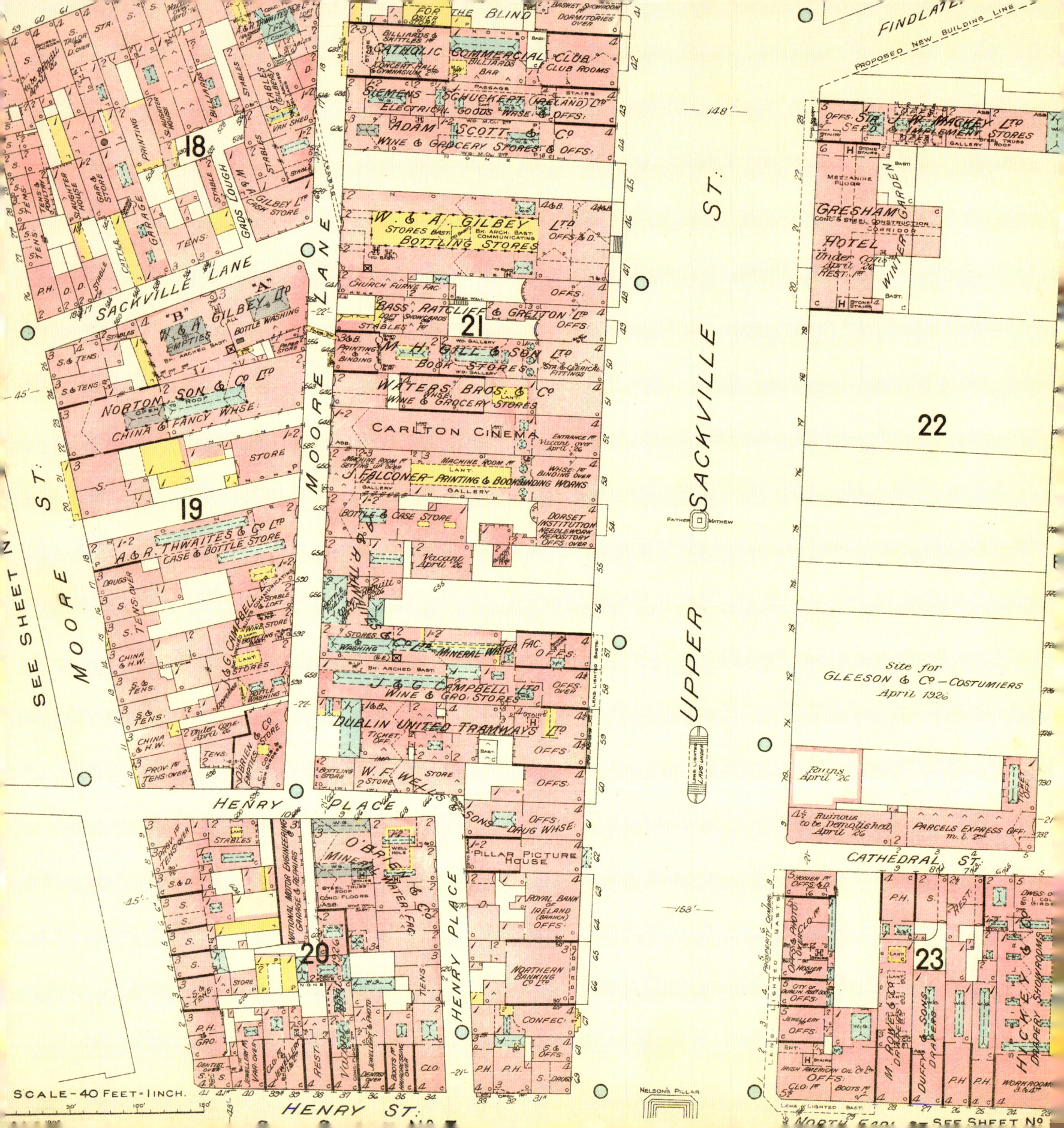

# 1926

## Destruction and renewal: Sackville/O'Connell Street

The Goad fire insurance plan (discussed later) for Upper Sackville Street in 1926 showed a street still in need of redevelopment, with large gaps in the top right of the sheet, between the Gresham Hotel and Cathedral Street. In contrast, the lower left-hand quadrant showed a complete and regular streetscape along Henry Street and its environs. While the destruction of the 1916 Rising had largely been dealt with by 1926, there was still a legacy from the Civil War and it would be some time before the urban landscape was complete. Indeed the Goad plan was wrong when it showed a proposed building line along Findlater Place – Cathal Brugha Street would soon occupy that.

The Rising began on 24 April 1916, and by the end of the week much of the city centre in the vicinity of Sackville Street lay in ruins. The destruction was mainly the result of fires caused by shelling, or looting in some cases. The block containing the General Post Office (GPO) from Middle Abbey Street to Henry Street was almost completely gutted, with damage extending to the other side of each street. Most of the east side of Sackville Street, from Cathedral Street to the river at Eden Quay, was destroyed, and the damage along Eden Quay extended to its junction with Marlborough Street. The buildings lost were mainly commercial and two of Dublin's most exclusive hotels, the Metropole and the Imperial, were destroyed. Also lost was the very distinctive Dublin Bakery Company (DBC) building, which was instantly recognisable in any photograph of the city centre. There were some remarkable survivals. On the corner of Lower Sackville Street and Bachelor's Walk was the business of Kelly and Co., whose façade once proclaimed that it was a Gunpowder Office.

The question of reconstruction was addressed very quickly. The UK government decided to compensate owners from public funds but to manage restoration under legislation. Under the Dublin Reconstruction (Emergency Provisions) Act 1916,

---

Charles E. Goad Ltd, Sackville Street, extract from Goad fire insurance plan, sheet 4 (1926) [AUTH]

Charles E. Goad Ltd, Sackville Street, extract from Goad fire insurance plan, sheet 4, showing destruction in pink following the 1916 Rising (1916) [AUTH]

Dublin Corporation was given powers to assist reconstruction by making funds available at a low rate of interest. However, they also had power to insist that new building did not injure the 'amenity' of the street and so could control in large measure what kind of urban environment emerged from the ruins. They also had power to acquire compulsorily any property by means of an order submitted to the Local Government Board. This was intended to speed things up and get the project going as quickly as possible. All this was managed by a Reconstruction Committee, comprising local architects but with the assistance of Raymond Unwin. Despite the difficulties caused by wartime shortages of workers and materials, the City Architect was able to report progress to the Reconstruction Committee in October 1917. He noted that 87 sets of plans had been submitted and 70 approved. Thirteen buildings had been completed and 44 were under construction. There was good progress on Henry Street, Earl Street and along the quays, but while there were plans for Sackville Street, building had yet to commence.

The issue for Dublin Corporation was the degree to which it was going to influence the nature of reconstruction. The damage offered the opportunity to implement some of Abercrombie's ideas as set out in *Dublin of the Future* (not yet published but widely known). That would have been radical and would have been possible, if expensive, but the business community wanted the street restored immediately. This was understood by *The Builder* in its 12 July 1918 issue which included the plan that is discussed here. In their view, it was a pity that the authorities had not worked to ensure a 'greater architectural unity', but they acknowledged that more had been done than would have been possible 20 years previously. They hoped that they might yet see 'the day when the general appearance and harmonious arrangement of a street as a whole will be the paramount consideration, instead of the individual treatment of the buildings in it'.

The opportunity was taken to widen North Earl Street by taking 14 feet from the southern side as far as Earl Place. The shop frontages on Henry Street were regulated so that the street had a uniform width. Across the area, blocks were required to be one of three heights, 60 feet in Sackville Street and lower in less important streets. Where the frontage of individual premises exceeded 40 feet in width, the architect was allowed to treat his building as a separate unit and adopt his own height, as in Clery's premises on Sackville Street. Where a group of architects co-operated, they were allowed to formulate their own standard for a block, provided it did not interfere with the general appearance of the street. A Georgian style was required but details were left to the architect, though there were controls on the materials to be used in the façades. The corner buildings of important intersections were to have all-stone frontages. The drawing included with the magazine compared what was being built with what had been there

# 1926

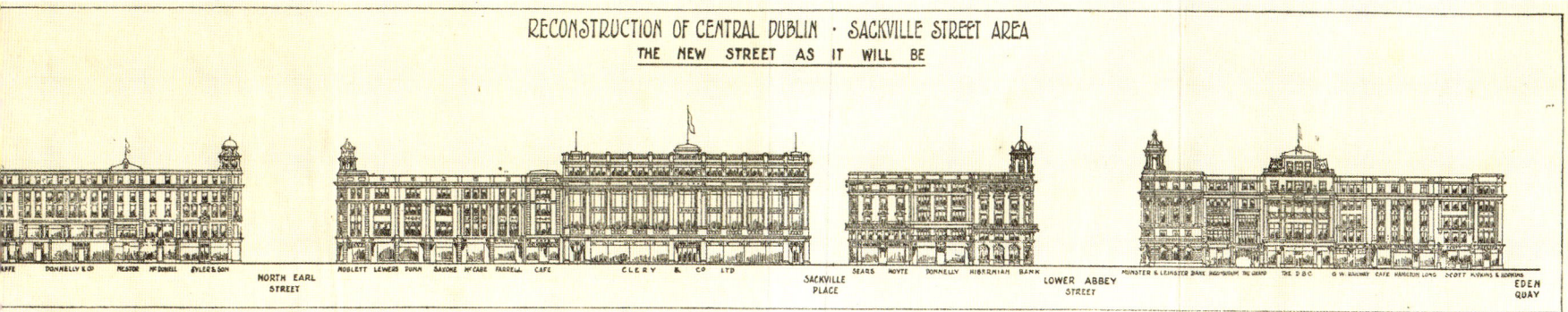

Plans for Sackville Street, *The Builder*, 12 July 1918 [AUTH]

previously. It can be seen that the new buildings were more substantial, though still in proportion to the width of the street. There was an overall coherence to the designs, more so than had been there previously, but more could have been done to ensure uniformity. The flamboyance of what had been the DBC was not restored. A similar approach was taken on the western side of the street and it was reported that the rebuilding of the Scala Theatre and Eason's premises were well underway. However, as with the housing programme generally, the effects of the First World War slowed down the project. It was difficult to raise money, and workers and materials were in short supply with price inflation rampant.

Unfortunately, just as the restoration of the lower part of Sackville Street was nearing completion, the Civil War ensured that the same approach was needed for much of the upper part of the street. A fierce bombardment of the eastern side of Sackville Street over the previous night, resulted in a fire on Wednesday 5 July 1922 which gutted the street from Cathedral Street to Findlater Place. This caused the loss of two of the remaining high-class hotels in the city – the Hammam and the Gresham. The Hammam was an institution in the city, famous for its Turkish Baths. Also lost in the flames were the premises of the Dublin United Tramways Company, Mackeys (the seed merchants) and the high-class grocery business of Findlater. The western side of the street fared a little better in that not all buildings were destroyed; however, serious damage was done to the Edinburgh Hotel, the YMCA and the mineral water plant of Thwaites. The entire shop front of Messrs William Laird and Co., chemists, at the corner with Henry Street, was blown away. This was another city landmark, a long-term feature of postcards of Sackville Street.

One of the casualties was St Thomas's Church on Findlater Place. Though the shell survived, it was decided to cut a new straight street which would link O'Connell Street with the parallel Marlborough Street; this would complement the preexisting Findlater's Lane which ran at an angle. In the space between the two streets, a new church was built and was consecrated on 22 December 1931. The new street was called Cathal Brugha Street long before it was officially named, a phenomenon not all that uncommon in Dublin. Its construction left a vacant site, the space later used for the St Mary's College of Domestic Science, which opened in 1941.

The rebuilding process was long drawn out but was very similar to what had happened earlier with the City Architect having oversight of any plans. The issue was that it was generally felt that the compensation was not adequate to rebuild. The City Commissioners had the power of compulsory purchase but that seemed pointless if there was no interest in building. It was reported in the *Irish Times* in August 1926 that the Hammam Hotel site had been bought and a six-storey building planned, which would see the hotel re-opened. Though the building was completed, the hotel did not return and the only reference to its existence was in the name given to the new block – Hammam Buildings.

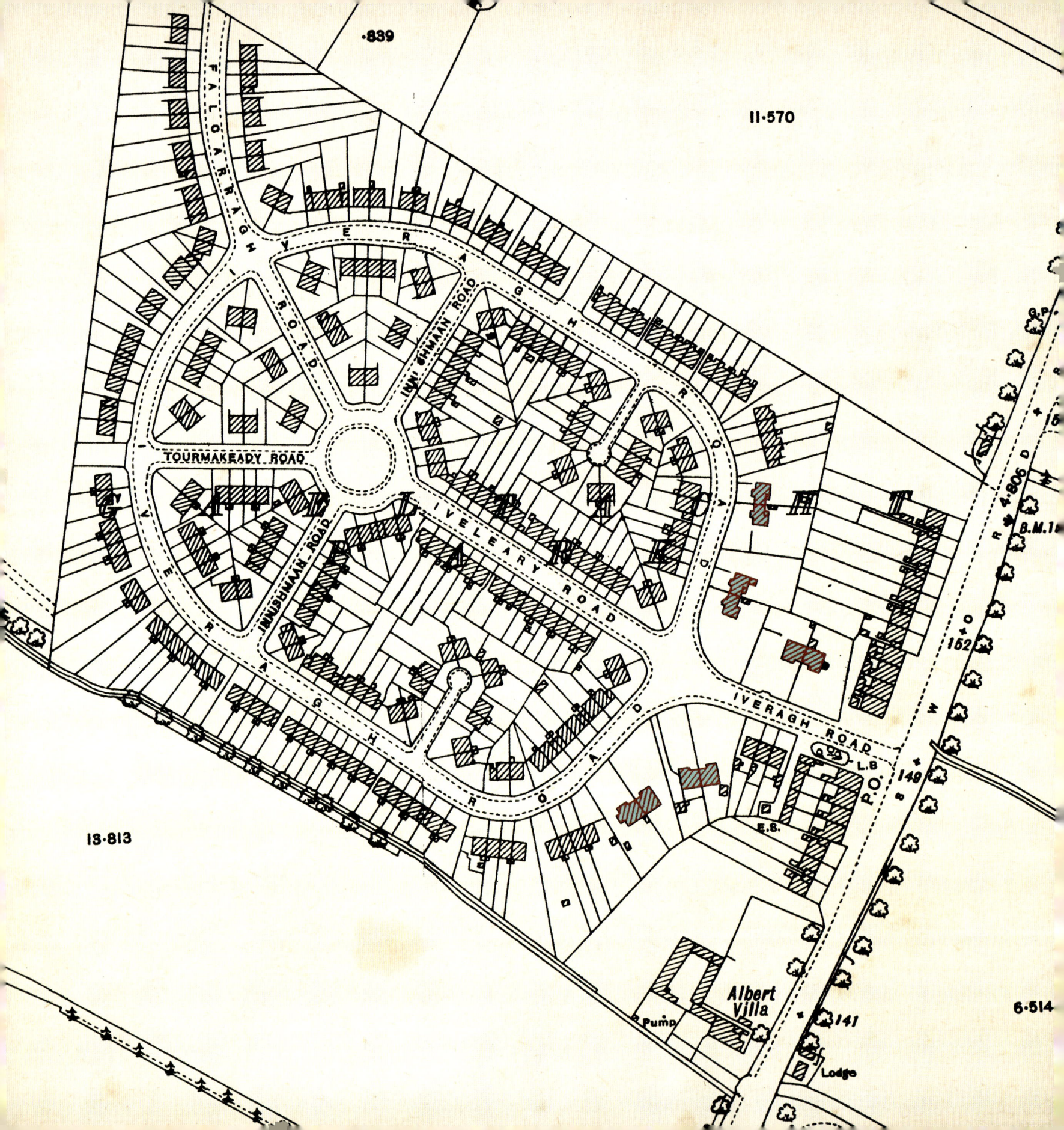

# 1929

## An Irish-speaking colony, Gaeltacht Park

Dublin has a few housing developments that have a shape which allows them to be easily identified on a map or from the air; Marino and Crumlin are two which come to mind. Gaeltacht Park is another which not only has a distinctive layout but, had it been developed as originally anticipated, would have had a very distinctive character too.

One of the concerns of the new state in the 1920s was the restoration of the Irish language. The creation of an Irish-speaking colony was the initiative of An Nua Ghaeltacht Teoranta, supported enthusiastically by General Richard Mulcahy, the Minister for Local Government, and by Dublin Corporation. In 1924, an organisation called An Ghaeltacht sought land from Dublin Corporation on which they would build houses for their members. What they wanted was a low-density development at about five per acre (Marino was 11), which would give the houses very large plots. This suggested a suburban location to Dublin Corporation and they considered a plot of about 39 acres, part of the city estate, located beside the Albert College in Glasnevin. This was being used for farming and could easily be turned to housing – it led directly onto the main road to Swords and there was good connectivity to the city with a tramline nearby. At the same time, the site was separate from other housing areas and the nature of land holding in the area meant it was unlikely to be encroached upon for a while.

It took some time for the necessary arrangements to be made and a public utility society called An Nua Ghaeltacht Teoranta was seen as the most efficient vehicle from a legal perspective, as well as being able to maximise the support funding available. In 1927 the commissioners running Dublin Corporation decided to test the concept and to offer a 2-acre plot for ten houses (Report 100/1927) of two types, costing

---

Gaeltacht Park, Ordnance Survey plan, 1:2,500, sheets 14 (XVI) and 14 (XVII) (1939) [AUTH] The original Gaeltacht houses are shaded in red.

£834 for nine rooms and £634 for seven rooms, about standard for large public utility society houses in the locality.

The foundation stone for Páirc na Gealtachta (Gaeltacht Park) was laid on St Patrick's Day 1928 by General Mulcahy, and a plaque marks the occasion. Work began immediately and it was expected that this would be but the first phase of building. The ambition was for at least 80 houses (with even more in the future), playing fields, two schools and a church. It was going to be a 'colony' and there was a sense of exclusivity about it. People buying houses there had to have shown their 'fealty' to the language and it was expected that most would have originally have come from Gaeltacht areas. Only those who spoke Irish would be entitled to use the playing fields and the residents would do business only with those who could conduct it in Irish. As the *Irish Independent* newspaper put it on 19 March 1928, 'Although the colony will be almost in the city, the residents hope to keep the place as Gaelic in speech and outlook as if it was in the centre of the Gaelteacht in Connemara, Tirconell or Kerry.'

However, this first phase of ten houses was as far as it got. By early 1929 it was clear that the enthusiasm for the project had not translated into funding and the society found themselves unable to build the roads, which was part of their original lease. That might have been the end of the matter but Dublin Corporation decided on a bail-out. In return for the surrender of the original lease, the Corporation gave individual leases to each of ten members of the society at an increased ground rent to cover the costs of the roads. The remainder of the plot was still available but there was no obligation on Dublin Corporation to do anything about it. However, the commissioners decided that they would continue with a housing scheme there and they advertised for tenders in 1930, but got no interest initially. The development could not be said to be rural but it was still somewhat detached from the nearest development on Griffith Avenue. These days, the five-minute walking distance would be seen as trivial but the sense of separation which the colony sought was a barrier to others. This proved to be only a temporary problem and in 1932 the Civil Service Housing Association was persuaded to take the north-east portion of the site where they would build 56 houses at prices of £600 or £650. The Corporation felt it had a good deal since the ground rent was at the same level as the best streets in their nearby Drumcondra scheme, clearly their baseline. Other parties became interested and W.H. Goulding, a builder from 37 Old Cabra Road, applied for a similar size plot on which he would also build 56 houses. Terms were agreed, which included the use of natural slates, and a similar deal was then completed with the Post Office Public Utility Society for a plot on the southern side of the site on which they would build 50 'better class' houses.

So, after a pause, the scheme was completed with its distinctive oval shape and its roads focusing on a small circular park. The original Nua Gaeltacht houses formed the entrance onto the main Swords Road. The extract from the Ordnance Survey 6-inch plan dates from 1938 and shows the completed scheme with the name Gaeltacht Park. The survival of the name to the present day required local knowledge and interest because none of the internal roads carry the name. Instead they make reference to Iveragh, Mayo and Donegal, but it would be difficult to infer any particular association from the combination. It seems that the geographical coherence of the scheme and the involvement of public service utility societies was enough to give a social cohesion to the area. From the beginning, the Gaeltacht Park Residents' Association was active in promotion and defending the area, and over the years they campaigned for a library, better bus services, better lighting and traffic lights as well as expressing concern about suggestions that Albert College might be developed for housing. In recent times, the fashion for identifying the 'villages' within Dublin has seen Gaeltacht Park once again highlighted and now marked with a nameplate at the entrance.

*Opposite top.* Gaeltacht Park in its geographical context, Ordnance Survey plan, 1:10,560, sheet 14 (1936) [AUTH]

*Opposite bottom.* Gaeltacht Park retains its distinctive footprint to the present day, despite the developments around it.

# 1930

## The new suburbs: Bacon's maps

George Washington Bacon was born in the United States in 1830, but moved to the UK where he came a successful publisher of atlases and maps of the UK and Ireland, London and the world. His company, G.W. Bacon, merged with the Scottish company of W. & A.K. Johnston around 1900, but they continued to use the Bacon imprint until the mid-1960s. One of their publications for Dublin was a nicely designed map of the city and its immediate environs. This was published from the 1880s into the 1930s and was intended for visitors, folding down to a convenient pocket size.

As with most such visitor-oriented maps, they were never dated so their age must be inferred from the features that appear. The map shown here was at a scale of 6 inches to 1 mile (1:10,560) on a sheet of about 32 inches square. It folded down to about 7 × 4 inches and there was an index to the streets enclosed within the cover. This size made it convenient to carry but did not help longevity, and many examples are found torn along folds. Of course, this did not bother the Bacon company greatly since the maps were designed to be ephemeral, but they did offer an edition which was backed on linen. It covered an area from Drumcondra to Clonskeagh and east from Crumlin – most but not all of the built-up area at the time. It was nicely produced with main roads in yellow and the buildings as narrow blocks in brown. Roads were named in a clear font and some editions, such as this one, showed the bus and tram stops with the route number indicated.

This map is from the early 1930s and it shows the transformation of the north city. Over the previous decade, Dublin Corporation had built a number of distinctive suburbs. While they had been busy on the southside, this was less obvious on the map and it would be well into the 1930s before building would commence at Crumlin. Marino was the most distinctive of these northside developments and the map shows its geometry very clearly. It was organised around two circular

G.W. Bacon and Co. Ltd, *Bacon's plan of Dublin and Suburbs* (1930) [AUTH]

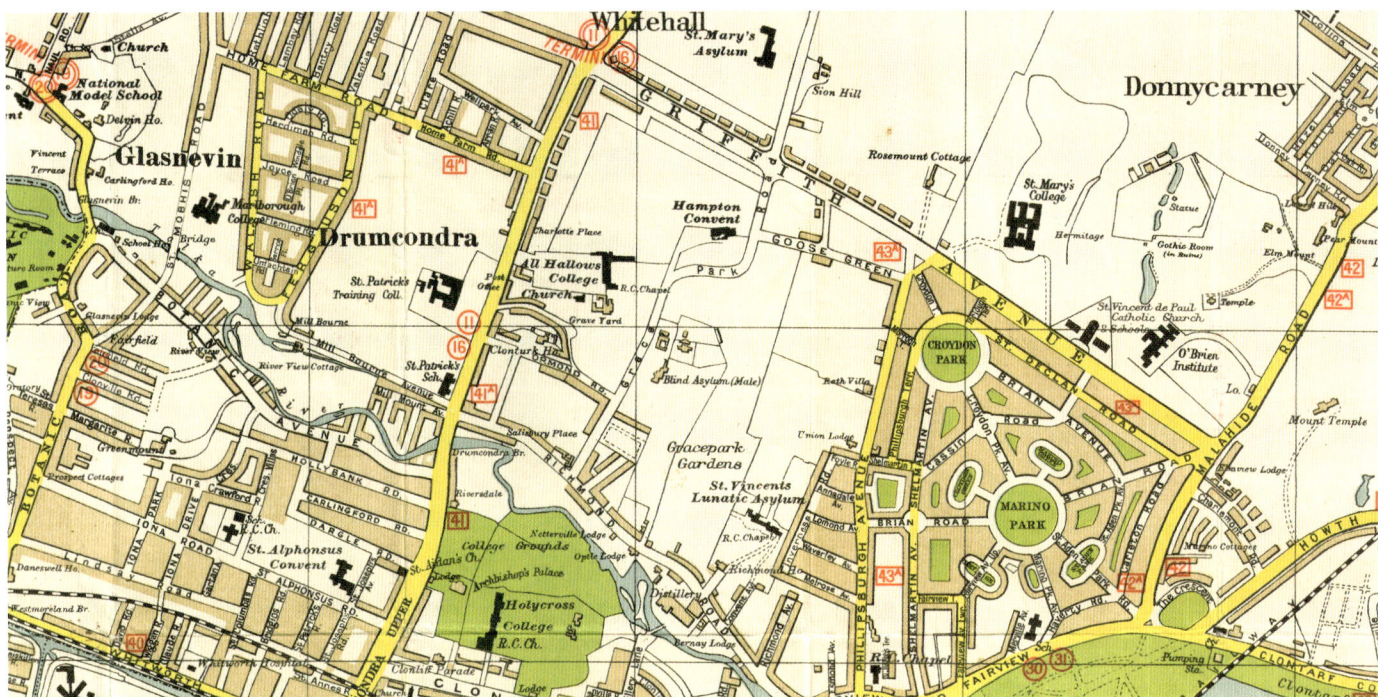

G.W. Bacon and Co. Ltd, Marino and Drumcondra, extract from *Bacon's plan of Dublin and Suburbs* (1930) [AUTH]

parks with four smaller parks radiating from the more southern of the two – Marino Park. The completed scheme in Marino was quite different to that originally proposed in 1919 and discussed here earlier. A larger site had been obtained and they built a total of 1,283 houses there. All were five-roomed and, eventually, all had fully fitted bathrooms. It was decided that the two parks would be managed in co-operation with a local committee. Marino Park was not open to the public. The local committee controlled access within the general framework of the regulations, though the Corporation could permit other groups to use it. Croydon Park (the more northerly of the two) was open to the general public but not before noon or later than sunset. It had a particular focus on tennis, and football was not permitted there. The design values in Marino were high: a number of housing styles was used in short terraces of different lengths with variation in the street line reducing any sense of monotony. Density was kept at 11 houses to the acre and this ensured a great deal of open space. In addition to the community parks, people had pocket-gardens to the front but large gardens and space for allotments to the rear. For those who could afford them, the value was excellent. They got a well-finished three-bedroomed house for about £450, compared to about £750 for a similarly sized house built by a speculative builder in the locality.

While they were building Marino, the council began work on their Drumcondra scheme. This can be seen at the top of the map between Glasnevin and Drumcondra. Again a formal geometric shape was favoured on a southwards-sloping site. Within an outer parabola there were three cross-streets. Culs-de-sac ran at right angles to two of these cross streets while the one to the north had an elliptical feature. As in Marino, the houses were in short terraces with different finishes and setbacks diminishing any sense of uniformity. These too were for tenant purchase but the Corporation provided two- and three-bedroomed houses in different configurations, reflecting their unease with tenant purchase.

Directly above the Drumcondra development, the map shows a series of roads running in an approximate north–south direction – Rathlin Road to Valentia Road. These are of interest because they are the best examples in Dublin of the Corporation's reserved area policy. The Corporation had the idea in the design of Marino of finishing the edges of the scheme with better-quality housing; something that would improve the overall impression. To conform with state policy with regard to maximum housing costs, they found it necessary to do this via a public–private partnership with private builders. This was the beginning of the reserved area policy whereby a portion of a Corporation housing scheme would be reserved for more expensive private houses. These houses attracted significant subsidies and were a major feature of Marino and Drumcondra. The houses in the roads mentioned above were built largely by public utility societies (co-partnership societies) for their members and are distinctive for their significantly lower densities.

The Donnycarney development may be seen at the top right-hand corner. This was smaller than either of the two mentioned above but both private and social housing would be built along Collins Avenue and its environs over the next 30 years. The big scheme towards the western edge was in Cabra. Enclosed by the loops of the Great Southern Railways, it was not quite as geometric in design as Marino or Drumcondra but the same approach to housing was taken – short terraces with setbacks. The big change came with Beggsboro, north of Fassaugh Avenue on the map where the layout was shown but the roads were yet to be named. This marked a shift from tenant purchase to rental. There was always an unease with the relatively exclusive nature of tenant purchase at both state and corporation levels but it reflected the parlous state of the country's finances and the gap between building costs and what could be afforded by tenants. Matters had improved by the beginning of the 1930s, however, and housing legislation shifted the balance from purchase towards rental. Dublin Corporation's reaction was swift and

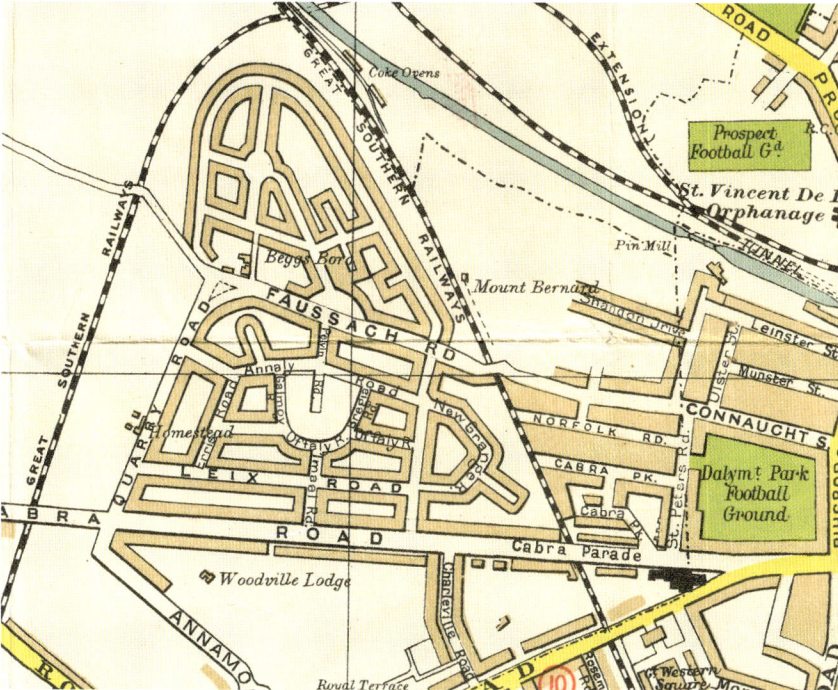

G.W. Bacon and Co. Ltd, Cabra, extract from *Bacon's plan of Dublin and Suburbs* (1930) [AUTH]

immediate. Tenant purchase was abandoned in its entirety and would not return until the 1950s, while a much reduced reserved area policy would continue to operate.

South of the river, the Corporation's housing schemes were more integrated into the urban environment and did not stand out as clearly from the map. Fairbrothers' Fields, between Cork Street and the South Circular Road, was complete. This was where the shift from rental to tenant purchase had first been applied. Further to the west, just below Kilmainham, the map showed Kehoe Square, formerly Richmond Barracks. To the east was the scheme of the same name. This was almost on the same scale as Marino but not as distinctive in its layout.

Private builders were also active during this time but there was a significant difference in scale between them and the Corporation. The latter were now building on a large scale whereas the former still followed an earlier approach of building in small tranches over an extended period.

# MAP OF PHOENIX PARK

**INDEX**
- First Aid Station — X
- Caterer's Tent — ▢
- Male Latrine — ⊙
- Female " — ⊙

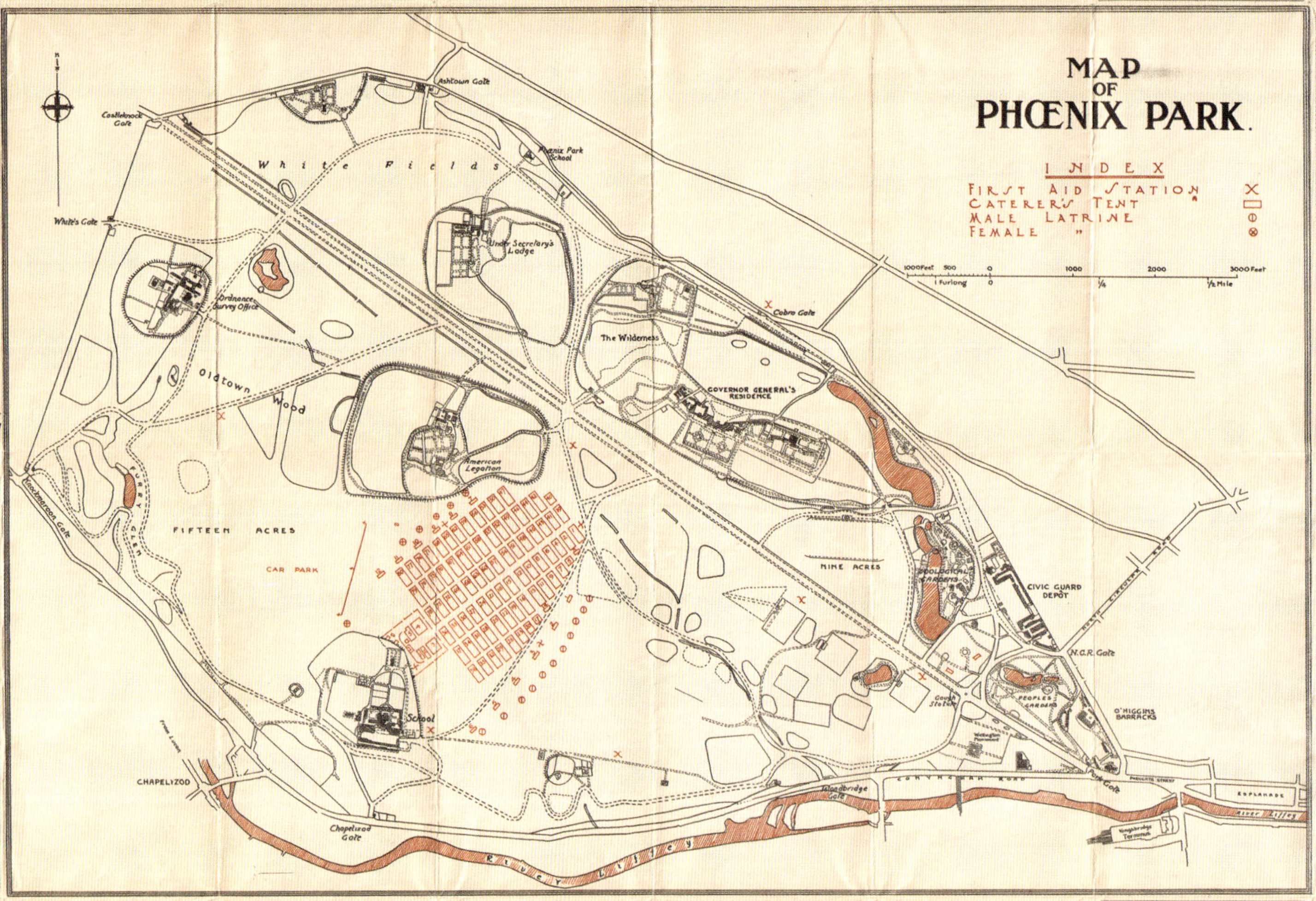

# 1932

## Public piety: the Phoenix Park

The Phoenix Park was opened for public use by Lord Chesterfield in 1747, after whom the main avenue in the park is named, and he installed the Phoenix Monument, an ornamental column on which a stone phoenix sits. By the early years of the nineteenth century, the park had become neglected but was revived in a 20-year programme of renewal under the British architect Decimus Burton and captured on the various directory maps. This resulted in the present landscape of wide, open green spaces, punctuated by small knolls of forest. Its open landscape facilitated its development as the location of many sporting activities, even motor racing on its roads. The park had been suggested as a suitable location for international exhibitions and even as an airport.

Dublin is a rather compact city and has few public spaces that can accommodate large crowds. College Green became a favoured location for mass meetings, while Smithfield, which could hold larger crowds, was rather neglected. As the public transport network evolved, the geography of the park suggested itself as a place to hold activities that attracted large crowds. Located 4km west of the city centre, it was reasonably well connected to the major routes into the city and by tram. Its 712 hectares (1,762 acres) could not only hold huge numbers, it also allowed for the well-managed dispersal of crowds. It was the perfect location for the Eucharistic Congress planned for 1932. Dublin had been chosen late in 1929 but expectations had been high for some time that a suitable celebration would take place during the 1,500-year anniversary of the arrival of St Patrick. The successful use of the Phoenix Park for the commemoration of the centenary of Catholic Emancipation earlier in the year had shown that (a) the park was a suitable location and (b) the organisational structures were in place.

The Congress was Ireland's opportunity to demonstrate its Catholic character on the international stage. It was also

---

Souvenir Map, Committee of the 31st International Eucharistic Congress (1932) [AUTH]

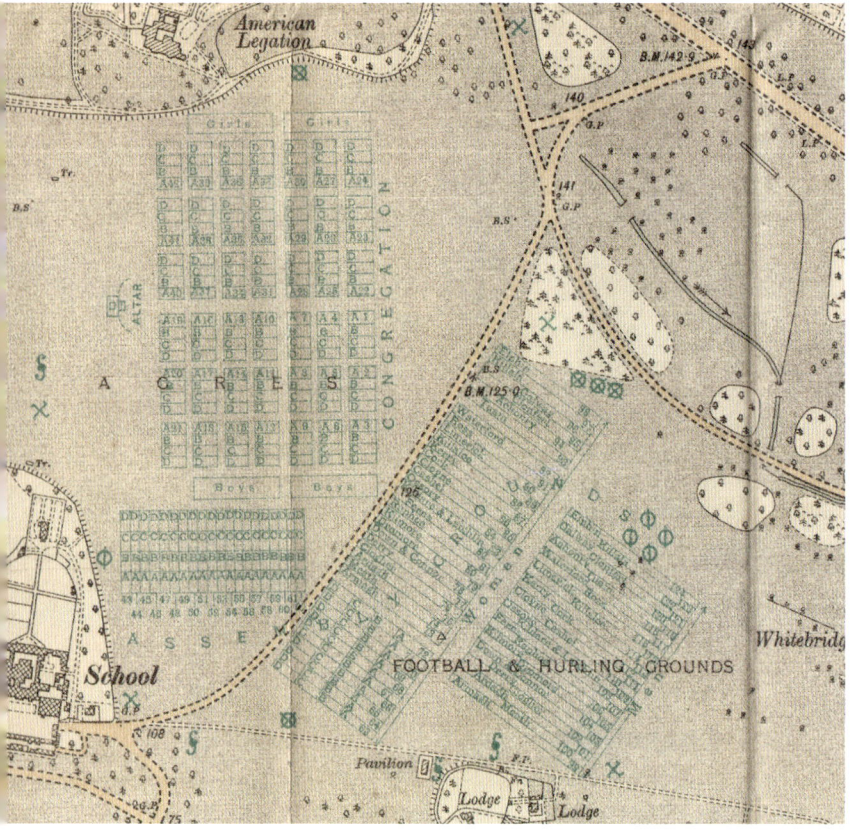

Ordnance Survey, *Map of Phoenix Park*, extract showing layout for Mass, Catholic Emancipation Centenary (1929) [AUTH]

the opportunity to show its organisational ability, the sophistication and modern nature of the State. Events were held from 12 June onwards, but the official programme began on Wednesday, 22 June in the Pro-Cathedral and was due to end with a solemn Benediction on O'Connell Bridge on the evening of the following Sunday. While many large-scale events were scheduled, the centrepiece would be a pontifical Mass in the Phoenix Park for which an attendance of 1 million people was expected.

The dress rehearsal for this event had been a similar Mass to celebrate the centenary of Catholic Emancipation. Mass was celebrated on 23 June 1929 in the Fifteen Acres in the park and it was estimated that at least 300,000 people had attended (the *Longford Leader* suggested 500,000). It had all gone well and organisers had found that the amplification systems had worked, even in such a vast open area. Despite the religious nature of the event, there was a range of souvenirs available. The Ordnance Survey produced a high-quality map at a scale of 8 inches to the mile (1:7,920). It was a detailed picture of the entire park and all of the various elements within it. Its immediate practical purpose was that it set out the arrangements for the Mass. It showed that the altar faced west and that the congregation was arranged in an east–west grid, with children on either side of the adults. Because many of the congregation would later be participants in a Eucharistic procession back to the city centre, the map also showed the marshalling arrangements for that. Arrangements were sufficiently finalised by the time of printing that the location of each diocesan group could be indicated. The map also showed the arrangements for car parking and the direction that cars were to face to facilitate their exit. Cars were still a relative novelty, but it was assumed that many who had a car would choose to use it and provision was made for significant parking.

The Pontifical Mass for the International Eucharistic Congress was scheduled for the morning of 26 June 1932, followed by a solemn procession towards the city centre. This time the arrangements were even more elaborate and complex than in 1929, involving street closures and specific routeways. They had been tested on the previous Thursday and Friday, when separate services were held in the park for men and women. The merchandising was also more sophisticated, and books, pamphlets, souvenir badges, plates and cups, all suitably decorated, were produced in large numbers. Newspapers sold special supplements, which had to be reprinted, such was the demand. It is not known why the Ordnance Survey did not produce a revised edition of their 1929 map, but the map shown here was an official souvenir produced by the committee organising the Congress. It was double-sided, with dimensions of 42 × 32 inches, and was intended to meet all information needs for visitor and local alike. Each side contained a map surrounded by advertisements for goods and services.

The map of Dublin city covered the area from Glasnevin

in the north to Rathmines in the south and east from the Phoenix Park. It showed the tram and bus routes, but not with the kind of detail that such maps often included. Hostels were located but there was no mention of hotels (presumably because these had been booked solid for months). These were not named on the map, but separate panels around the edge give their map square and the tram to take. Similar information was provided for the churches. The enquiry bureaux were located, and for those out and about the location of public toilets and catering facilities was given. The advertisements around the map were many and varied, and a number had a particular Congress focus. The Gramophone Store on Grafton Street offered recordings of a selection of approved hymns. Others were just normal advertising for the shops or services indicated. There was nothing relating to the Congress in the advertisement taken by the South of Ireland Asphalt Company, though it took the opportunity to enjoin people to 'ensure the retention of your money in your country'.

On the reverse side was the detailed map of the park. It was not as well drawn or as detailed as that produced in 1929, and it concentrated on the practicalities. The location for Mass was similar to 1929, immediately south of the American Legation and across the road from the Governor General's House. This time the altar was oriented towards the south-west, with the congregation in a grid formation towards the north-east. There were catering tents and first-aid stations as previously. The assumption was that most of the attendance would be male, at least if the number and distribution of latrines was anything to go by. They were provided on either side of the seating area, just in case there could be any confusion, with 12 for males compared to five for women.

Chesterfield Avenue was intended as the main pedestrian route and to that end the Phoenix Monument was moved to the side and the main gates and pillars removed. Though cars were more actively managed than in 1929, parking was provided on site in the same location and estimates suggested that 25,000 cars were accommodated.

Mass was celebrated at 1 p.m., but the crowds assembled from early dawn. The newspaper reports over the following

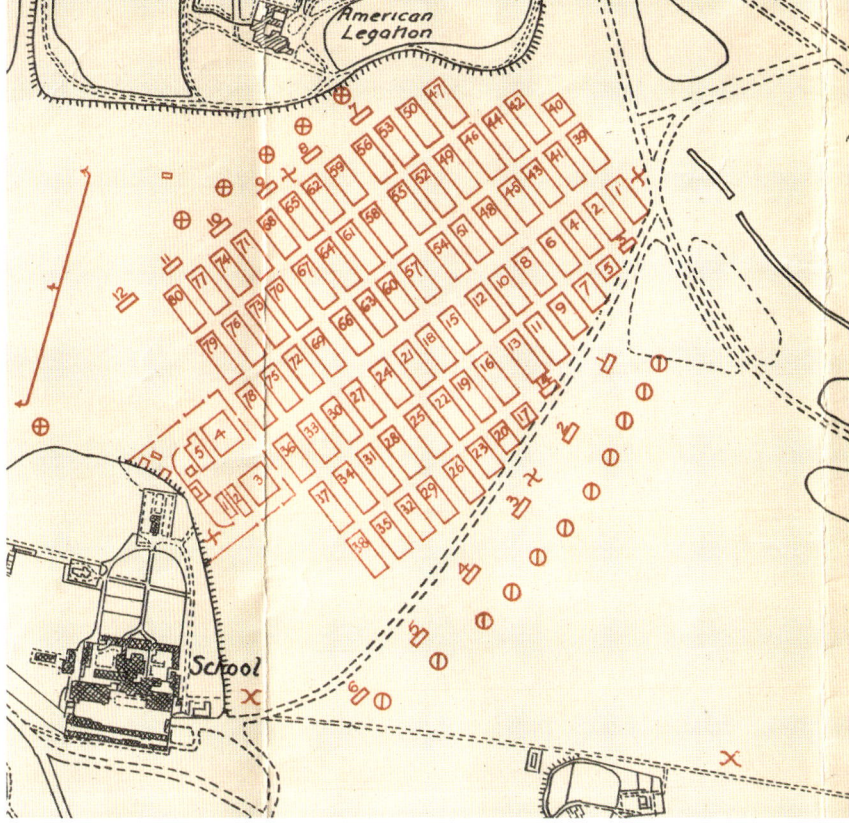

*Map for International Eucharistic Congress*, advertising showing layout for Mass (1932) [AUTH]

days spoke of the success of the event, with estimates of attendance at Mass varying from 750,000 to over 1 million. It was a long day because, following Mass, the greater part of the congregation made its way towards O'Connell Bridge, where the Congress was closed with great pomp and ceremony with a Benediction that evening. The city was effectively closed for the event, with no other activities possible.

Neither event left any permanent mark on the landscape – the altars were taken down. Neither was there any rush to restore the Phoenix Park to its pre-Congress state. The entrance pillars survived in storage until they were reinstated in 1986. No trace, however, could be found of the gates. The Phoenix Monument was restored to its central position at the junction of Chesterfield Avenue and Acres Road in 1989.

# 1935

## The best shopping and Goad's fire insurance plans

By the middle of the nineteenth century, there were many companies in Dublin who offered life insurance, marine insurance, annuities and fire insurance. As the Liverpool and London and Globe insurance company, with premises at 51 Dame Street, noted in their 1868 advertisement in Thom's directory:

> the beneficial effects of insurance against loss by Fire are no less obvious and important than the consequence of neglecting such a precaution is oftentimes ruinous ... When all these consequences can be guarded against by an annual outlay so small as to be within the reach of all, no thinking man would remain uninsured.

The business was competitive, and companies offered incentives such as no-claims discounts and professed themselves open to considering any circumstances that might reduce the risk and thereby the premium. While determining the premium for a domestic dwelling might be relatively straightforward, it was more complicated for a business in an urban setting. Here, the form and construction of the building, the nature of neighbouring buildings as well as the layout of streets were important considerations.

Charles Edward Goad (1848–1910) was an English cartographer and civil engineer who saw a business opportunity in providing this information, and established a worldwide company that specialised in fire insurance maps. The business was most active in Canada and the United Kingdom, but also operated in Turkey, Egypt, South Africa, the West Indies, Mexico, Venezuela and Chile. There was nothing quite like a Goad plan and each sheet contained a huge amount of information, as well as being attractive.

He produced atlases for all major towns and cities and no insurance company would be without a copy. He was a

---

Charles E. Goad, Grafton Street and environs, Goad fire insurance plan, sheet 19 (1935) [AUTH]

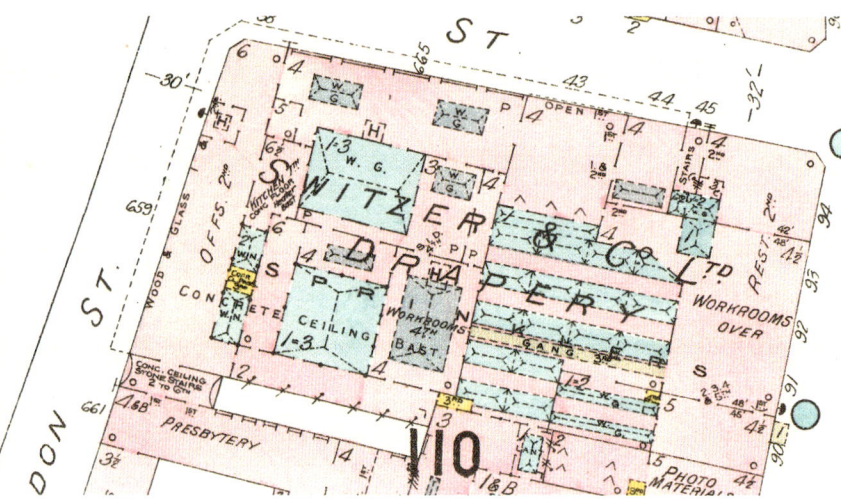

Charles E. Goad, Switzers, extract from Grafton Street and environs, Goad fire insurance plan, sheet 19 (1935) [AUTH]

shrewd businessman and he provided the atlases on a subscription basis. This involved producing paste-on modifications reflecting change in an area. From time to time, a new edition of the atlas would be produced but the older ones would be retrieved. Goad thus ensured that the availability of his product was limited to his subscribers. Unfortunately, this also meant that Goad plans had a poor survival rate. Individual sheets were either modified or replaced on a regular basis but no older sheets were retained locally. The only sheets that survived were those that were put away for whatever reason at Goad's headquarters. The plans continued to be produced at high quality into the 1960s and were still available during the 1980s, but both the coverage and the quality had deteriorated significantly because by then the nature of the market had changed and other sources of information were more important.

The series of maps for Dublin was usually organised into two volumes, one for each side of the river. The index map showed the coverage for each sheet which was numbered and dated by month and year. That index map also provided for revisions to be noted. The coverage was limited to the main business areas but it also included the docks where fire insurance was particularly important.

The information in maps was conveyed by a standard system of colour, symbols and text. It was the use of pink for standard buildings of brick stone or concrete that gave the maps their distinctive colour. Yellow was used for wooden buildings, grey for metal buildings and various shades of blue for skylights of different types. Symbols indicated chimneys, door, water hydrants of all kinds while walls were shown by different thicknesses of lines. A long list of abbreviations was used to indicate materials used, such as asbestos, concrete or corrugated iron, as well as whether skylights had wired glass or wired netting over glass. Various types of motors were identified as were building uses such as P.H. for public house or S. for shop and TENS for tenements. The number of storeys was shown, with basements identified, together with the width of streets and often the actual height of the building. It made the maps very 'busy', but ensured that they were very, very useful.

A *Guide to Dublin* was produced by the Advertising Company Ltd in 1904 and this 12-page guide explained that Grafton Street was 'the most fashionable business thoroughfare in Dublin. Here are many of the smartest shops, and the wealth, fashion, and beauty of Dublin may be seen in the season, in the forenoons, engaged in "shopping" and in the afternoons in promenade.' Sheet 19 shows part of Grafton Street and Nassau Street in 1935. The character of the street was little changed since that 1904 guide and the *Irish Times* commented in 1935: 'Grafton Street may be little credit to Dublin from an architectural point of view, but as a shopping centre it has no peers.' Two large department stores facing each other across the street ensured a regular flow of customers. Both claimed an air of exclusivity and Switzers made no apologies for expecting their customers to dress appropriately, with staff available to escort those who did not fit in to other establishments. Switzers were careful to strike the balance between price and quality. In 1933 they made the point explicitly by advertising in the *Irish Times* that: 'At a time when the myth of "cheapness" is almost everywhere emblazoned in inviting golden letters, the House of Switzer maintains unswervingly its policy of Quality First. Switzers

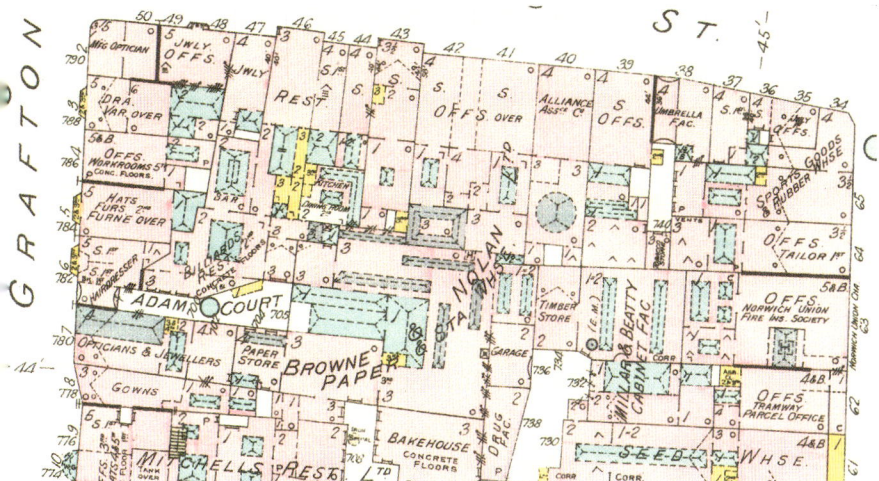

Charles E. Goad, Millar and Beatty, and Browne and Nolan, extract from Grafton Street and environs, Goad fire insurance plan, sheet 19 (1935) [AUTH]

Grafton Street, extract from postcard showing Switzers, Millar and Beatty and Brown Thomas [AUTH]

stake their reputation on the goods they sell.' Switzers occupied half the block bounded by Wicklow Street, Clarendon Street and Grafton Street itself. Natural light entered the building through quite a number of large rooflights. The restaurant faced onto Grafton Street on the second floor while the offices faced onto the less interesting Clarendon Street. Workrooms were an important element but these were located on the third and fourth floors, far away from the main retail space but mainly towards Grafton Street. Having workrooms on site was typical of large undertakings and there was still a significant amount of manufacturing in the centre.

Across the road, Brown Thomas faced Switzers and, having once been part of the Selfridge group, sought to maintain that kind of image. In addition to a wide range of high-quality clothing, it had a beauty salon, photographic studio and a fabric department. The plan noted a restaurant on the second floor and extensive workrooms. Both stores held fashion weeks. In March 1936 Miss Jeanne Kent, described as 'the famous fashion commentator', spoke to a packed Fabric Hall on the new season's line and the colours which would be most worn. The range of clothing on offer included street wear, afternoon and evening dresses, sports clothes and outfits for cruising. Such was the demand that the talks were given at 3 p.m. and 4.30 p.m. each day. Switzers Fashion week in October 1938 was described in the *Irish Times* as revealing 'the secrets of the well-dressed woman – fur-trimmed hats, the right tweeds, dresses and coats, full-skirted-evening gowns and, above all, the right accessories with the right garments'. Millar and Beatty were the kind of store chosen by upmarket estate agents to furnish their show houses. They had been on Grafton Street since the middle of the nineteenth century and they were a quality house. They offered 'outstanding quality at competitive prices', sold furniture and furnishings with an emphasis on traditional styles and would send a catalogue on request. They also offered a service whereby carpets were 'shaken and cleaned' in their works.

The passer-by would not have noticed that Browne and Nolan, the paper manufacturers, printers and stationers, had a large operation in the inner part of the block bounded by Nassau Street and Dawson Street. Millar and Beatty's furniture store faced onto Grafton Street but it had a factory wedged between offices which faced onto Dawson Street and Browne and Nolan. Around these units were found a wood store, a garage and a drugs factory. The density of development is apparent and it must have been somewhat of a fire hazard.

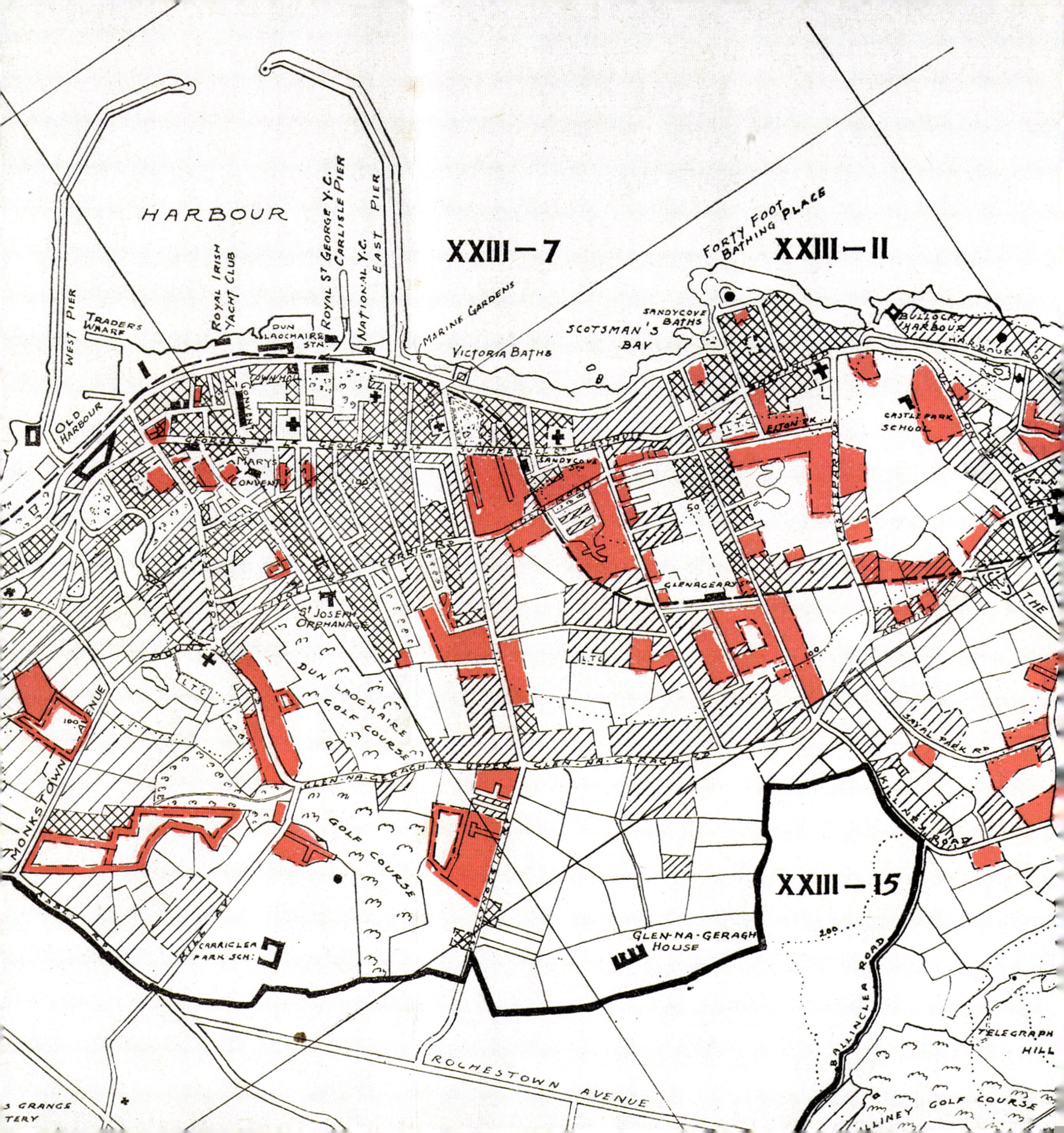

# 1936

## Dún Laoghaire civic survey

Dún Laoghaire had successfully resisted attempts to incorporate it into the city in 1930 and looked forward to its future as an independent borough. It had the Local Government (Dublin) Tribunal to deal with but it was confident that its status would not change. Indeed, its expectation was that there would be an increase to the existing borough area. That confidence resulted in it adopting the Town and Regional Planning Act 1934, which required that it make a town plan.

Just as in the city, it was believed that a necessary first step was to undertake a civic survey. As they put it 'to be in a position to guide development alright, one must know what are the conditions which exist'. Manning Robertson prepared *Dún Laoghaire – The History, Scenery and Development of the District* as part of that process. Robertson was an architect and town planner, and a commentator and critic on architectural matters. Throughout the 1930s and early 1940s he was a consultant to Limerick and Cork corporations and town planning adviser to Dún Laoghaire borough corporation. He was also a member of the team that produced with Patrick Abercrombie and Sydney Kelly, the *Sketch Development Plan* for Dublin, which was published in 1941.

The study of Dún Laoghaire was published in 1936 and followed much the same format as Dublin's 1925 study. Robertson discussed geology, climate, vegetation, antiquities and the history of the place. There was an examination of general administration which took in population, religion, housing and public health. Other aspects examined were recreation, traffic, industry and amenity. As well as a small general map of the neighbourhood, there were three more detailed ones which examined geology, development and housing, and public spaces.

The maps were approximately 44 × 28cm, slightly narrower

Manning Robertson, extract from 'Development' map, Dún Laoghaire civic survey,
showing, in pink, housing built since 1908 in the central area (1936) [AUTH]

Manning Robertson, 'Development' map from *Dún Laoghaire – The History, Scenery and Development of the District* (1936) [AUTH]

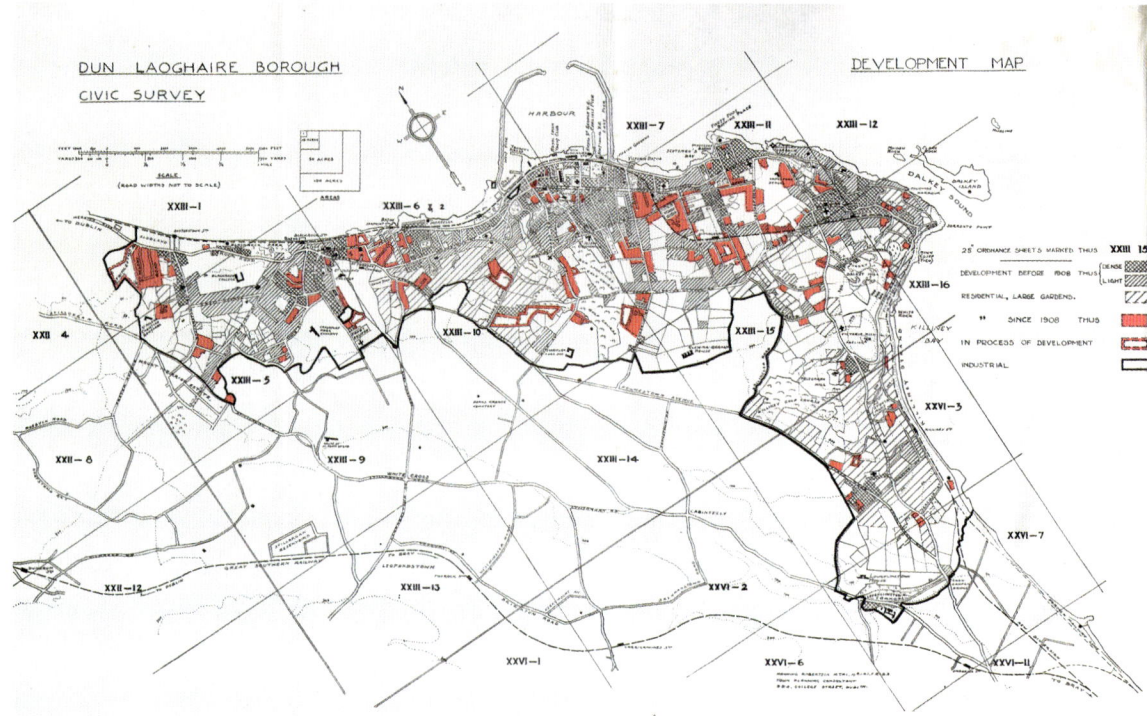

than A3, and were folded into the hardbacked report. They were outline maps, not to the same standard as the Ordnance Survey, but with colour. While the main roads were named, a reader would need to know Dún Laoghaire to get the maximum from the maps. The 'Geology' map showed a very simple picture. As Robertson put it: 'A line drawn from Dundrum Station to Blackrock Station divides the Carboniferous Limestone floor on the west from the Granite on the East.'

The 'Development' map was the most detailed of the three. It showed the historical growth of the borough, with the period before 1908 being divided into dense and light development. Residential development post-1908 with large gardens was given its own category. A pink shading was used in a solid block to indicate other post-1908 housing development while a block outlined in pink indicated that development was current. The map showed that a lot of new building was infill and spread across the borough. Coastal locations showed the greatest extent of building and the largest blocks of unbuilt land were inland, towards the borough boundary.

Unlike the city of Dublin, the housing situation was said to be satisfactory. The municipal authorities had built 1,574 houses and flats since 1897. In the period since 1930, private enterprise was responsible for 652 houses while the council had built 458. The report noted as a matter of pride that no flats had been erected since 1914, 'thus keeping abreast with the policy everywhere adopted, except where circumstances make flats an unfortunate necessity. Such circumstances do not exist in Dún Laoghaire.' They did exist, however, in Dublin city where an intensive flat building programme in the city centre was then underway. It was instructive to look at the third map, the housing and public open spaces map in the context of these numbers because it gave the location of each of the social housing schemes, and the number of houses in them. While they were found throughout the borough, there were concentrations in the city centre but the most noticeable was a cluster south of Glen na Geragh Road Upper (Glenageary Road) in Monkstown (then under construction), Carriglea Gardens and Sallynoggin.

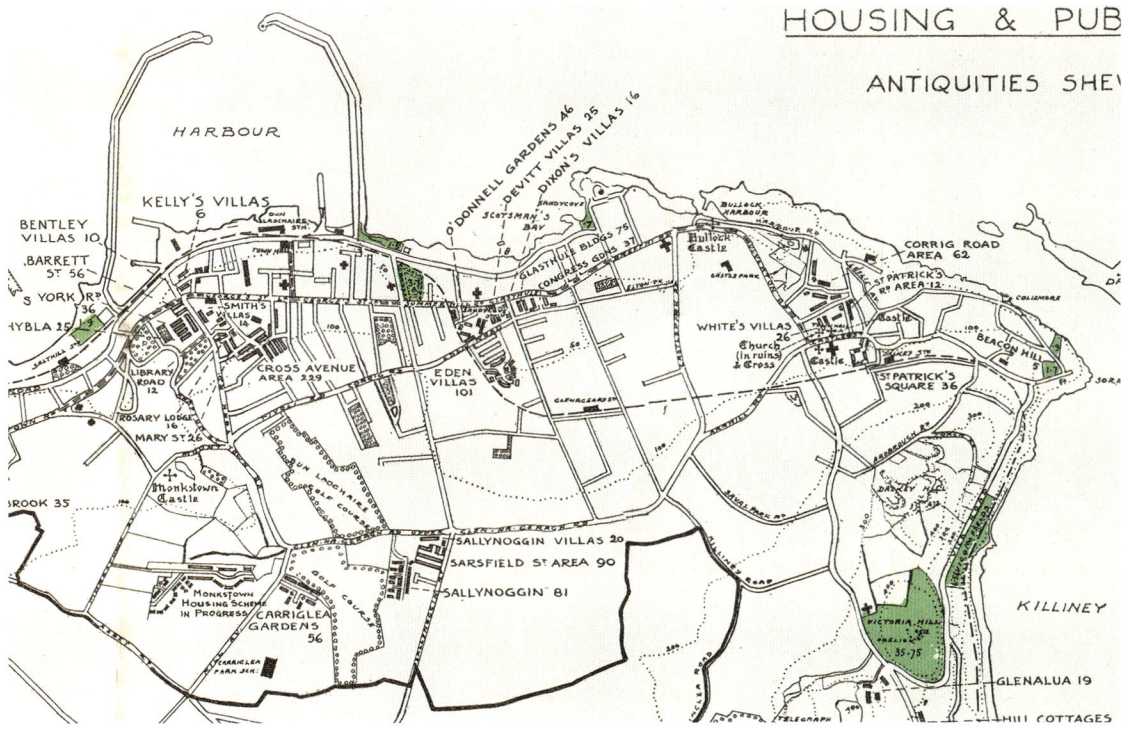

Manning Robertson, extract from 'Housing and Public Open Spaces' map, Dún Laoghaire civic survey, showing housing schemes, public (in green) and private open spaces (1936) [AUTH]

The borough was growing steadily and they expected that to increase the pressure on development land. There were two areas of concern though. The first was that there could be an influx of people from 'outside' seeking social housing and the second was the possibility of some of the multi-storey over basement houses becoming tenements.

Amenity was very important to Robertson and having arrived at the chapter entitled 'Parks, Open Spaces and Public Playing Fields', he noted that 'we have now come to what is, in many ways, the most important function of a town plan – the provision of sufficient public open spaces'. This was not because housing was unimportant, rather it had become so important that there was a danger that open spaces might be neglected, and these were essential to the 'physical welfare and contentment of an urban community'. To prevent this happening there were long sections on public open spaces and on playgrounds. 'Bathing, Yachting, Music, etc' were given a separate chapter and noted on map No. 3. The public parks were highlighted in green but the map also showed the location of golfcourses, clubs and private squares. There was a surprising number of these, distributed across the borough, some of which appeared quite large.

As expected, there was a chapter on traffic, but it seemed to be less of an issue in Dún Laoghaire than in the city generally. There was some congestion near the seafront and a need for parking spaces to accommodate day-trippers, and it would also be good if a one-way system could be instituted on the front since this would allow the width of pavements to be increased. On the whole therefore, the problems were relatively minor and did not require major reconfiguration of the road system, which Robertson felt was self-contained. What did bother him, and what he called a 'new menace', was the arrival of the motor bus. These burned heavy oil and omitted noxious fumes which were not only poisonous and unpleasant, they obscured the view of traffic following and encouraged dangerous overtaking. All in all, it was a picture of a prosperous and well-provided town which was growing rapidly, but where the issues seemed within the capacity of the authorities.

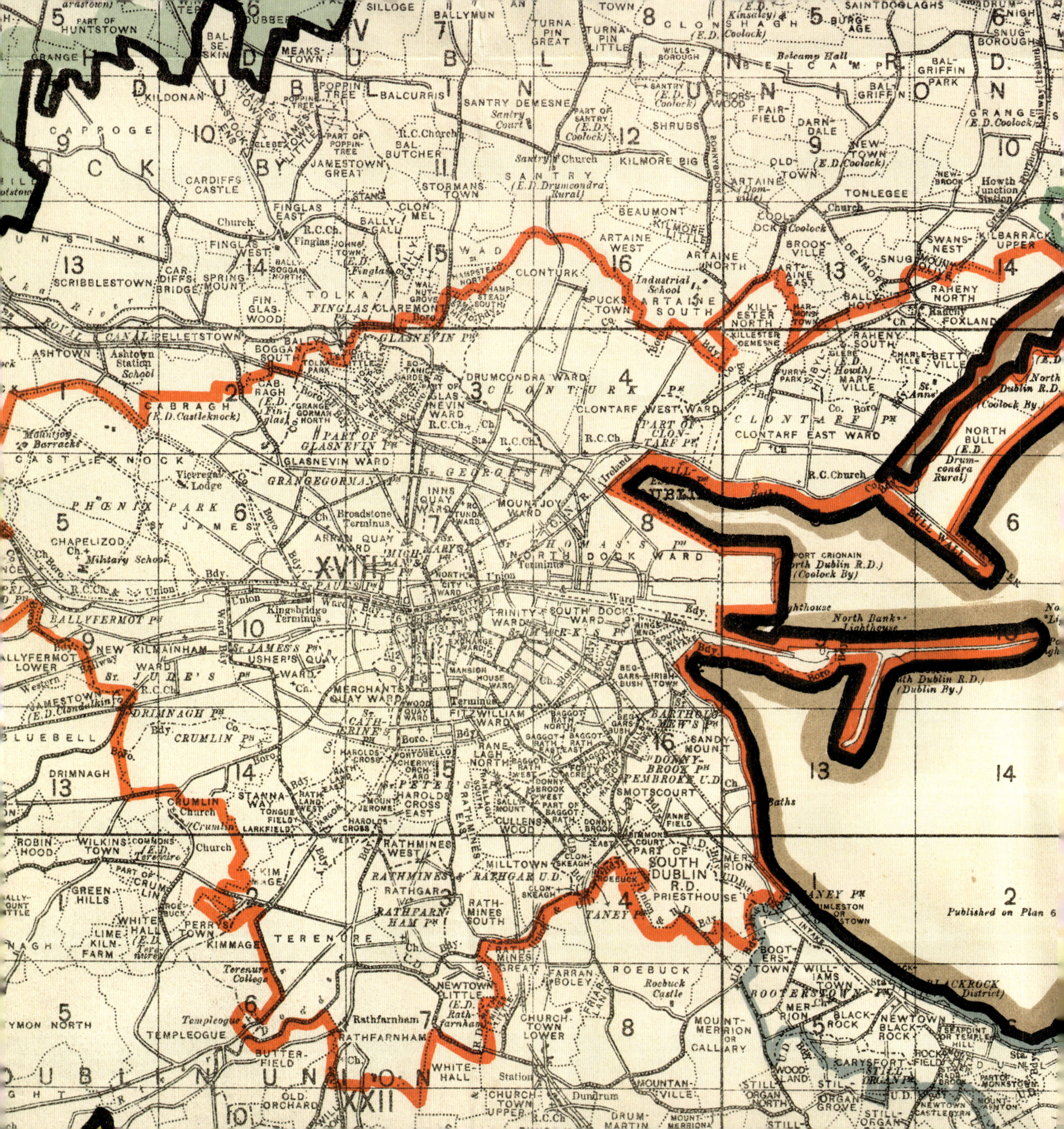

# 1938

## Governing Dublin

For a relatively small urban area, Dublin city has always had complicated governance. Until the middle of the nineteenth century, Dublin Corporation had control over a rather ill-defined area with a Grand Jury holding sway in the remainder of the county. The mid-century reforms, discussed earlier, resulted in the urban area being contained within the canals, roughly speaking, and the remainder under the control of a county council. This binary arrangement was complicated by the creation after mid-century of a significant number of townships, independent urban districts. Within a very short period of time Dublin had a complex and fractured system of governance with no overall control on growth or development. An additional element in that mix was that Dublin port was governed not by Dublin Corporation, but by the Port and Docks Board which also had a role in bridge building and river-channel maintenance.

It was soon recognised that this made no sense and in 1881 the Municipal Boundaries Commission (Ireland) reported on Dublin and recommended that the area of Dublin city be extended to include the townships of Rathmines, Pembroke, Kilmainham, Drumcondra and Clontarf, and other significant areas of the then county of Dublin. They did not recommend the inclusion of the coastal townships of Kingstown, Blackrock or Dalkey within Dublin but rather that the latter two should be annexed to Kingstown. Nothing happened until 1900 when the townships of Kilmainham, Glasnevin, Drumcondra and Clontarf were annexed to the city, but Pembroke and Rathmines managed to escape, despite the best efforts of Dublin Corporation and a vote (later overturned) in favour in the House of Commons.

There were some small boundary additions to the city between 1900 and 1922 but it was not until the Greater

---

Local Government Tribunal, County Dublin, alteration of
administrative area, extract showing Dublin county borough (1938) [AUTH]

Dublin Commission of Inquiry reported in 1926 that the issue was revisited. The commission was asked to examine the administration of local and public utility services in the city and county and it recommended significant amalgamations. Dublin was to be extended by the annexation of Rathmines and Pembroke but also by the addition of Howth and some rural areas. A single 'coastal borough' was to be formed by the amalgamation of Dún Laoghaire, Blackrock, Dalkey and Killiney (and Ballybrack) with a significant rural element. Common services were to be administered by a Great Council in both the new county borough and the expanded coastal borough with local councils to deal with local services. It also recommended that a 'city manager' be appointed to the whole joint administrative area. The Great Council idea never came to much but some amalgamations took place as a result of the Local Government (Dublin) Act which was passed in 1930.

This Act finally saw the absorption of Pembroke and Rathmines into the city. By then the enthusiasm within Dublin Corporation for this had cooled somewhat and it was a reluctant marriage on both sides. There was also a significant extension to the boundary of the city by the inclusion of Rathfarnham, Terenure, Kimmage and Crumlin, as well as a swathe of land from Cabra to Killester and along the coast to the boundary with the Howth urban district at Kilbarrack. The effect of the boundary extension for the city was to increase its area from 8,172 acres to 18,776 acres. The Howth Urban District survived as an autonomous unit for a little while longer.

This left the question of what to do with Dún Laoghaire. Certainly it made sense to them to amalgamate the coastal townships but there was no way that they were prepared to be absorbed into the city. During a fractious debate in the Dáil, it was explained that they were 'different' and that it made no sense to merge them into the city. What was meant, and understood by all, was that they were of a higher social status and they did not want that diluted in any way. They had enough political power in the Oireachtas to ensure victory and Dún Laoghaire emerged from the process as a separate administrative entity. With the new county borough of Dublin came the city manager (decried as an 'anti-democratic' concept by some in the Oireachtas) who took on direct responsibility for certain functions: finance, housing, public health, streets and waterworks, leaving only a rump set of reserved functions for the councillors.

One of the features of the 1930 Act was a compromise which contained a provision for a review. This resulted in the Local Government (Dublin) Tribunal which began its work late in 1935 and their report was published in 1938. The optimism with which the inquiry was undertaken is impressive given what had happened only five years previously. The report makes for interesting reading and the simple and basic conclusion was that the three strands of governance – Dublin Corporation, Dún Laoghaire Borough and Dublin County Council – each managed their affairs as if the others did not exist. Nobody exercised oversight over the entire urban area. Indeed, Dublin Corporation had chosen not to take up the opportunity provided under the 1934 Town and Regional Planning Act to become the regional planning authority for Dublin city and district.

The Tribunal's report recommended the creation of a Dublin metropolitan corporation which would encompass the whole of Dublin County. There would be an 'urban' and a 'rural' portion, of which the rural part would be in two zones; one to the north of the city and one to the south. The urban part would comprise the existing Dublin Corporation area and the Dún Laoghaire Borough would reach to the foothills of the Dublin mountains and encompass Lucan, Clondalkin, Tallaght, Dundrum and Ballinteer. On the northside it would encompass Hartstown, Finglas North, Ballymun and then strike north to include Swords, Malahide, Portmarnock and Howth. It would have been very similar to the contiguous urban area of today. The entire county would be under the control of an elected metropolitan council and an appointed metropolitan manager, on the lines of the city manager. They felt that it would be best to keep the council at no more than 40 persons with only a small number of electoral areas. The proposals were radical but they were largely ignored, with the

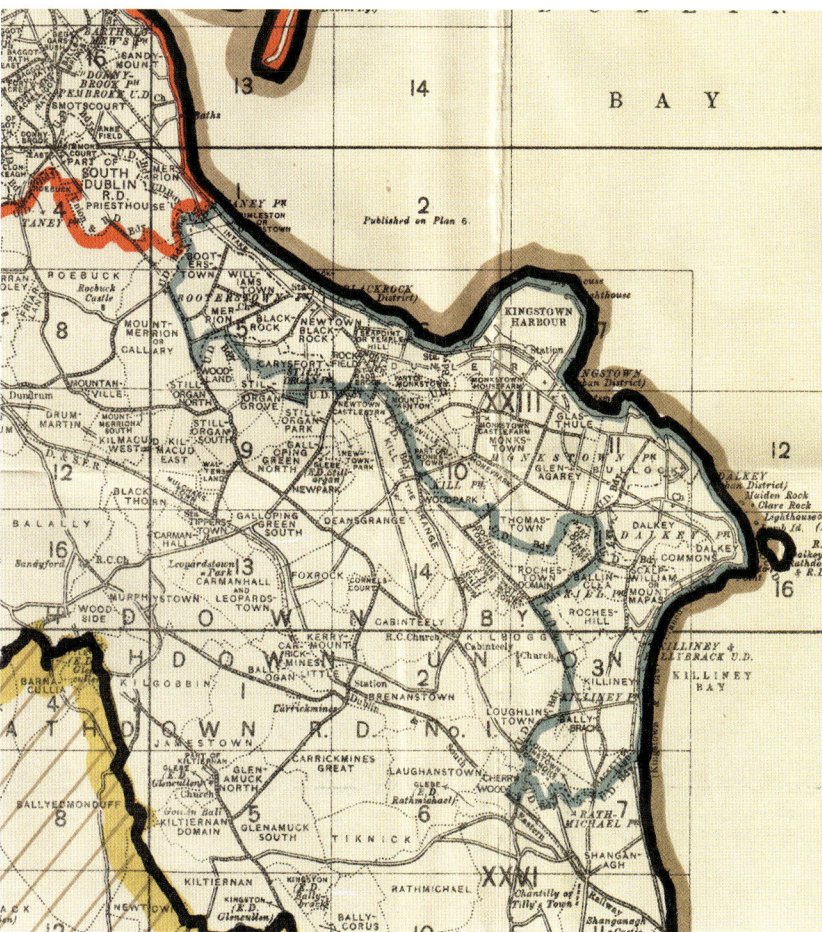

*Left.* Local Government Tribunal, County Dublin, alteration of administrative areas (1938) [AUTH]

*Above.* Local Government Tribunal, extract showing Dún Laoghaire borough (1938) [AUTH]

only tangible outcome being the creation of a combined post of city and county manager and the absorption of Howth into the city in 1942. The official view was that the time was not right for additional action but that if ever such a time came, then the report would be useful. Nothing much changed in the governance of the city until a brief period in the 1970s when the Fianna Fáil government decided to tackle the administration of local government nationally. That soon fizzled out and it was not until the 1980s that any kind of reform was undertaken. However, even that failed to produce a structure that managed the city as a single entity and Dublin continues to be governed by four separate, though elected, councils.

# 1941a

## Sketch Development Plan

The map shown here was published in 1941 and that probably goes a long way to explaining why it is not better known. It was also printed on very flimsy glassine paper which did nothing for its longevity, yet it is a very important map because it set out in considerable detail the town planning projects and possibilities for the next 30 or so years. It was not intended to have that degree of importance, being just a step on the road to a full development plan. That it became a surrogate for that plan was the result of a strange combination of circumstances.

The Town Planning Act of 1934 was an important step in the approach to urban development in Ireland. This promoted town planning and the production of planning schemes but did not require any local authority to do so. Dublin Corporation was an early and enthusiastic supporter of the idea of town planning and they decided on 6 January 1936 to prepare a planning scheme. This gave them considerable interim control over development but they now had the obligation to prepare the plan with 'all convenient speed'. The plan did not appear and, as the years passed, the exercise of that interim development control became more opaque. Dissatisfaction with how Dublin Corporation was behaving resulted in High Court and Supreme Court proceedings in the early 1950s. A plan of sorts was published in 1955 but it was never formally adopted because of flaws in the process. It was to be the end of the 1960s before the first approved planning scheme arrived.

In 1936, though, the aim was to produce a good comprehensive plan. As an important step, Dublin Corporation commissioned Professor Patrick Abercrombie to produce a *Sketch Development Plan*. This, however, would not be the plan, as its title indicated it would outline the priorities and the potential solutions. The city had had considerable experience of Patrick Abercrombie: he had won the international

*Sketch Development Plan*, extract showing port and bay reclamations (1941) [AUTH]

# ḃAILe ĀṪA CLIAṪ : SKETCH DEVELOPMENT PLAN : METRO

CONSULTANTS: PATRICK ABERCROMBIE
SYDNEY A. KELLY
MANNING ROBERTSON

ALD. E. E. BENSON, T.D., CHAIRMAN, TOWN PLANNING COMMITTEE

PRINTED BY DOLLARD, LIMITED, DUBLIN.

MAP PREPARED BY DUBLIN CORPORATION TOWN PLANNING DEPARTMENT AND BAS

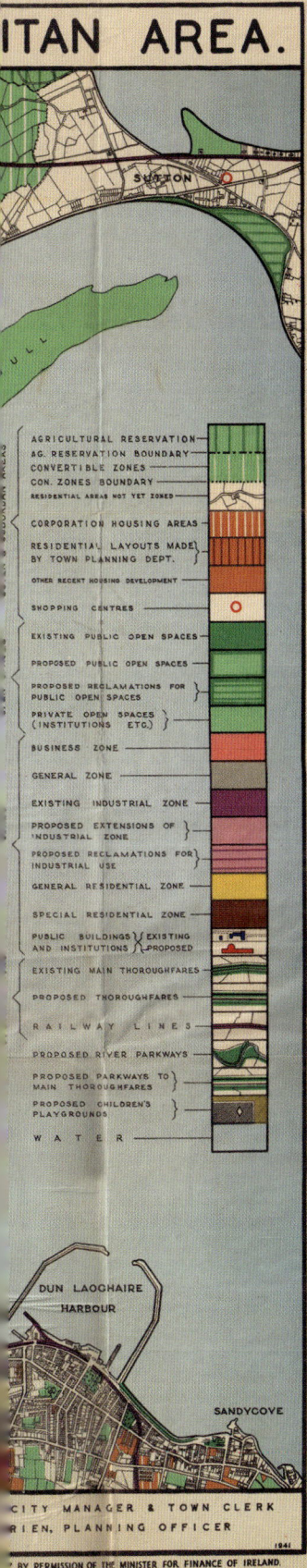

*Left.* Patrick Abercrombie, Sydney Kelly and Manning Robertson, *Baile Átha Cliath – Sketch Development Plan – Metropolitan Area* (1941) [AUTH] This was the detailed map to accompany the printed report.

*Right. Sketch Development Plan*, extract showing the classification system (1941) [AUTH]

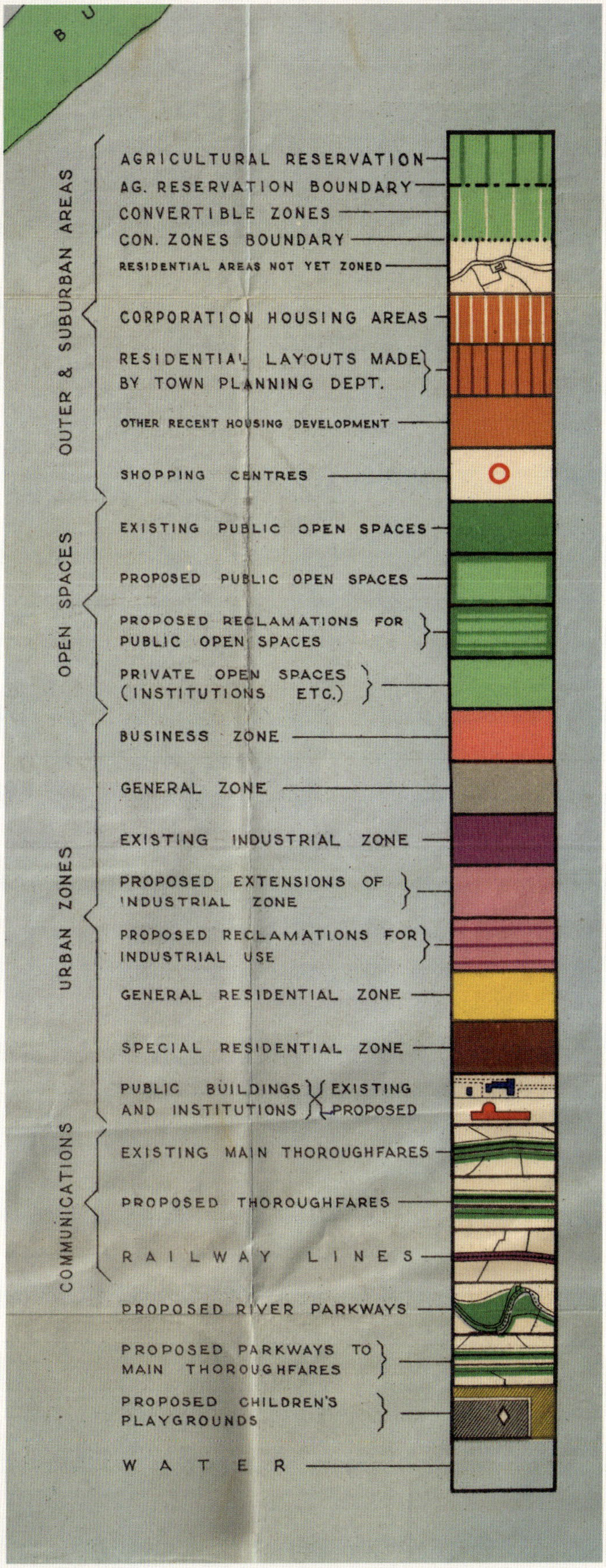

competition for a town plan in 1914 and had become an important consultant on planning issues to the city. Produced by Patrick Abercrombie, Sydney Kelly and Manning Robertson, the *Sketch Development Plan* for Dublin and its environs was presented to the Corporation on 20 October 1939, and published in 1941. The plan dealt with housing, civic improvements, transportation and regional development. There were relatively few graphics in the plan but this important map was produced as a separate publication. Because of its size, it was possible to see the detail of many of the proposals.

The map is a large size, about 97 × 75cm and at a scale of 1:20,000. It encompassed an area just to the north of Howth and the south of Dún Laoghaire and east from Castleknock. It is a highly coloured map with strong differentiation of the various categories used. It included a land-use zonation, an important element in any plan since it allowed the authority to determine what uses were compatible with others and what were not, as well as fixing the best locations for all activities. This was a generalised scheme which identified the business zone, the existing and proposed industrial zones, the general residential area and the special residential area – essentially one where residences had to co-exist with a variety of commercial uses. There was an area which defied classification and was referred as the 'general' area. Another novel feature on the map was the provision of shopping centres. These were spaced regularly across the map in the undeveloped areas of the city as a way of trying to ensure services within a reasonable distance of inhabitants. While planned shopping centres had been a feature of American cities for quite some time, they had not arrived in Ireland and it would not be until the later 1960s before they finally did. Until then, there would be nothing to challenge the dominance of the city centre and any new shopping provision in housing developments would serve only daily needs. Abercrombie was a strong advocate of providing green spaces in the city, and this was also reflected on this map: existing and proposed public parks were indicated together with zones along river valleys which would be preserved as amenities. The plan suggested that the growth of Dublin would be contained within a green belt and the future expansion of the city would be accommodated in satellite cities. Abercrombie was greatly influenced by Ebenezer Howard's garden city movement and his plans for the reconstruction of London after the Second World War would set in motion the new-towns programme in the UK.

Dublin still had a perennial problem with traffic congestion, as discussed earlier in various maps. For historical reasons, the road system focused on the city centre and the work of the Wide Streets Commission, despite being marvellous in many ways, had created a number of major pinch points. Abercrombie's plan involved the addition of circular elements to the existing radials so that traffic could travel around the city without engaging with the centre unless this was necessary. He had done this in his 1914 plan and this was a logical restatement of those ideas. The map showed a system of circular roads at increasing distance from the centre. On the northside, the innermost one would be based on Griffith Avenue and would run from the Phoenix Park to the Malahide Road. The middle one would be based on Collins Avenue and go as far west as Ashtown, while the outer ring would run from Dollymount Avenue in the east to Finglas North. The southern city would also have three rings. The innermost would link Nutley Avenue via Rathfarnham before heading north to Chapelizod. The middle road would have travelled along Foster's Avenue through Dundrum and Templeogue to link up with the Naas Road. The outer road was called a scenic route, because it would have been in the green belt, and would have linked Stillorgan with Tallaght. These would not have been sufficient in themselves to solve the problem and enhanced radial roads were proposed for city-bound traffic. In the centre, the roads would be realigned to permit easier north–south flows. This is where Abercrombie was able to combine his transport planning with his civic projects.

Sandymount Strand once more came under pressure. Abercrombie, in *Dublin of the Future*, was tempted by the ease with which a large stretch of the bay could be absorbed into the city. It could be done by simply building a sea wall between the Great South Wall and Booterstown or even Blackrock, and a great arc of the bay would be encompassed.

*Sketch Development Plan*, extract showing northern ring roads and shopping centres (1941) AUTH]

This would provide land for all sorts of projects including housing and industry. The idea was taken up by Desmond McAteer in a paper published in *Studies* in 1935. He suggested that the Municipal Airport, then in contemplation, could with advantage be located there. He believed that the future lay with flying boats. Abercrombie returned to the idea of using Sandymount Strand in this plan but this time a more modest reclamation to Merrion Gates was proposed, which would be used to provide a park with some space parallel to the South Wall being developed for industrial uses. This might not be the end of the matter though, because he indicated a much wider zone for future reclamation.

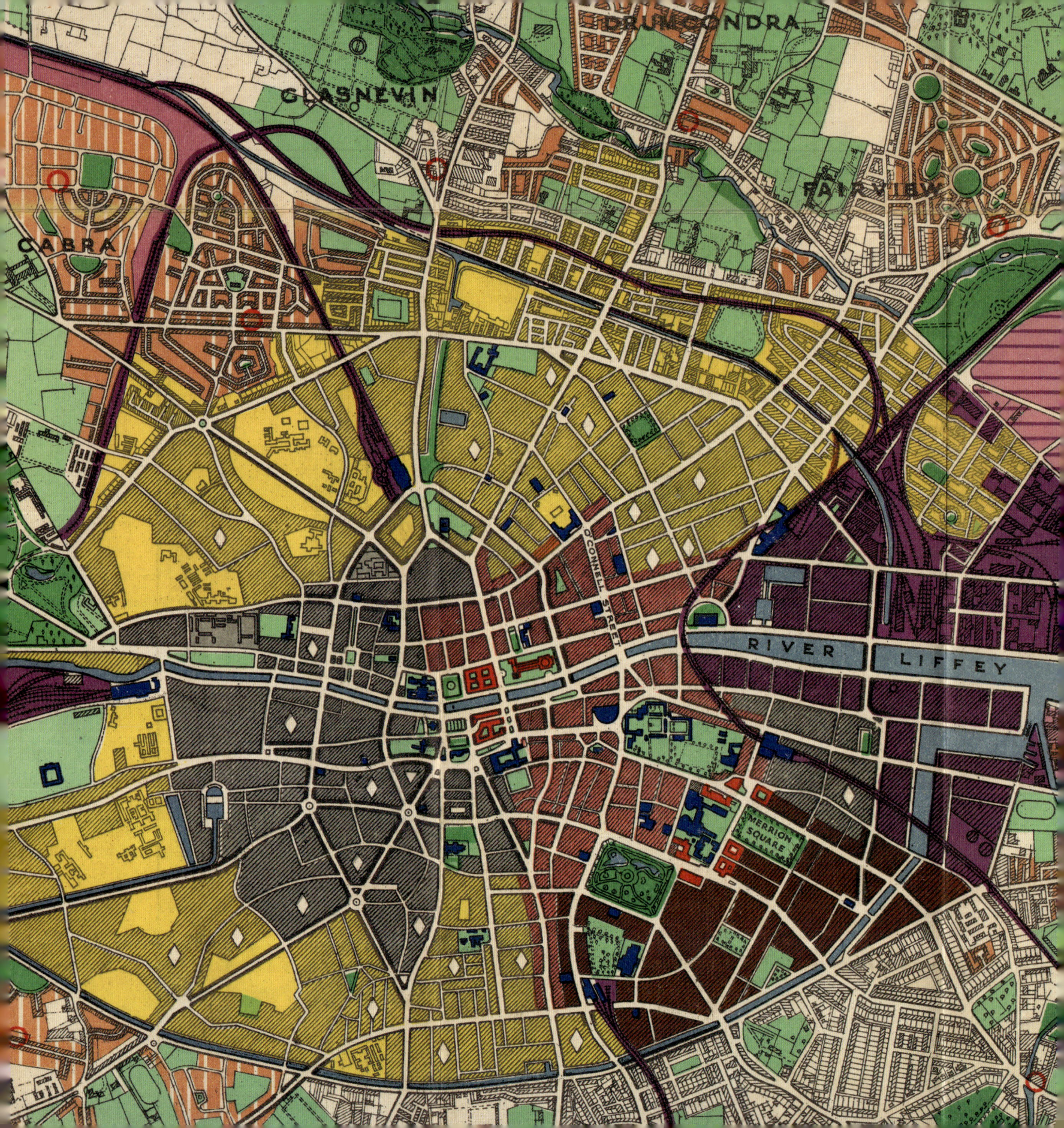

# 1941b

## A new city centre

There was also consideration of civic projects in Abercrombie's plan, but they were somewhat less ambitious than had been included in *Dublin of the Future* and they get relatively little discussion time in the report, where the focus was on transportation and communications. Nevertheless, the report retained the idea of resetting the central focus of the city closer to the medieval city and so help invigorate a declining part of the city.

The axonometric view of the city centre provided in this plan was focused on this area and shows the proposed transformation. There was no mention of the dramatic 'traffic centre' of the 1914 plan but the opening up of the streetscape between Christ Church Cathedral and the new cathedral would have given the opportunity to optimise the traffic circulation system. In his view:

> There is no one who does not deplore the loss of the Custom House view from O'Connell Bridge due to the loop line railway bridge. We envisage a similar possibility of view for the new Cathedral up river from the bridge with fine buildings of quiet dignity connecting to the Custom House on the North bank, while the City Hall with a riverside arcade, would dominate the South bank.

The space around Christ Church Cathedral would be opened up, providing a vista not seen since medieval times: a park would lead down towards the river. However, such was the height difference between the ridge on which the cathedral sat and the river that it would be possible to insert a bus station at the quay level which would be shielded from view. Another park on the left bank of the Liffey would open up the landscape around the Four Courts complex. He envisaged clearing the quayside up to the Ha'penny Bridge. In his view, the city offered a wonderful vista along the river and this

---

*Sketch Development Plan*, Central City showing road system, cathedral and other civic projects (1941) [AUTH]

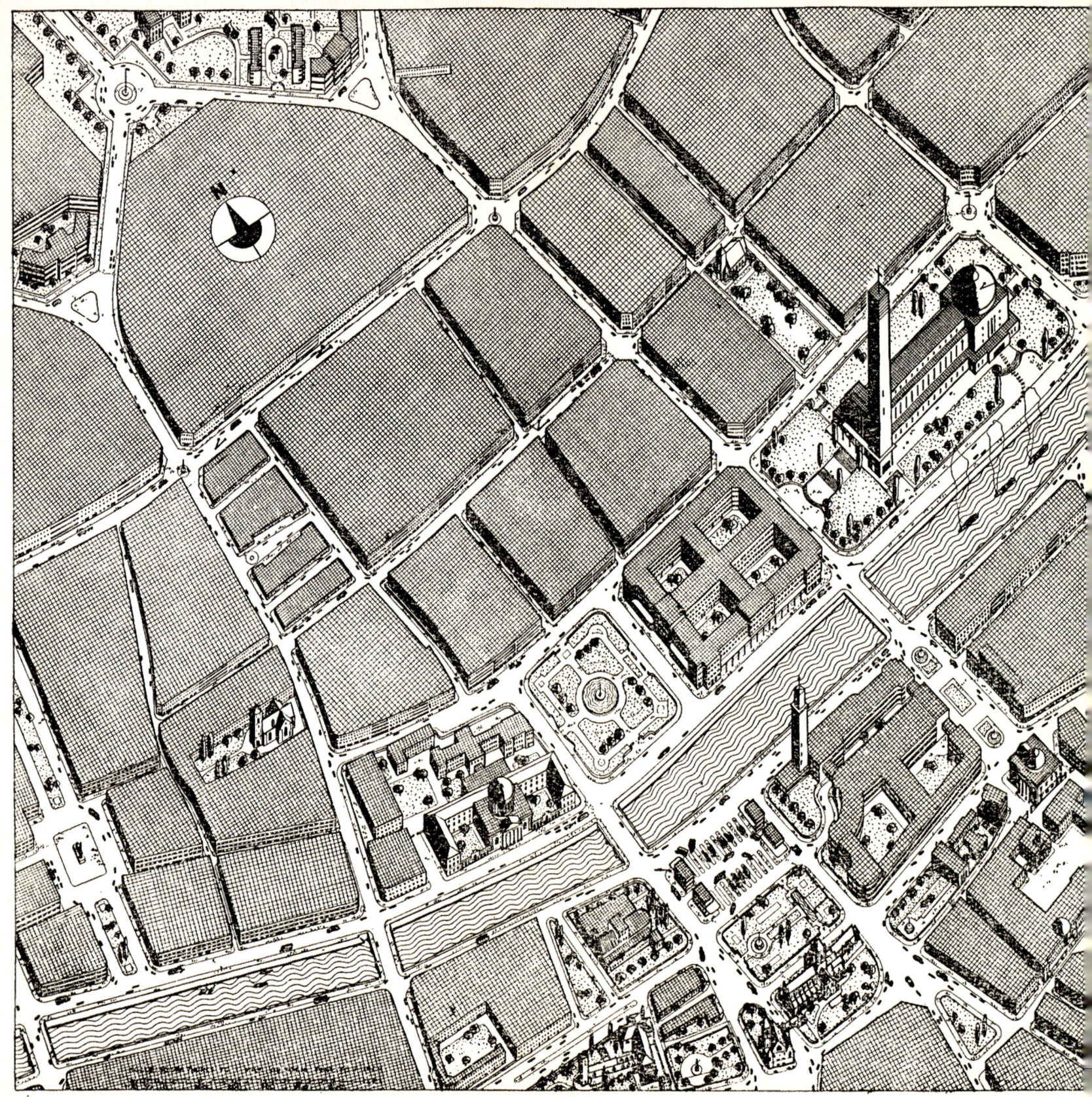

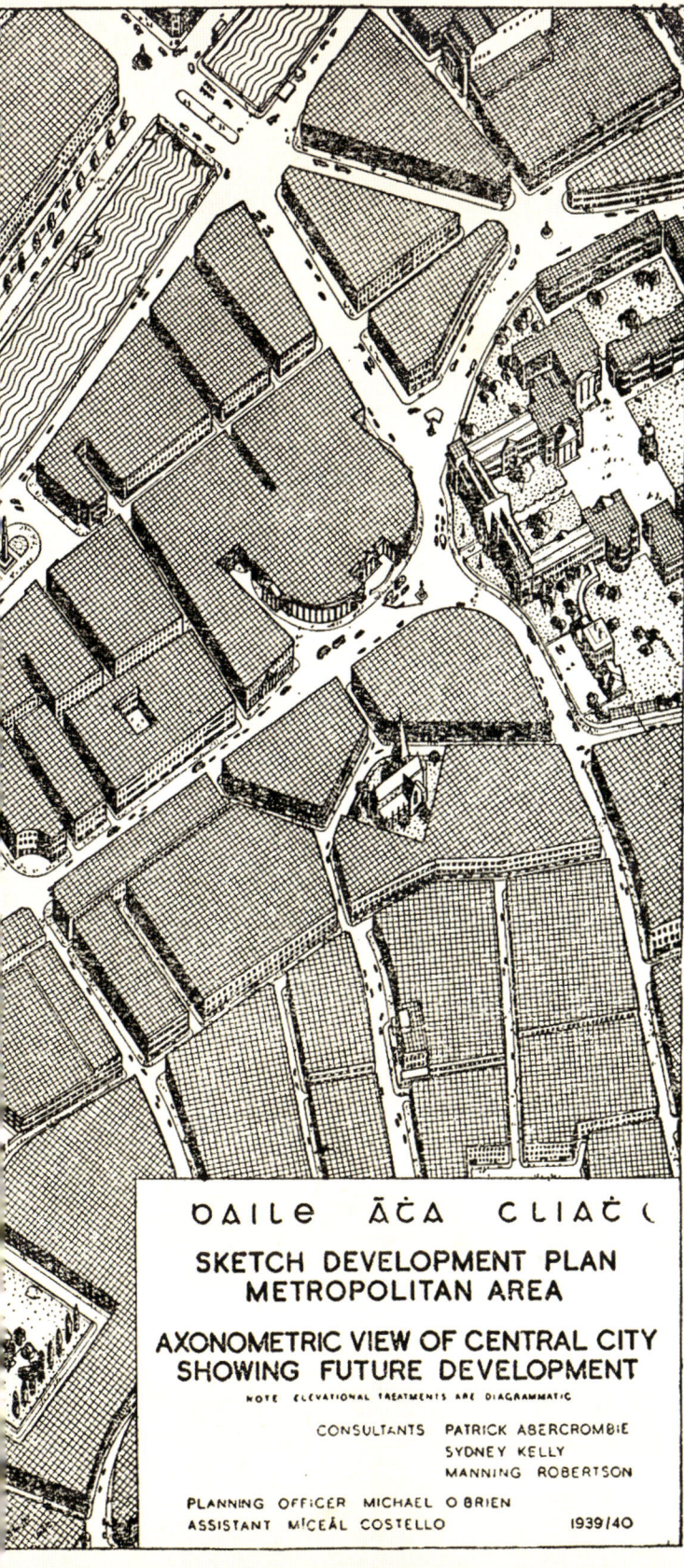

would be a great site. The area bounded by Lower Ormond Quay, Lower Liffey Street, Upper Abbey Street and Capel Street, and the quays frontages, was not being used to any adequate advantage. Here was the opportunity to replace the current admixture of buildings with something special on both sides in 'appropriate gardens and oases of green'. The centrepiece would be the Roman Catholic cathedral, the 'greatest building and noblest monument of the future Dublin'.

At a more practical level, this would also provide the opportunity to realign the road system and improve circulation. This would permit an inner tangent route that would link South Great George's Street with Liffey Street and onwards to Dominick Street where it would link to the orbital roads. It would not be a direct link but rather one which took in a block of buildings around Eustace Street and Temple Lane. This would have involved significant demolition as well if the road was to have any capacity, but this area was also peripheral to the main business of the city. Certainly a clearance would have been needed to create the small traffic circus on Wellington Quay, and the Ha'penny Bridge would have to go since a new bridge of much higher capacity would be needed. Though the bridge is now part of the iconic image of the city, it was not universally admired during this period. Its disappearance would have been lamented by some, but it would not have generated any controversy. These changes would open up a vista on the cathedral from the southside.

The new district would be completed by the provision of the civic offices. Dublin Corporation had long sought a central location in which it could bring together its various departments that were spread across the city centre. It was part of Abercrombie's brief to suggest a solution and he thought it would be a good idea to retain the straight line of Parliament Street, with its almost perfect focus on City Hall, and build there. However, the Wide Streets Commission proportions would be abandoned, and the street widened to 150 feet. This

Patrick Abercrombie, Sydney Kelly and Manning Robertson, 'Axonometric View of Central City Showing Future Development', frontispiece, *Sketch Development Plan* (1941) [AUTH]

The environs of the Four Courts at the end of the 1940s, the suggested location for the new cathedral and a range of civic functions [AUTH]

would both improve the vista of City Hall and permit the remodelled street to become a dual carriageway with grass borders, tree planting and a taxi rank. The Civic Offices would occupy the western side of the street and extend along the quays in an L-shaped configuration. Abercrombie suggested retaining the medieval Essex Street within this block by building a bridge which would connect the upper floors. This approach could also be used further up Parliament Street over an entrance to an internal court.

The traffic proposals generated little commentary in the general press. Likewise the proposal for the Civic Offices was of the greatest interest to the Corporation but not to anyone else. That is because the proposal for the cathedral generated an unholy row. The problem was the Roman Catholic archdiocese had already decided that the metropolitan cathedral was going to be on Merrion Square. It was incongruous that the capital city of a predominantly Roman Catholic country did not have a Roman Catholic cathedral. The building of a metropolitan cathedral had been a long-held ambition of the Church, which felt that St Mary's pro-Cathedral, though a fine building, was neither imposing enough nor suitably located. After much debate and not a little plotting during the late 1920s, an option to build on Merrion Square had been obtained in September 1930 from the Pembroke/Fitzwilliam

*Above.* View of Christ Church Cathedral from the quays (1960s) [AUTH]

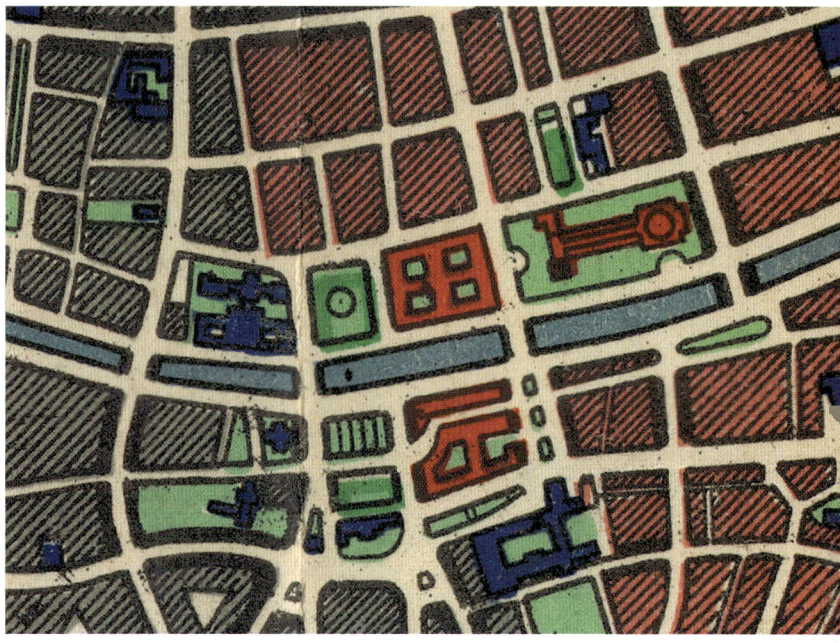

*Right.* Central City, showing redesigned road system around South Great George's Street and the new cathedral, extract from *Sketch Development Plan* (1941) [AUTH] Along the north quays, the Four Courts are in blue, and the cathedral is outlined in red on the east side of Capel Street.

estate. The square would become available in 1938 and the archdiocese was given a lease which allowed 60 years to complete the project or see the property revert to the Pembrokes. After a great deal of initial enthusiasm, augmented by the Eucharistic Congress of 1932, very little happened to make progress with the plan. No significant design work was undertaken and enquiries about future plans tended to be met with very terse responses. This may have given Abercrombie and his co-authors the impression that the Church was not heavily invested in Merrion Square and would welcome an alternative suggestion for a more imposing location. If they did, they were wrong. The proposal produced a furious response from the archdiocese along the lines that it was not the business of Dublin Corporation (the sponsors of the plan) to determine where the cathedral would be built. The Corporation (or at least most of its members) did not back down and the matter continued to simmer for some time. What took the heat out of it was that neither archdiocese nor Corporation had any intention of proceeding with the project.

While the main objection of the archdiocese was one of interference, there was a practical concern too. Dublin Corporation had become the planning authority for the county borough under the 1934 planning legislation. Under that Act it was obligated to produce a town plan with 'all convenient speed'. The archbishop's advisers were concerned that if Abercrombie's proposal was incorporated into that plan, it would be impossible for the Merrion Square project to succeed because, once adopted, the plan would have the force of law. The concern about the town plan abated because Dublin Corporation did not get around to making its plan until 1955 and had to be dragged through both High and Supreme Courts before they met their obligation. By this time, neither party had any enthusiasm for a cathedral project and another Abercrombie idea of a government quarter in the vicinity had similarly evaporated. The archdiocese never deviated from its position that the cathedral would be built on Merrion Square but never did anything to advance the project either. Dublin Corporation never revisited the project, concentrating instead on the practical problem of getting their civic offices built. Ultimately, Merrion Square passed quietly into the hands of Dublin Corporation and was transformed into the public park which Dubliners and visitors alike enjoy today.

# 1957

## Henry and Moore Streets

While the maps produced by the Ordnance Survey were the gold standard in terms of accuracy and quality, there was a significant time lapse between their various editions. The 1:1,250 series for Dublin showed the city in very useful detail but most sheets were not updated after the mid-1940s. Changes were noted on maps used by local authorities, but these were not generally available. Therefore, having a Goad plan for a later period was extremely useful because of the information which it contained.

Map 3 (1957) covered an area very close to O'Connell Street and in the heart of the northside shopping district. The northern side of Henry Street was at the bottom of the map and the shops would have been well known. On the corner was Egan and Co., drapers, and beside them was a small branch of Burton, tailors (a much larger one was on the corner of O'Connell Street and Earl Street). Another well-known store on this block was Slowey's. This was a women's clothing store which offered a full range of styles but specialised in clothing for the fuller figure, ensuring a clientele drawn from a wide area. The map captured something very new – a supermarket. H. Williams, described as a 'Grocery and Provisions' store was at No. 47. Supermarkets were a brand new concept in Dublin, and while self-service was commonplace in the United States before the Second World War, it arrived in Dublin only in 1956. Cummiskey's Self-Service opened at 45 Grafton Street and so radical was the concept that they felt it necessary to take advertisements in the newspapers to explain how the system worked. The potential impact of the concept was immediately apparent, particularly the capacity for larger stores selling more lines to be managed by a staff no bigger than needed for the traditional counter service. However, it was recognised that they would have to deal with another phenomenon – pilfering. The H. Williams store was their first supermarket in town and they followed with another in 1963

---

Charles E. Goad, central Dublin, extract from Goad fire insurance plan, sheet 3 (1957) [AUTH]

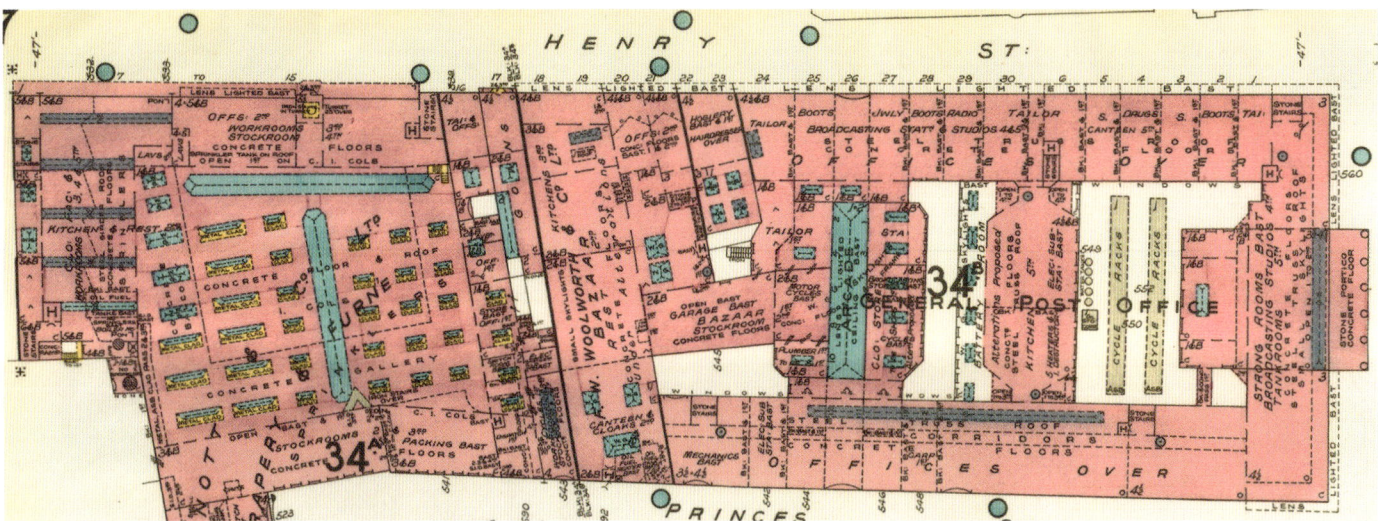
Charles E. Goad, Henry Street, extract from Goad fire insurance plan, sheet 7 (1961) [AUTH]

in part of the Pim's department store on South Great George's Street. Other companies, both local and international, followed quickly and by the mid-1960s there was an oversupply of supermarkets in the city centre. Companies came to the realisation that stand-alone supermarkets were better suited to the suburbs from both the retailer and shopper perspective, and began to invest in larger and larger units there. So, the map captured a moment of significant change.

The middle block on Henry Street/Mary Street was occupied entirely by Roches Stores. Department stores were seen as crucial to the success of shopping on a main street – they generated the footfall and the other shops benefitted from the passing trade. On Grafton Street, it was the presence of Brown Thomas and Switzers facing each other across the street that provided that draw, but Henry Street had three major draws. There had been a 'warehouse' there since 1867 which became the Henry Street Warehouse in 1881, and was taken over by the Roche family in the late 1920s, becoming as much an institution as Arnotts across the road.

The main northside shopping district was linear and continued along Mary Street. It was a more mixed offering here and never had quite the same status as Henry Street. What drew people along it was the presence of another department store at the junction of Jervis Street. Todd Burns focused on fashion, as they all did, but there was a particular emphasis on making goods available on credit and they marketed themselves heavily to engaged couples with offers on home furnishing. By the end of the 1950s they needed a new business model and the old style department store was replaced with a 'supermarket' style approach to clothes selling.

The area to the north of Henry Street was very different. By the early 1930s the area bounded by Parnell Street, Moore Street, Henry Street and Chapel Lane was a warren of narrow lanes and courts. Tenements were ubiquitous and second-hand shops were the norm. It was able to support businesses that needed a lot of space because nobody else wanted it. Decline continued into the late 1950s and it remained a warren of lanes and passages but there had also been a lot of demolition. The main central block was largely vacant and even Hendron Bros, who had replaced Hely's envelope business, were able to use much of the site as no more than a yard; this within a short walk of O'Connell Street and Henry Street. Tenements were still present and the focus on second-hand selling still dominated the service sector. Moore Lane had been characterised by wooden buildings but there were now a lot more wooden sheds and stalls, an indication of the general decline

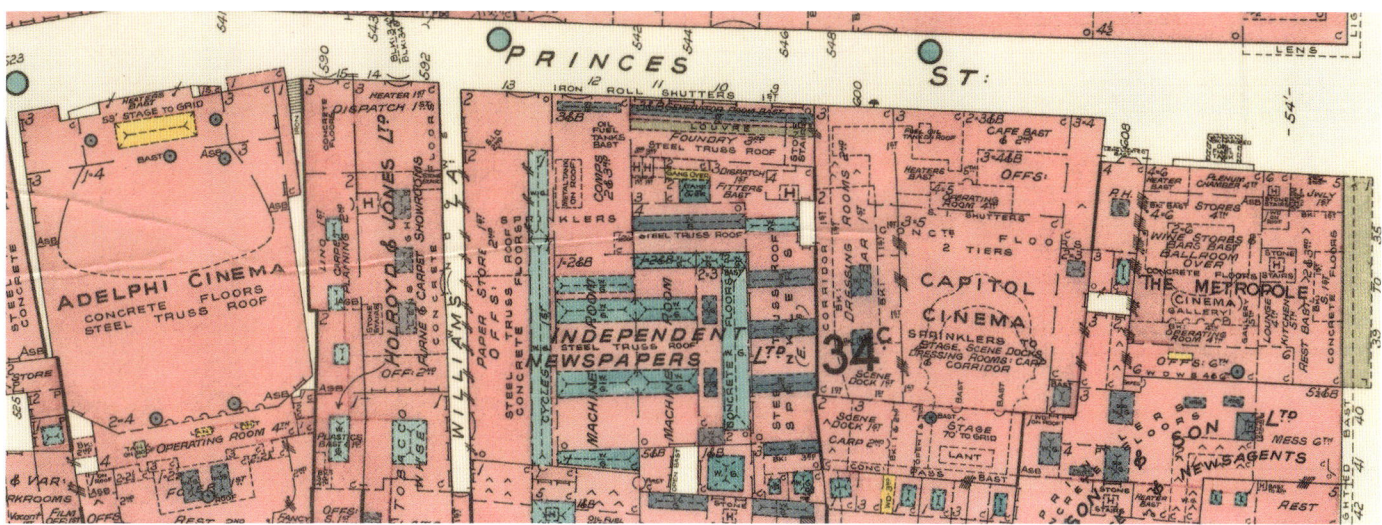

Charles E. Goad, Princes Street environs, extract from Goad fire insurance plan, sheet 7 (1961) [AUTH]

and unwillingness to invest in anything more permanent. Despite Dublin Corporation's promotion and involvement in renewal from the late 1960s, it would take another decade before the dereliction in the area would begin to be addressed with the opening of the ILAC shopping centre.

The picture of the other side of Henry Street comes from 1961. Sheet 7 was dominated by the Arnotts department store, which occupied most of the block and extended almost to Middle Abbey Street. It was another Dublin institution and had managed to survive the 1916 Rising by a hair's breadth. It aimed to give the same kind of quality shopping experience as was available on Grafton Street. It sold the full range of goods with a focus on women's clothes and on furniture. The plan shows that the operation was concentrated within the building with stockrooms on the top floor and warehousing in the inner block. There were also workrooms on the third, fourth and fifth floors. Woolworth's two main stores were on Grafton Street and on Henry Street. The Henry Street premises had a rather narrow frontage but, as the plan shows, there was considerable depth as well as a restaurant on the second floor. It also had an entrance onto the Henry Street arcade, whose glass roof was marked on the map. Woolworth's was just as much an institution in the city as any of the department stores, selling all manner of objects, toys, cosmetics, costume jewellery and ice cream. The remainder of the shops towards O'Connell Street were occupied by a range of activities that included a jewellers, a shoe shop, a tailor and a branch of Hayes, Conyngham and Robinson, the chemists. At the O'Connell Street end was the block occupied by the rebuilt General Post Office. Much of the upper floor space was given over to offices as part of the post office operation. However, this was also the location of the broadcasting studios of Radio Éireann. Television had yet to arrive in Ireland so the radio was a vital medium for information and entertainment. The plan noted the studios on the fourth floor facing O'Connell Street and on the fourth and fifth floors facing Henry Street. Another feature that made the area distinctive and which was captured here was the cluster of cinemas. O'Connell Street, though regarded as the main street in the city, had become an entertainment centre with four cinemas and many restaurants. The map located the Adelphi, which opened onto Abbey Street, the Capitol, which opened onto Princes Street, and the Metropole, directly on O'Connell Street. The Metropole had been one of the city's prestige hotels prior to the 1916 Rising, and though rebuilt, was never a hotel again but became a restaurant, cinema and ballroom.

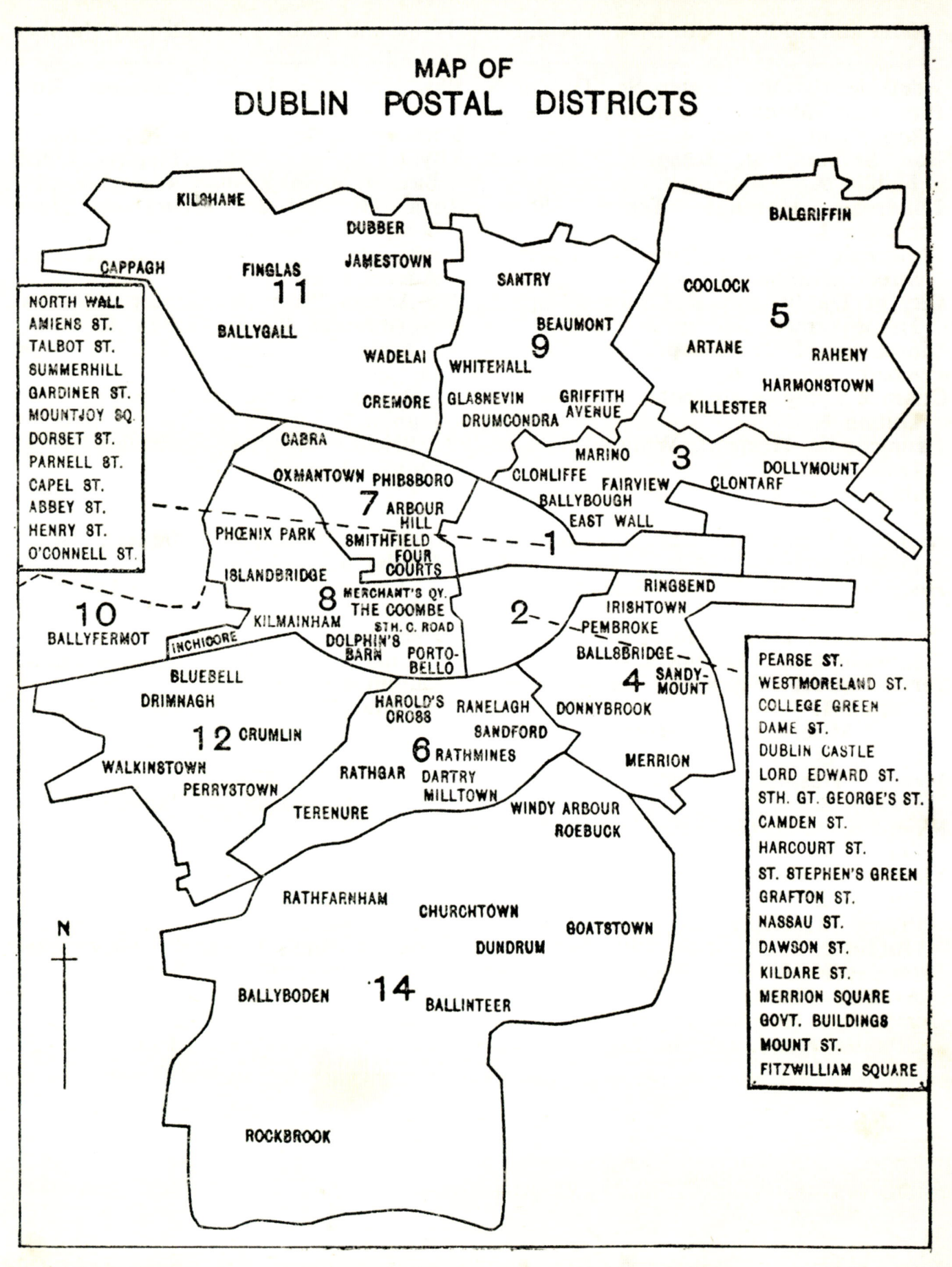

# 1961

## Dublin postal districts

There was an announcement in the *Irish Times* of 5 September 1963 that the Ordnance Survey had published a new edition of their map of Dublin. This map, at a scale of 2.5 inches to the mile (1:25,000), was the fourth Popular Edition, the edition that showed individual streets. This, it was explained, was the first to show the bus routes and the postal districts. Though this might have been the first time that the Ordnance Survey showed bus routes, these had previously appeared on Bacon's maps of the city, but the appearance of postal districts was significant. Despite having been introduced simply to speed the delivery of post, they were quickly becoming an important factor in differentiating 'good' residential areas from the 'not so good'. Having them on a large-scale map now removed any doubt about where a property was located, though it took a little effort to find the boundaries on the map, so light was the printing.

The current system had been introduced in January 1961 when all householders received a card in the post telling them in which of the 13 postal districts they lived. Odd-numbered districts were on the north side of the Liffey, even-numbered on the south side. The names given were as follows: 1 Pearse Street (north), 3 Fairview, 5 Killester, 7 Phibsborough, 9 Whitehall and 11 Finglas; and 2 Pearse Street (south), 4 Ballsbridge, 6 Rathmines, 8 James's Street, 10 Ballyfermot, 12 Crumlin and 14 Churchtown. It was explained that there was no district 13 because there was no need for it at the present and not because of some concern about ill-luck. It would be introduced if and when needed.

This was not the first time that postal districts had been used in Dublin. In the nineteenth century Dublin was divided into six districts – C, N, S, W, SE and NE – though it does not seem that these were either widely known or used by the general public. It was only in 1927 that a significant reform of the system was attempted together with a notification to

---

The new postal districts (1961) [AUTH] Map distributed as part of the publicity campaign.

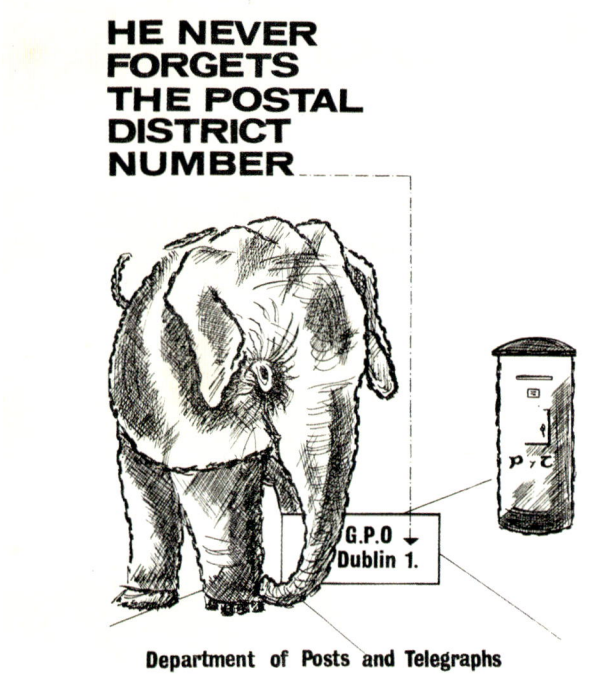

'Reminder to use the code', advertisement from Thom's *Dublin Directory* (1965) [AUTH]

all households of the new system. Each location would be identified by a combination of letters and numbers – 10 Cabra Road NW1 – very similar to what was being used in London. Customers, both domestic and commercial, were encouraged to make use of the system but it seems to have made little impact. Advertisers generally ignored it, though official and government communications tended to persevere with it, even in some cases after the new system was introduced in 1961.

This time, there was hope for greater success and the Department of Posts and Telegraphs reported satisfaction in September 1961 with the degree of take-up of the idea. They reported that 30 per cent of households and businesses were making use of the system. They intended further information campaigns and hoped eventually to get usage up to 90 per cent, though it would remain voluntary. This was assisted by the inclusion of a general map of postal districts in the telephone directory from 1961, though phones were still a relative rarity in the domestic world. The postal district of every address was now included and there would be 166,000 copies distributed.

Even though the purpose of the postal districts was to ensure the efficient distribution of the mail, the potential of using the zones for classification purposes was obvious. The city was divided into a number of electoral areas and these had been used for years for census and statistical purposes, but they had long and confusing names. A simple numerical system had many advantages, especially as numerical (computer) processing of data was becoming more common. Superficially it seemed that these areas might have some degree of social homogeneity and that they might divide the city into social areas. This might have made some sense within the county borough south of the Liffey because of the way in which Dublin developed historically. Even there, the reality was more complex and certainly the idea did not make sense when looking at the city holistically. This did not stop people using the postal districts in this way and they came to be associated with what was generally perceived to be the dominant socio-economic character of the area. Dublin 8 crosses the Liffey, breaking the pattern of odd numbers on the northside, even numbers on the southside. Many will assert that this was done to ensure that Áras an Uachtaráin and the US Ambassador's residence, got a more desirable 'southside' post code. When Micheál Mac Gréil published his groundbreaking research on educational disadvantage in the city in 1974, he felt it useful to classify areas using postal districts and included a map showing the percentage of those with national school education only based on these areas. Likewise, property owners began to understand the value of being in the same postal district as desirable residential areas and of the difficulties posed by the opposite. It took only a little while for the desirable characteristics of 'Dublin 4' and 'Dublin 6' to be recognised as a postal address. In fact, Dublin 4 (or D4) came to be more than just a location, it was seen to epitomise a certain kind of upmarket lifestyle and attitude. Matters were crystallised in 1988 when An Post (the post office services of the Department of Posts and Telegraphs) decided that it

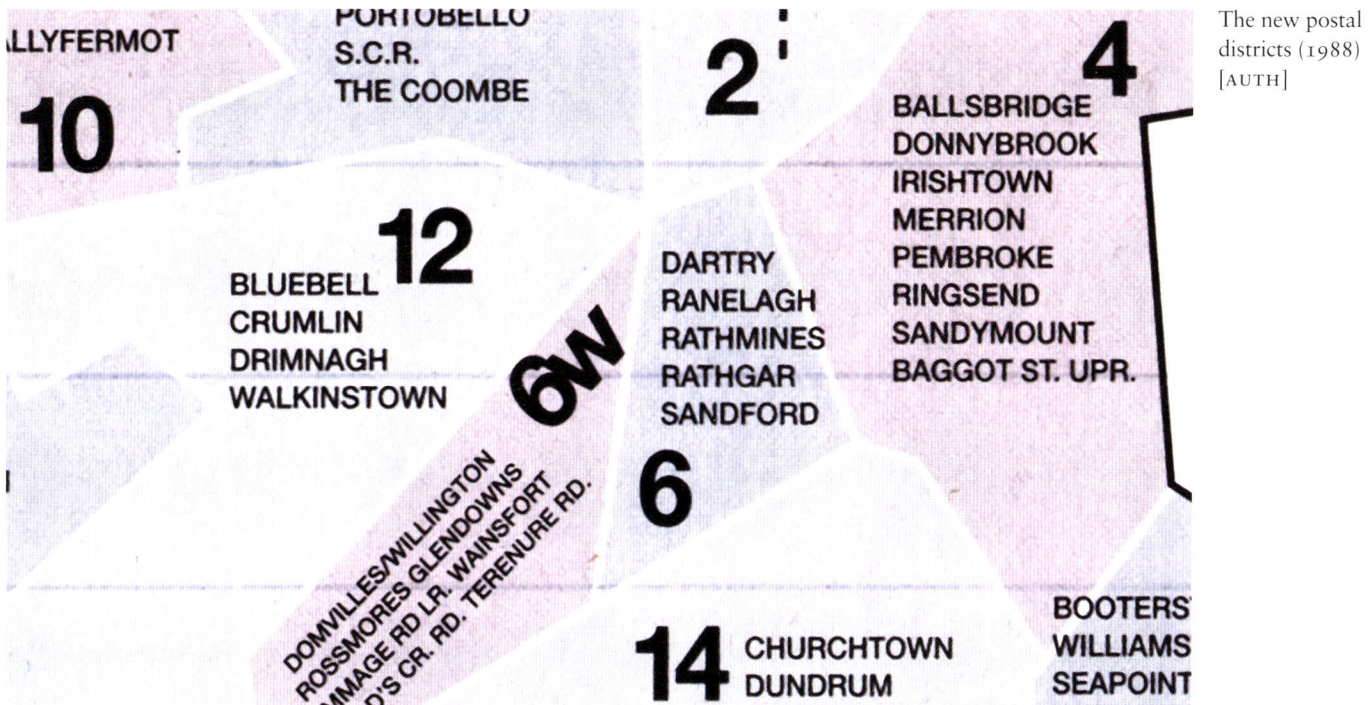

The new postal districts (1988) [AUTH]

needed to make a significant adjustment to Dublin 6 and Dublin 12 to improve efficiency by taking parts of each to create a new Dublin 26. The suggestion was met with a furious response in Dublin 6, with public meetings and action committees established and threats of legal action. The reason was simple enough but never stated; the loss of the cachet of a Dublin 6 address. An Post was asked to come up with a creative solution and it found one in the arrival of Dublin 6W, which allowed An Post to reorganise their routing and met, to some degree at least, the concerns of the residents. Postal districts continued to take on far more significance than merely aids to postal sorting and could generate intense emotions, and the occasional court case.

By the early years of the twenty-first century it was clear that the system was no longer fit for purpose and that a system of individual postcodes was needed. It took a long time before the Eircode system was developed, but the previous experience with change was remembered and the existing postal codes were retained as the first part of the Eircode. Thus all residential Eircodes in Dublin 4 now begin D04. In the UK, the postcode system has long been used for demographic profiling of areas. Each code identifies a block of properties, not an individual house, and it has proved possible to associate postcodes with socio-demographic/health/crime data. Since they are also spatially contiguous, it is possible to aggregate them easily into larger units. This has not always been to the advantage of residents as they can become associated with undesirable traits by virtue of their postcode. In Ireland, the Eircode system was designed to identify individual properties but the second part of the code – the individual identifier – is a random allocation of characters. So knowing the Eircode of one house does not allow someone to the predict the Eircode of houses around. This makes the kind of profiling that has become so prevalent in the UK more difficult, but it does not stop it. Companies which use Eircodes to schedule deliveries have the ability to find and group spatially contiguous properties, so properties can be grouped and classified. It is just not as easy to do as in the UK, and not as widely accessible.

*Top.* Parnell Monument (1911)

*Middle.* Nelson Pillar (1809)

*Above.* O'Connell Monument (1883)

Twentieth-century postcards [AUTH]

# 1966

## The monumental landscape

At 1.30 a.m. on 8 March 1966 an explosion blew off the top portion of Nelson Pillar, destroying the statue of Admiral Horatio Nelson. The Pillar, as it was known, had been a landmark in the city since it was completed in 1809. In fact, it was regarded as the city centre and distances were measured from there. It was also an important attraction – for a payment of 6d and a breathless ascent of 168 steps, a panoramic view was available in all directions over the city.

Nelson Pillar, however, had proved controversial since the beginning. For almost as long as cities have existed, those in power have recognised the value of commemoration in prominent locations of what they believe important. Dublin Corporation recognised this and Sackville Street, with its noble proportions, became an important place to assert its more nationalist orientation following the local government reforms of 1840. It was the obvious choice for the O'Connell monument, whose foundation stone was laid in 1864 in the presence of vast throngs, and later for the monument to Charles Stuart Parnell (1911).

The Pillar was out of place among these monuments to Irish nationalism and there were regular calls for its removal, but these were little more than ritualistic. It was well known too that the Pillar was managed by a Trust which had no power to demolish it, even if they wanted to. It would have taken an Act of the Oireachtas to remove it, and nobody was going to go that far. People liked the Pillar – it gave its name to a number of adjacent businesses – and it was a practical point of orientation and meeting. Referring to particular sides of the base gave precision to meeting locations – the Nile side or the Trafalgar side. There was a genuine surprise therefore when the explosion occurred on the eve of the 50th anniversary of the Easter Rising. However, the fate of the Pillar was decided with unusual haste. Having been damaged but perhaps repairable, it raised the thorny question of who might

---

Monuments on O'Connell Street, extract from Ordnance Survey plan, 1:2,500, sheets 18 (VII) and 18 (XI) (1939) [AUTH]

Nelson Pillar in the immediate aftermath of the explosion (March 1966) [AUTH]

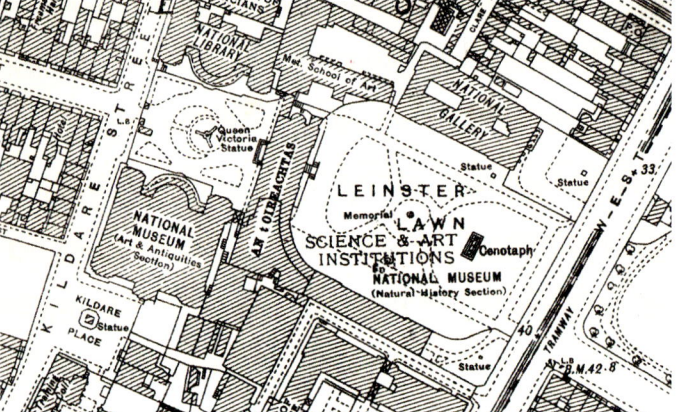

Leinster House and lawn, placement of monuments, extract from Ordnance Survey plan, 1:2,500, sheet 18 (IX) (1939) [AUTH]

sit on top. Rather than address that, the state, without consulting Dublin Corporation, decided that the simplest solution was to remove the pillar entirely and this was done before the Easter Week celebrations, leaving a major gap in the street and the city without a central focus.

While Dublin Corporation had been active in adding new monuments to the city's streets, it had never been particularly iconoclastic and most of the time it was content to leave imperial monuments as they were. Most of these though were removed following an act of vandalism, with the Corporation taking the opportunity to remove rather than repair.

That subtlety of approach was evident in the question of how to commemorate Prince Albert and Queen Victoria. Following Prince Albert's death in 1861, the Queen indicated that she expected a suitable monument to Albert in the city. A committee was established to raise money by public subscription and a sufficient amount was collected. An initial decision by Dublin Corporation to place it on College Green proved not to have public support and, following some diplomatic gymnastics, an alternative site was found in the grounds of the Royal Dublin Society. After all, it was rationalised, this had been site of the first International Exhibition in 1853 and the Prince had been hugely important in its success. Dublin Corporation could now take comfort in the view that this was not public land and they were merely facilitating it. In any event, it took some time for the statue to be completed but it was in place by the middle of the 1870s.

This offered a solution when it came to commemorating Queen Victoria. She had made it clear that she felt that Dublin had been lax in failing to honour her. Her visit in 1900 generated new enthusiasm and following a public meeting in the Royal Dublin Society (RDS) in May 1900, a committee was formed and subscriptions raised. A statue was commissioned in 1903 which was ready in 1908. It fell to the Lord Lieutenant, Lord Aberdeen, to do the unveiling and the ceremony was fully reported in the *Irish Times* of 17 February 1908. It described a large and enthusiastic crowd, over 1,000 soldiers in 'review order' and gave a full account of the speeches, but there was no mention of Dublin Corporation. It was a major political event but, at the same time, it was a private ceremony. The land on which the memorial stood was private space, behind

Memorial to Prince Albert on Leinster Lawn looking out over Merrion Square, late nineteenth-century postcard [AUTH]

Victoria Statue outside Leinster House, postcard from early twentieth century [AUTH]

substantial railings and whether or not it was a public monument could be left to the observer to determine.

The tricky question of how to commemorate Queen Victoria and Prince Albert in Dublin had been answered with some finesse, but the problem was not destined to go away. With Irish independence in 1922 came the question as to where the parliament would be located. Various locations, including the Royal Hospital, were examined and during this process in 1924 the RDS building was taken as a temporary solution, which became permanent. This meant that the parliament building had a statue commemorating Queen Victoria at its entrance and another in memory of Prince Albert on its lawn. The matter was raised occasionally in the Oireachtas, but the prevailing view was that it was best to leave things as they were.

Following their untimely deaths, it was felt appropriate to place a monument to Michael Collins and Arthur Griffith on Leinster Lawn which was unveiled on 13 August 1923. This was done by the simple expedient of putting a cenotaph between the Albert Memorial and Merrion Square. Since the new monument was in the form of a Celtic cross, 12 metres high with a large base, it effectively obscured the Albert monument. This is the layout which is seen on the map here, an extract from the 25-inch sheet for 1939. The monument to Prince Albert is described simply as 'memorial'. When the time came in 1949 to replace the cenotaph (which suffered from dry rot) with the present monument, Albert was not removed, rather he was moved to a peripheral location near the Museum of Natural History where he currently remains, largely unnoticed behind a hedge.

The question of the more substantial monument to Queen Victoria rumbled on over the decades but it was not until 1947 that it was moved. Political motives were denied, and instead, the need to provide parking for the members of the Oireachtas (now that petrol was available again) and the need to improve direct access to the main door meant that it was no longer possible to accommodate the statue. It was now simply in the way.

Of course, this raised yet another question: what to do with her? It was suggested that she might be moved to another site in Dublin but that cannot have been suggested seriously. Without much ceremony, she was put on a truck on 23 July 1948 and put in storage in the Garda Headquarters in Kilmainham (Royal Hospital). After some decades in storage, the statue was given on permanent loan to Sydney, Australia, in 1988. She sits on her plinth outside the Queen Victoria Building, a five-level shopping centre in a restored nineteenth-century building. However, the three smaller bronzes which were on the plinth – Hibernia at War, Hibernia at Peace and Fame – did not travel with her.

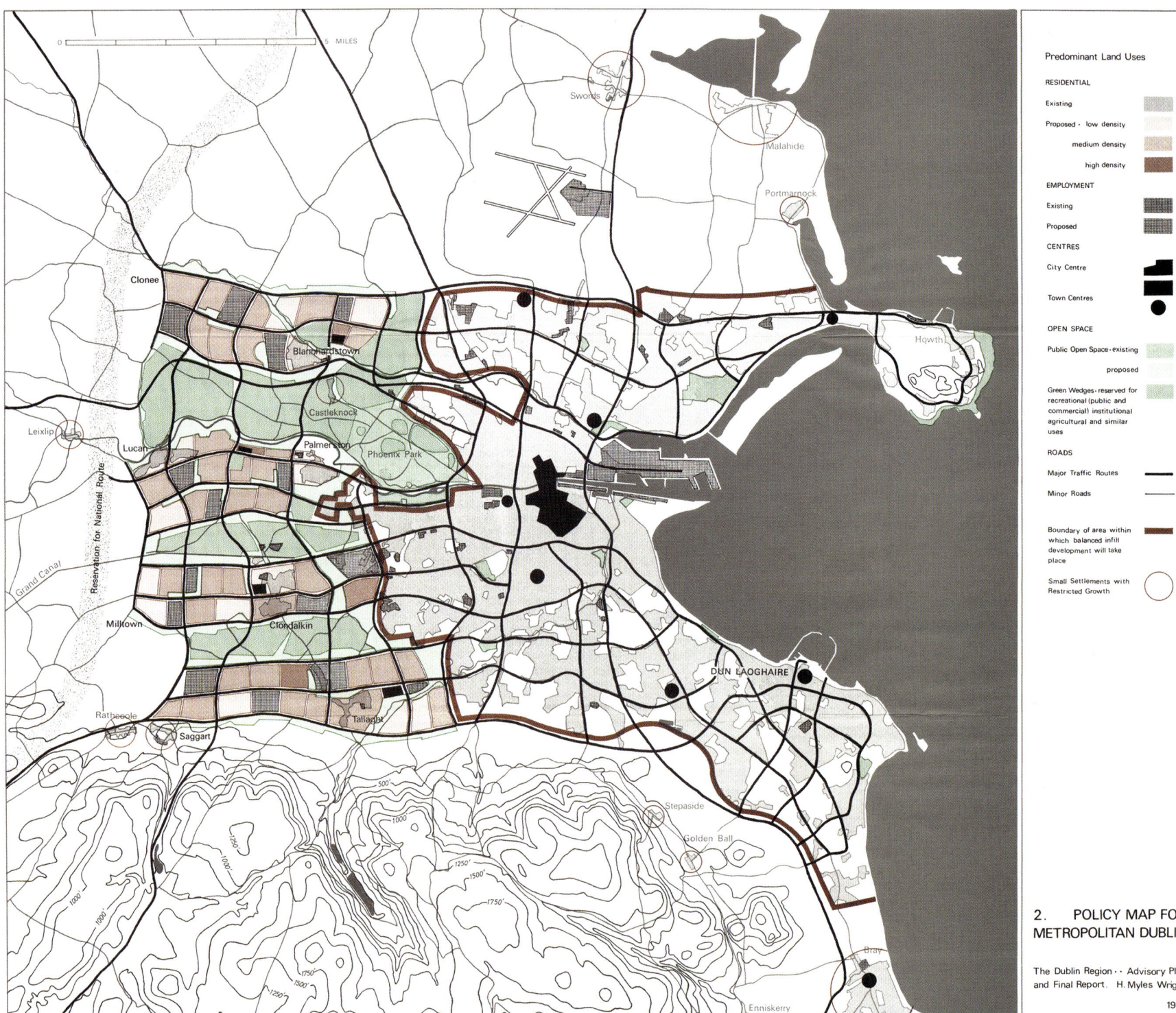

# 1967

## Planning the Dublin region

Despite the early enthusiasm in Dublin for town planning, the city was slow to set up the formal processes for the production of a town plan. One of the features of the 1934 Town and Regional Planning Act was that Dublin Corporation had the power to act as the regional planning authority for Dublin, Kildare, Meath and Wicklow. It chose not to do so, indeed it developed a reluctance to produce even its plan for the city. Having been forced to produce a plan in 1955, the Corporation and the state found that the planning process set out in the 1934 Act (as amended) was unworkable and the plan was dropped. This led to the 1963 Local Government (Planning and Development) Act, which among other things, set out the process for the development of a town plan. It also provided the legislative basis underpinning the decision of the Minister for Local Government, Neil Blaney, to commission an advisory report for the Dublin region, defined as Dublin, Kildare, Meath and Wicklow, with quite comprehensive terms of reference. It set out a land-use pattern that would deal with future growth and the settlement network, roads, communication, public services and utilities.

The minister did not look far for his chosen consultant. Abercrombie had been a force in Dublin's planning for most of the century until his death in 1957, long after his retirement from Liverpool University. Myles Wright was now the Lever Professor of Civic Design in Liverpool University and he was chosen to lead the team of experts.

The first and summary report was received in 1966 with the more detailed technical report arriving in 1967. The future options for Dublin were limited by its geography. Despite suggestions being made in previous decades, there was no appetite for filling-in the bay and neither was it seen as reasonable or feasible to extend development into the Dublin

---

Myles Wright, 'Policy Map for Metropolitan Dublin', *The Dublin Region* (1967) [AUTH]
This shows the new linear towns separated from each other and the existing city by green belts.

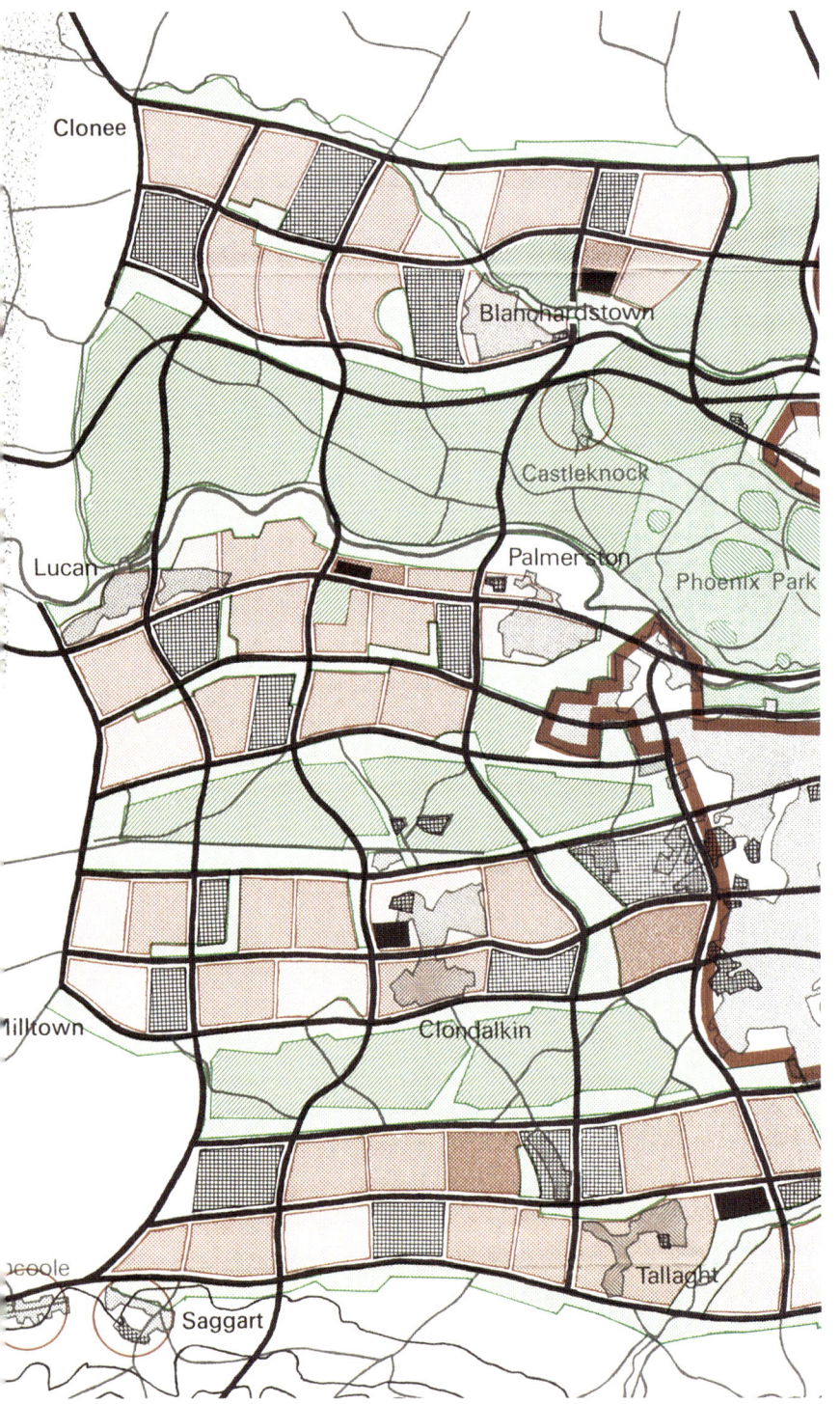

mountains given their recreational potential. Development could take place to the north but it would have to skip over the airport zone. There were possibilities along the coast to both the north and south but Wright felt that the problems caused for infrastructure did not make this a good option. That just left the western edge of the city and, unsurprisingly, it is here that Wright suggested that growth should be accommodated.

The question now moved to the practicalities of how to design the expansion of the city. Wright's suggestion was that the city's growth should be organised into new settlements – Abercrombie had said the same in 1941. These would be laid out on a rectangular transport network which would connect with a new high-capacity 'national' route to the west and to the existing city via a new routeway around its edge.

The 'new towns' programme was coming to an end in the UK, but Wright was influenced by the approaches taken in Cumbernauld and Runcorn, with their compact and highly structured planning, which had more than a hint of early *sotsgorod* planning in the Soviet Union. He proposed four quite narrow linear towns, oriented on a west–east axis which would permit growth to the west. Accepting the value of green belts and in line with UK new towns, the new Dublin towns would be separated from each other and from the city by amenity spaces. However, these could not be new and independent towns in the UK sense because their proximity to the city ensured that they would be closely linked to the capital in all aspects. The locations were Blanchardstown, Clondalkin, Lucan and Tallaght, with Blanchardstown suggested as the first to be developed.

While the importance of the car was understood, the provision of good public transport was seen as essential. The main spine roads would run on the long axis of each town with perpendicular routes dividing the town into zones. Within these would be residential areas and industrial areas. Express

Myles Wright, 'Policy Map for Metropolitan Dublin', extract showing the zonation of the region and layout of the new towns (1967) [AUTH]

bus routes would operate along the spine roads but with a local bus service within the town. The width of each town would be between 1 and 2 miles, depending on whether one or two bus routes were provided, to minimise walking distances to the buses. A town centre would be provided with a wide range of shops and services, including colleges, clinics and local administration, with the aim of encouraging people to stay local. While the level of service provision would be comprehensive, Wright did not envisage the town centre being in competition with the range of goods and services available in the city centre. The local bus service would link with the town centre, where there would also be linkages to the express bus service to the other towns and the existing city.

The residential areas would be enclosed by the main traffic routes with the access roads gradually filtering traffic towards the individual dwellings. There would be very little through traffic, it being directed outwards towards the enclosing routes. A pedestrian route through the centre of the residential area would provide easy access to the church, local shops and primary schools. This would be quite high density, with open space limited to the areas immediately around the dwellings and along the pedestrian route. The aim would be to ensure that no house was more half a mile from the local services, with small shops located no more than a quarter of a mile away. Wright did not specify what kinds of dwellings might be provided but high rise would seem to be inevitable in this kind of configuration. The pedestrian system would link to other zones and ultimately to the town centre, though Wright did not envisage an entirely separate pedestrian circulation system. Any demand for additional open space would be met in the green belt, which would be within reasonable walking distance.

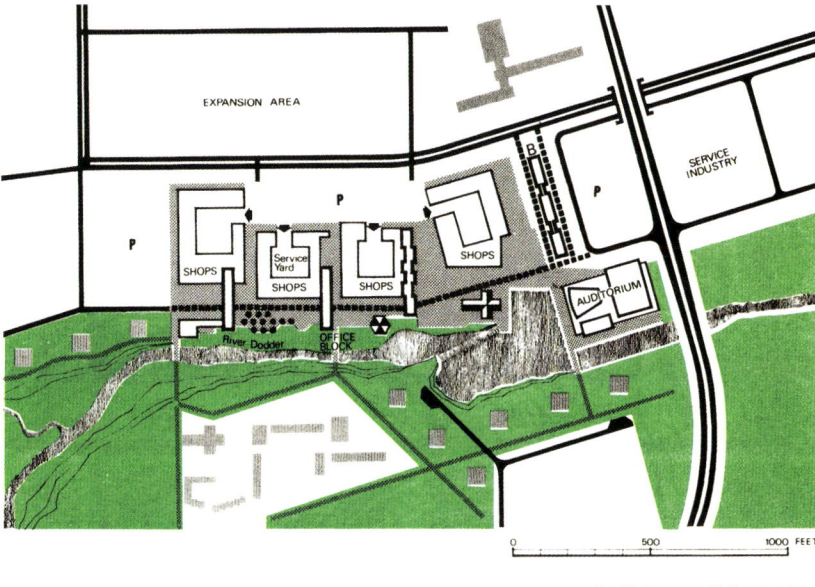

Myles Wright, 'General Layout for the Tallaght Centre', extract from *The Dublin Region*, 29 (1967) [AUTH] Although not explicitly mentioned, the principles outlined and the layout suggested the use of high-rise blocks.

The general layout for Tallaght showed that Wright envisaged the town centre bounded by the main roadways and expanding along the route of the river. The shopping area would be pedestrianised but provided with significant car parking. Wright suggested that it might be possible to run a local bus service through the pedestrian zone. The intention was to provide as much local employment as possible and so minimise commuting. Industrial estates would be integrated into the town, adjacent to the residential areas, and their scale could be adjusted to suit the production needs of the businesses located there. Employees would either be within walking distance or be able to make use of the local buses. The plans would permit the towns to grow to populations of between 60–100,000 or about 350,000 people in all.

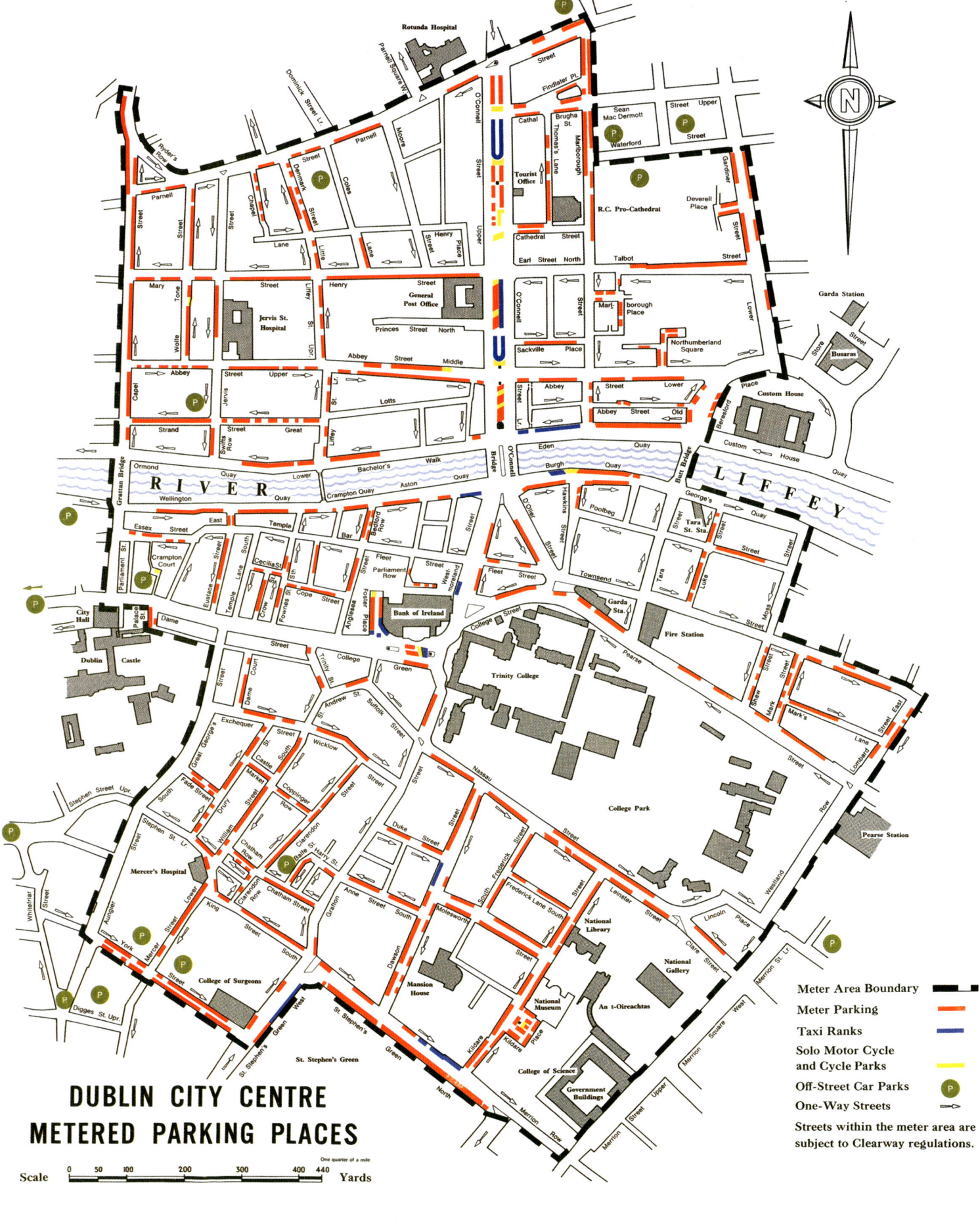

# 1969

## Bringing motoring under control

Traffic congestion in the city centre had long been a problem. Dubliners are often surprised to find out that traffic has never really moved freely in Dublin centre and that even by the 1920s there were calls for controls on motorists. One important contribution to the problem was parking. Photographs of Dublin from the 1920s show cars parked in every possible location, with much of O'Connell Street taken up with parked cars.

The approach up to the 1960s had involved bans and prohibitions, but a new solution was under consideration which might have the effect of both controlling parking and generating some revenue. Up to this point the only revenue generated went to the unofficial but ubiquitous parking attendants or 'lock hards' who, in return for a tip, would assist motorists to get in and out of impossible spaces. The approach decided upon was the installation of parking meters.

These finally arrived on 14 January 1970, after a decade of consultation and research, or dithering, depending on the point of view. This was partly because any decision on parking required a joint decision by An Garda Síochana, which had the power to make the necessary bye-laws, and Dublin Corporation. The power to control parking by meters was contained in the Dublin Parking Byelaws (S.I. 170) of 1969, and covered the County Borough, Dún Laoghaire, County Dublin and Balbriggan.

The map shows the 'control zone', the area within which meters would operate. It is dated 1969 but the system was introduced in stages early in 1970. The first phase covered an area which was encompassed by Parliament Street, Dame Street, College Green and Pearse Street, Lombard Street and Moss Street. This explicitly targeted commuters since maximum time was limited to two hours, for which the driver had to pay 1s per hour. However, compared to previous attempts at control from the 1930s, these were positively

Dublin Corporation, Parking Control Zones as approved in 1969 [TCD]

permissive, and meters operated only between 8.30 a.m. and 6.30 p.m. from Monday to Friday. Enforcement was by parking warden and a ticket could cost £1, though it was reduced to 10s if the excess time was less than 15 minutes, but it was not explained how this could be availed of. The sting was further diminished, though, by the fact that wardens did not come on duty until 10 a.m. It seems that most Dubliners did not see this map because it was widely believed that only this zone would be affected. There was criticism that civil servants and Corporation officials had engineered it in this way to avoid affecting the areas in which they parked. However, it soon became clear that much more was envisaged. Next was the area to the west of O'Connell Street and bounded by Parnell Street, Capel Street and the river, and meters arrived there on 16 February. On 9 March 1970, a further 500 meters were introduced into the area bounded by Parnell Street, Marlborough Street, Waterford Street, Lower Gardiner Street, Beresford Place and Eden Quay.

There was a brief respite during April but on 27 May meters arrived in the area immediately north of Dame Street and College Green and bounded by South Great George's Street, Aungier Street, York Street, St Stephen's Green West and Grafton Street. The final extension took place on 10 June 1970 with the zone further expanded to take in the area between Nassau Street, Leinster Street, Clare Street, St Stephen's Green North and Merrion Row as well as Westland Row and Lincoln Place. The only reason for phasing the process can have been the need to get the system up and running as fast as possible once the legal and administrative obstacles had been cleared. There was nothing novel in the area to be covered. It was basically the same control zone that had been in operation since controls were first introduced in 1937. At that time Assistant Commissioner Brennan was quoted in the *Irish Times* as saying 'the streets have been turned into garages'. So extensive and wide ranging were the controls that were introduced at that time that it was felt necessary by the Traffic and Safety First Association of Ireland to have a 40-page guide produced, which contained the second map shown here. The 1937 regulations limited 'waiting' on main streets to 20 minutes between 8.30 a.m. and 6.30 p.m. There was unilateral parking on some streets, with each side operating on alternate days and a host of limits of different kinds. The '20-minute' rule persisted into the 1950s and seems to have been enforced with some energy in the immediate post-war years. So, in many respects, parking meters were a gentler solution.

Very quickly, it was decided to extend the area covered by parking meters into the secondary shopping and commercial districts and, by the end of 1973, it was reported that there would be 3,600 meters in the city, a thousand more than the original plan. An extension announced in August 1971 included parts of Parnell Square, Merrion Square, Fitzwilliam Square, Castle Street, Cork Hill and Lord Edward Street. They reached Thomas Street on 27 November 1972.

The introduction went smoothly enough. An Taisce complained about the aesthetic effect on the Georgian landscape of the installation of meters around Fitzwilliam Square, and so did a group of residents and commercial interests, not just from Fitzwilliam Square but from Baggot Street, Leeson Street, Merrion Square and Herbert Street. Their spokesman was quoted as saying that 'the thousands of office workers and residents in the Georgian area must have somewhere to park their car for long periods'. This was, of course, to miss the point. The idea of parking meters was to prevent long-term parking and the clear message to those working and living in the inner city was that long-term parking was not in the interests of the city and was going to become much more difficult. The first prosecution noted was not for a failure to pay but for malicious damage. Three Dubliners were fined £10 each and had to compensate Dublin Corporation for wilful damage to four parking meters on Burgh Quay. The fact that the event took place at 12.15 a.m. might shed some light on the motive. Vandalism to the meters proved to be a significant problem, but whether this was motivated by annoyance at the concept or just because they presented an easy target was unclear. Burgh Quay and York Street were particularly targeted locations and while most damage was done by jamming match sticks or similar materials into the slot, it seems that some came prepared with sledgehammers and crowbars. The latter,

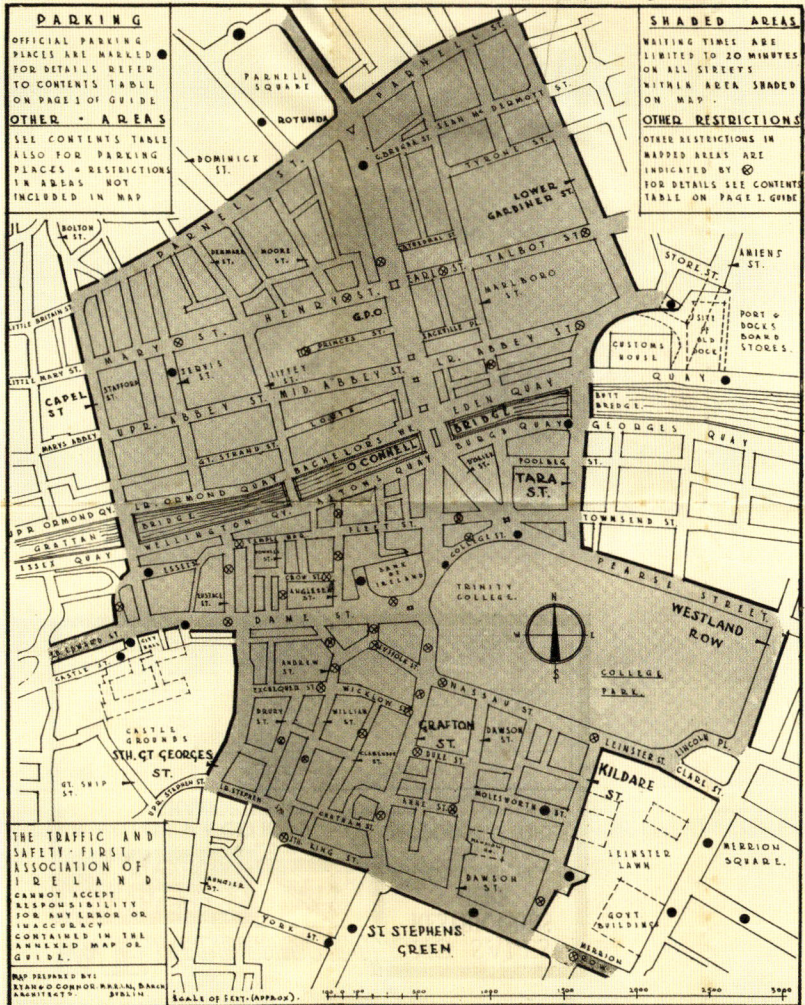

*Left.* Traffic and Safety First Association of Ireland, the 'original' control zone in 1937, *Dublin's New Traffic and Parking By-Laws* (1938) [AUTH]

*Above.* 'Parking for 20 minutes', *Dublin Opinion* (February 1955) [AUTH]

though, seemed to be more associated with theft, such as when hacksaws were used to behead forty meters in York Street allowing the thieves to make off with the contents. This led to a more robust form of parking meter being introduced and these survived, more or less, until the advent of the current electronic machines made them redundant.

Dublin Corporation was not yet hostile to the car, so attention was also given to accommodating those which would be displaced by the installation of meters. It was therefore fortunate (if unfortunate from an urban point of view) that the city centre had a great deal of brownfield sites. These were the result of slum clearances and the migration of industry and other commercial bodies to suburban locations. North of the Liffey, there was little or no demand for these sites as locations for new commercial building, despite a boom in office-block building in the late 1960s. These sites found an important use as off-street car parks as they waited for redevelopment. Since they were intended to be temporary (though temporary turned out to involve decades in some cases), there was little or no investment in their appearance – there was usually no more than a fence to prevent access and a rough surface. These did little to enhance the appearance of the city.

# 1971a

## Preservation and renewal: the development plan

The first development plan for Dublin was finally taking shape as the 1970s began. Dublin would have three plans, reflecting the fractured governance of the city – one each for the county borough (Dublin Corporation's area), one for the county and one for the borough of Dún Laoghaire. The long time it had taken to get a robust planning process in place ensured the plan was awaited with great interest. While Dublin Corporation's plan covered most of the built-up area at the time, even at this stage there were significant middle-class suburbs in the county area and most of the city's future growth would be concentrated there.

Making a plan involved producing a draft document, which then went for public consultation. Following consideration of observations and objections, the plan would be adopted and form the legal basis for development for the period of its validity, usually five years. Dublin's draft plan was felt to be sufficiently ready for the public in 1969 and it generated so much public interest and commentary that it was not ready for adoption until 29 November 1971. It probably got an easier passage than might have been expected because the political response to its proposals within Dublin Corporation was muted. The Corporation had been dissolved by ministerial order in April 1969 for its refusal to set a rate and it fell to the commissioner, Mr John Garvin, to accept the advice of the City Manager that the plan was ready, all objections having been considered.

The plan comprised a written statement which set out policy under a number of headings, housing and roads being two of the more significant. Important buildings and structures were listed for preservation and were categorised into one of three lists – A, B and C, with different levels of preservation applied to each. The plan was accompanied by a set of maps

Dublin Corporation, *Dublin Development Plan 1971*, Map 2, extract showing the
north city areas for preservation (red) and areas in need of redevelopment (purple) (1971) [AUTH]

and these contained the detail and the import of the more prosaic policy aspirations. Map 2 was particularly interesting with its emphasis on roads, preservation and urban renewal. Drawn on Ordnance Survey sheet 18 at a scale of 1:10,560 (6 inches to the mile) it showed the planned road developments, many of which would be within the central area. These were the tangent routes which the various consultants had suggested as being necessary to move traffic through the city. While the outline of routes was clear, Map 2 was unable to give any indication as to the real spatial impact that these would have. There was no suggestion as to the width of the roads or of the configuration of the junctions.

More dramatic was the highlighting of the problem of dereliction and the identification of areas for preservation. The boom in demand for commercial buildings had put the best-preserved part of Georgian Dublin under threat. It was precisely because the area was well preserved that the threat existed. The south-east quadrant of the city centre had retained its higher status and this is where businesses wanted to locate. At this time, there was no great love for Georgian architecture – its style was not appreciated by the general public and the configuration of single-family homes was not suited to modern business arrangements. Others, indeed, had a more negative view and saw it as the architecture of the oppressor and something to be rid of as quickly as possible. This was a minority view; the dominant view tended towards indifference in not recognising the cultural importance of built heritage. Visual education in Ireland had been limited and only the physical landscape was associated with culture and heritage. This attitude was beginning to change as the development plan was making its way through the approvals process – some ghastly modern buildings, the ESB building in Fitzwilliam Street for example, and the heavy-handed tactics of developers in Hume Street, had nurtured an increased awareness of the importance of preserving what was good in the urban landscape. This was reflected in Map 2 where the Corporation outlined the structures designated for preservation. The concentration in the south-east was very clear and this extended across the canal into then leafy Ballsbridge, which was becoming a much sought-after business location. In the northern city, the main focus was on an area around Gardiner Street from Parnell Square to Mountjoy Square, while Henrietta Street was singled out from its immediate surroundings. The map did not show that these areas were in a different state to those south of the river. Decay was evident throughout this part of the city and much effort would be required to bring these buildings back to acceptable levels. Nonetheless, Map 2 was an important statement of the new realisation that urban landscapes and not just individual buildings were worthy of, and in need of, preservation.

Of even more concern was the area in the various shades of purple. These were the areas badly in need of renewal, where obsolescence and decay were already obvious. The area between O'Connell Street and Capel Street, discussed above, was highlighted and the plan supported the idea of joining up the two main shopping districts. There were large areas in need of attention along and behind the quays to the west of the centre, a large area across the Liffey from the Custom House where the warehouse operations were no longer needed, the entire medieval city and much of the area bounded by Patrick Street, Thomas Street and the Coombe. The slum clearances of Dublin Corporation resulted in vacant sites and declining local businesses. Industries and businesses were moving to the suburbs where land was cheaper and transport more efficient and there was no demand for the sites that they left. Developers were unwilling to risk investment in anywhere other than the core area of the south-east, despite a boom in office demand. The development plan recognised the dangers and sought to discourage further intensification of the office sector in the south-east and encourage it elsewhere, though it was unclear how it might be done. It was this parlous state that caused the *Architectural Review* magazine in November 1974 to produce a lengthy feature on Dublin, later published as a stand-alone report, entitled *A Future for Dublin*. It would be the 1980s and 1990s until the problem was finally addressed.

*Opposite*. Dublin Corporation, *Dublin Development Plan 1971*, Map 2, extract showing the inner city and the south-east commercial area (1971) [AUTH]

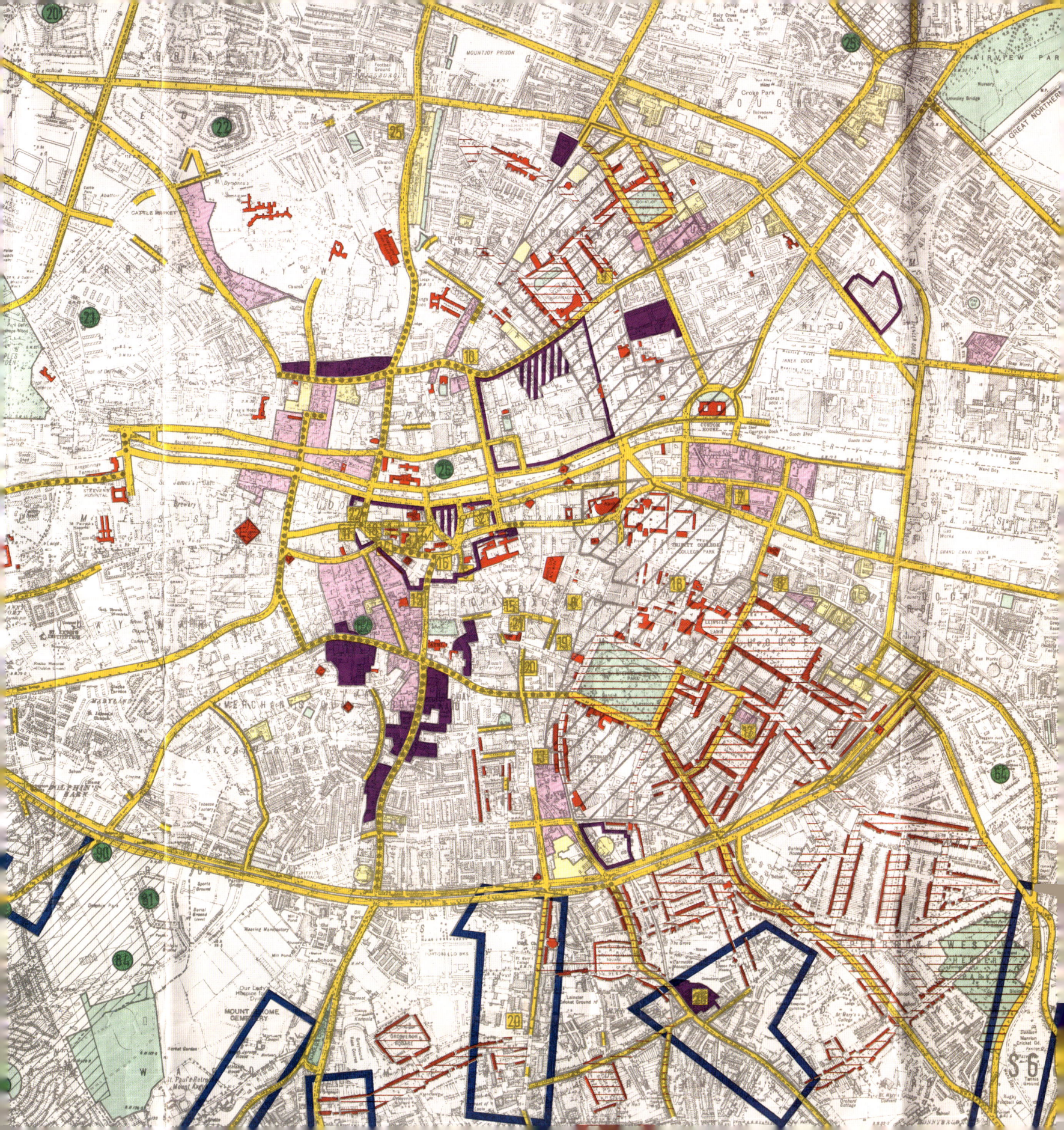

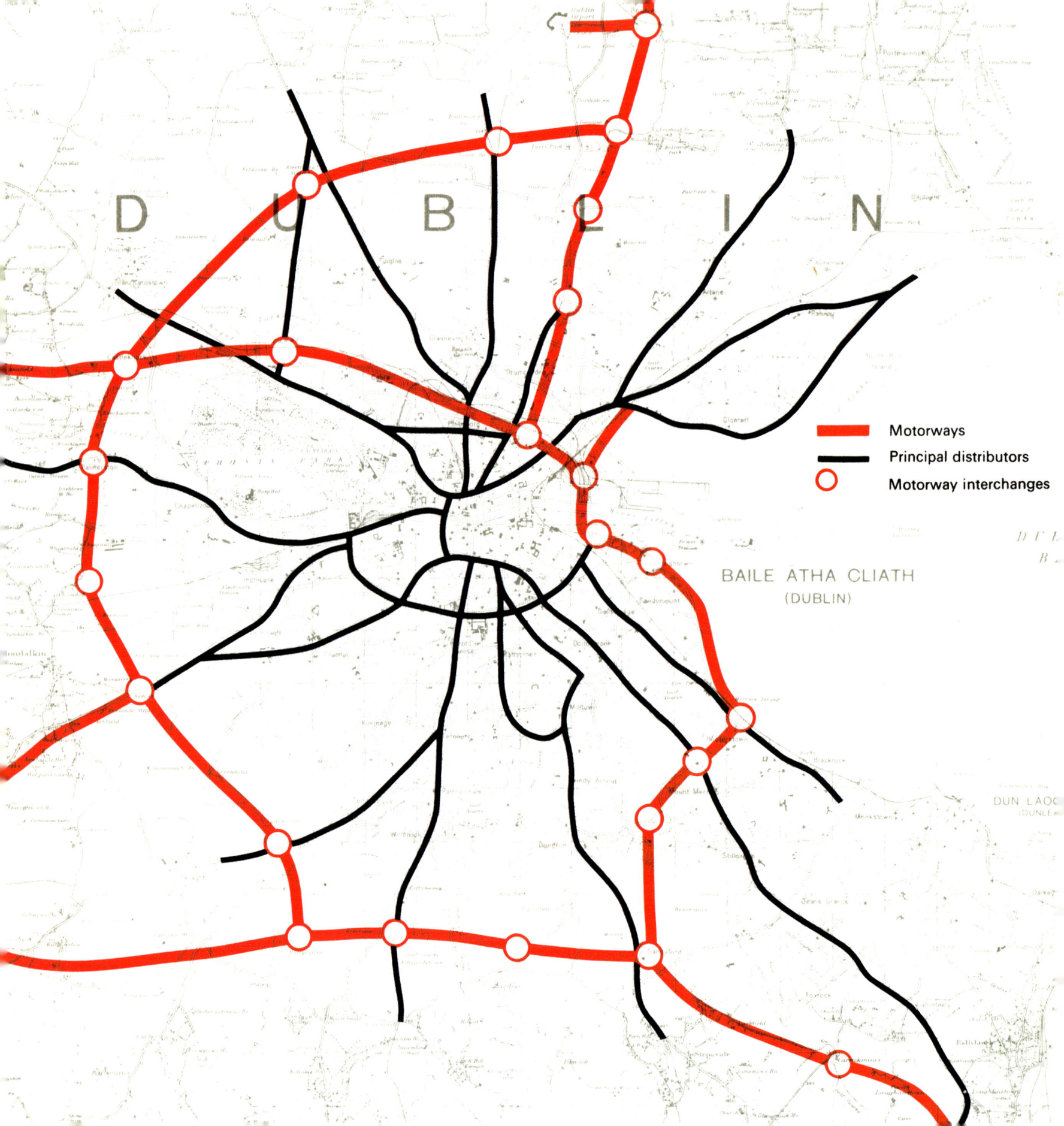

# 1971b

## Urban motorways and the city centre

As earlier maps have shown, Abercrombie had recognised the importance of planning to manage the traffic of the city even at a time when private car ownership was very low. It was still relatively low during the 1950s, but Ireland's economic success in the 1960s resulted in rising levels of home ownership in Dublin and an increase in private cars. While the approach to city planning was still focused on making the city more accessible to cars, it had long been accepted that some form of management was needed. The search for solutions came to involve traffic consultants and the decades following the 1950s were characterised by a variety of reports, though, fortunately perhaps, little was implemented.

The first of these comprehensive traffic plans was completed by Professor Schaechterle in 1965. One of the key elements of his plan was that traffic would enter the central area via 'express traffic' roads, essentially urban motorways, using a combination of the circular roads and the Royal and Grand Canals. This would lead traffic into a high-capacity 'tangent square' that would circulate traffic around the central area. The northern segment would be along Parnell Street and North King Street. This would link in the west with a southwards route along Bridgefoot Street and Pimlico to Cork Street. The southern link would join with Cork Street and run eastwards through the Coombe and break through to St Stephen's Green South. The final element in this square was a complicated linkage from St Stephen's Green East to Merrion Street into Westland Row and then across Butt Bridge to Gardiner Street where it would link with Parnell Street. These would be big roads of between four and six lanes and would require significant demolition of the existing fabric, but this did not generate a great deal of controversy, even in relation to the best-preserved Georgian district around Merrion

---

The Highway Network in 1991, proposals from the *Dublin Transportation Study*,
reprinted in R. Travers Morgan and Partners, *Central Dublin Traffic Plan*, 4 (1973) [AUTH]

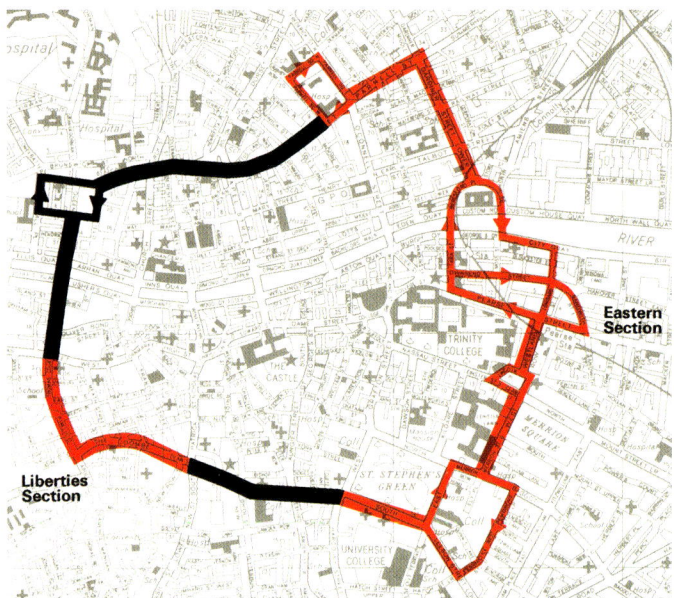

R. Travers Morgan and Partners, Corporation's suggested Inner City Tangent Route, *Central Dublin Traffic Plan*, 2 (1973) [AUTH]

Square. This was possibly because there were no maps or diagrams in the body of the report and people may not have had a sense of the physical impact of the plans.

What did generate controversy and ultimately a rethink of sorts was the Grand Canal proposal. The canals were not particularly valued in Dublin; the era of canal traffic had passed and the canals were in a spiral of decline. This was particularly so with the Royal Canal as it flowed unnoticed for much of its route through Dublin. The Grand Canal was different because it and the South Circular Road were in parallel for much of its route and accessible, but even so it was felt to belong to a bygone age. That was at least until Schaechterle proposed that one of the express traffic routes would run along the route of the Grand Canal. He did not explicitly suggest that the canal be drained but it would require either that or substantial demolition on its banks. Dublin Corporation, however, *did* suggest that they might drain the canal temporarily (perhaps) to facilitate a new sewer. This produced a surprising backlash from the middle classes who lived nearby. They did not believe that the canal would return

as they felt that the opportunity would be taken to build the road. A new love for the canal emerged, but more importantly a political storm ensued. Those in government understood what was at stake and the canal was saved. Subsequent plans for the Grand Canal were more modest but such modesty did not extend to other locations.

As Schaechterle's proposals were being digested, Myles Wright was preparing his plan for Dublin in its regional context. To manage connectivity between the new towns and the existing city, he proposed five major routeways, in essence a circular motorway around Dublin with the Royal Canal now providing the entry route into the city and with a coastal route linking the Belfast and Wexford roads. These proposals were refined further in the Dublin Transportation study, undertaken by An Foras Forbatha and which reported in 1971. They proposed an orbital motorway system, elements of which are recognisable today: the M1, the M50 and the M11. This system would also have a motorway along the line of the Royal Canal as well as one along the coast, including Sandymount Strand, then south-westwards along Foster's Avenue towards Sandyford where it would join with the outer ring.

Running a motorway along the Royal Canal route generated mostly positive reactions initially. The plan did not specifically mention filling in the canal; the reference was to the 'canal route' but anyone who knew the canal also knew that a motorway would occupy the entire cutting and more. This motorway also had very few interchanges and the one at Castleknock and Navan Road (Cabra) could be accommodated without much disruption. This was not the case with the proposed interchange at Drumcondra where the Royal Canal motorway would meet the extension of the northern arm of the network (the current M1 from Belfast). While much of the route would be through then undeveloped land this would change once the system got to Whitehall and then to Drumcondra. The canal motorway would be six lanes here and it would meet the airport arm at Jones' Road. As Technical Report 16 put it 'the section of the motorway between the interchange with Dorset Street and the New Liffey Bridge poses problems from the geometric design point of view. This is because there are two interchanges within 0.4 miles and the turning and weaving movements are large.' This too seemed to be initially acceptable, though it also seems likely that the local residents were largely unaware of what might happen.

Dublin Corporation decided that it needed another study to look at the practical land-use implications of the various proposals for the city centre and they commissioned a study from consultants R. Travers Morgan. Travers Morgan revisited the inner city tangent square and, reflecting growing public militancy, the more intrusive proposals for the best-preserved Georgian district were scaled back. What they proposed for the remainder of the route was dramatic and would have required considerable demolition. Their plans, overlain on the city streets, showed complex motorway junctions within a short distance of the city centre with grade separations.

The northern section of the tangent was similar to that previously suggested. From a link with the new Royal Canal motorway at Ballybough, it would go via Summerhill, Parnell Street and North King Street to join with the western section at Smithfield. Mostly dual carriageway but with three lanes at difficult junctions, there would be an underpass at the junction of Parnell Street and O'Connell Street and the Parnell monument would have to be moved. The western route would be much more impressive. It crossed the Liffey to the east of St Paul's Church in Arran Quay via an upper-level bridge and then continued between Patrick Street and Francis Street and from Kevin Street to Clanbrassil Street it followed the line of Blackpitts. This would comprise two three-lane carriageways with grade-separated interchanges. The southern element spared the Coombe by running on a more southerly route between Cork Street and Patrick Street and would cut through Cuffe Street feeding into a new system of one-way streets. While the effect of the plan can be gauged from the maps, the absence of visuals reduced its impact. The *Architectural Review* provided such an impression in its 1974 analysis of Dublin and it became clear that the road would cut a swathe through the urban environment and be a very significant visual and physical barrier. The proposal to run a motorway link along the coast at Sandymount also generated a furious

Figure 12.03 Recommended Route for the Western Section of the Tangent Ring Road

R. Travers Morgan and Partners, recommended route for western leg of Tangent Route, *Central Dublin Traffic Plan*, 79 (1973) [AUTH]

response from the local residents. They too had become better organised than previously and were militant, politically astute and well resourced. The road would have ruined the amenity of the bay and cut the city's population off from a much-loved promenade and strand.

The money did not exist for such schemes in the mid-1970s to be undertaken as large capital projects. Sections of road were built as opportunities arose and the intention to build the entire system (or something like it) continued into the 1980s. However, by then the public mood had swung against what were seen as intrusive developments and local opposition to urban motorways was much more organised. Also, there was greater understanding in planning circles that there would never be enough space for cars and that the focus had to shift to controlling their use.

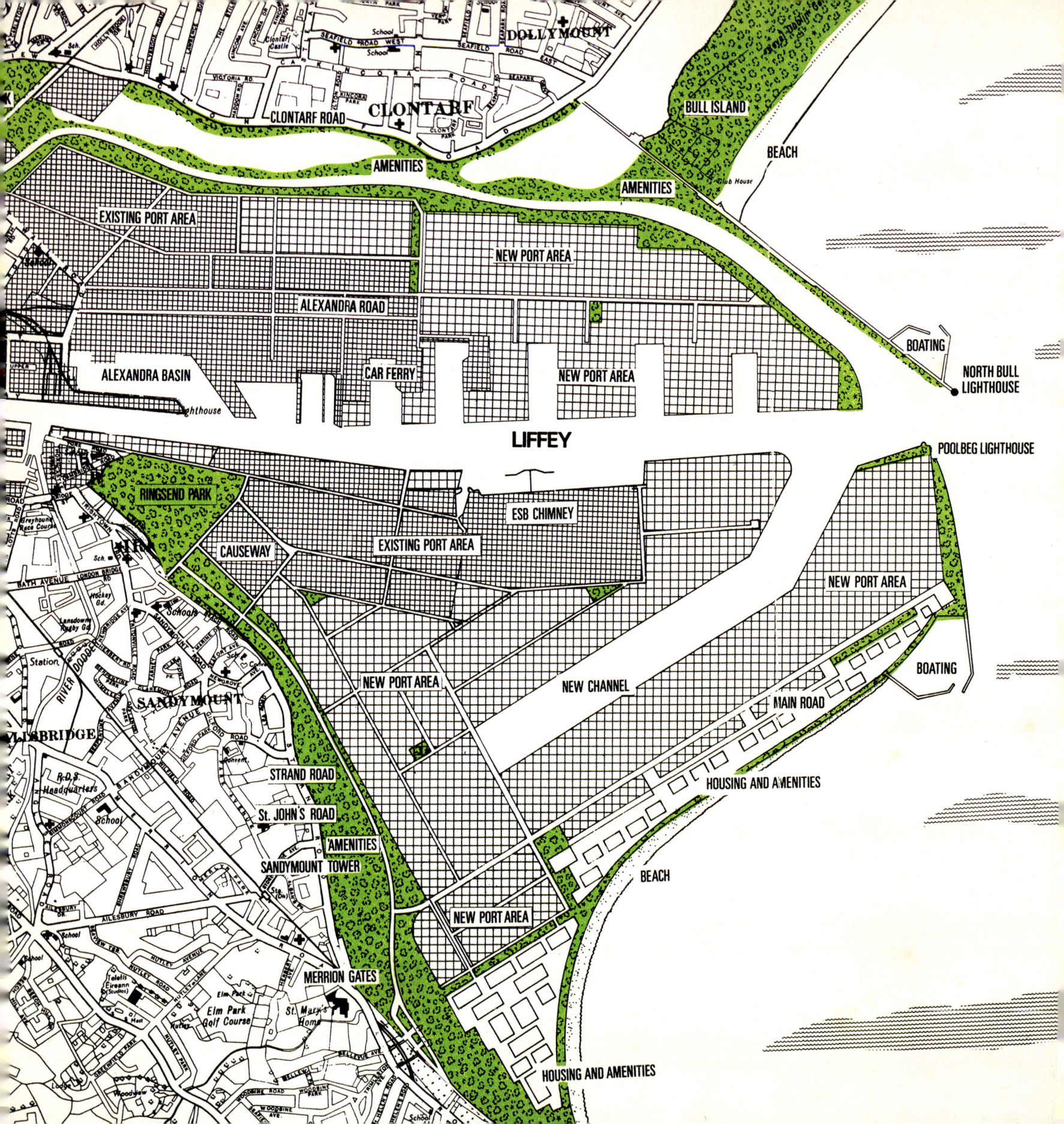

# 1972

## The port and the bay

Those tireless travellers and essayists, Mr and Mrs Hall, began their chapter on Dublin with the following:

> What a glorious impression of Ireland is conveyed to the eye and mind upon approaching the noble and beautiful bay of Dublin! It is, indeed, inexpressibly lovely; and on entering it after a weary voyage, the heart bounds with enthusiasm at the sight of its capacious bosom, enclosed by huge rocks, encompassed in turn by high and picturesque mountains.

Others saw the bay as a place with economic possibilities. Its shape and shallow configuration made reclamation a possibility. An arc could be drawn from a number of natural or manmade features which would enclose a considerable area of land. Abercrombie, as mentioned earlier, had seen the possibilities offered by the South Wall to enclose a large area from there to Blackrock. Another body which had a particular interest was the Dublin Port and Docks Board. Over the centuries the port had moved eastwards, gradually reclaiming land as needed, but further expansion now required much more. Consideration of future development led the board to publish a series of proposals in 1972. The *Studies in long term development of the Port of Dublin* was produced by a team of consultants which included two with much experience in the development of Rotterdam. It was emphasised at the time of publication that this document was not a plan, rather a series of possibilities which were presented for discussion and debate.

This was appropriate because the board was making suggestions for an area of the city which it did not control. The strange governance structure of Dublin gave the Port and Docks Board significant control over the port area and the River Liffey but Dublin Corporation was the planning

---

The proposed port, extract from DPA/635/010, *Studies in long term development of the Port of Dublin* (1972) [DP]

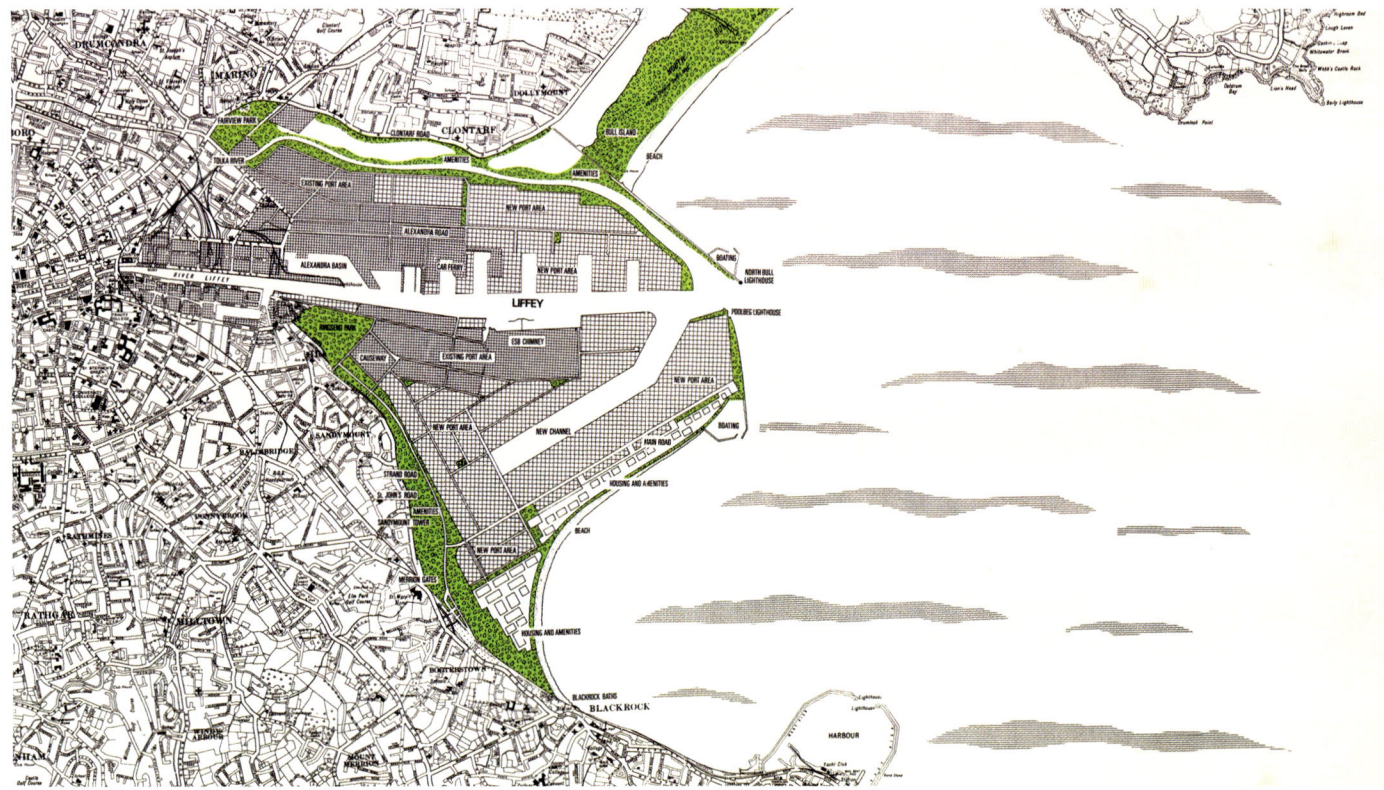

Dublin Port and Docks Board, proposed reclamation in Dublin Port, DPA/635/010, *Studies in long term development of the Port of Dublin* (1972) [DP]

authority for the city, with Dún Laoghaire Borough and Dublin County Council also having areas of interest. Dublin Corporation had very recently completed its development plan and there was no suggestion of significant reclamation in the bay. It was considering the possibility of a motorway along Sandymount Strand so it is fair to conclude that the amenity value of the bay would not have been foremost in their thinking. It had been known for some time that the Port and Docks Board was considering major reclamation and those suggestions had been important in the establishment of some of the groups who would now object. It was the scale of what was suggested that was breathtaking.

A number of proposals were up for consideration but at their core was reclamation of the area that was enclosed by the North and South walls together with a wide arc reaching from the Poolbeg lighthouse to Blackrock. The reclamation would involve 550 acres on the northside and would reduce the bay to a narrow inlet along Clontarf Road which would meet the sea at the North Bull lighthouse. The port activities running parallel to Clontarf Road would be screened from view by an amenity area. The reclamation on the southside would be more extensive and would involve 1,740 acres, with 1,150 acres devoted to port activities. The remainder would involve 220 acres for housing and an amenity area along the existing coastline. The housing proposal was particularly strange in a Dublin context, but perhaps made more sense in Rotterdam. The suggestion was that 5,000 dwellings could be provided by building high-rise blocks of from 14 to 18 storeys and apartment (spine) blocks of between 6 to 12 storeys. These were intended as a dramatic landscape feature and capable of being seen from as far away as Dún Laoghaire. The planners were somewhat dismissive of any concern that the residents along the Sandymount coast might have, noting that if it was not this kind of development, then they were likely to get a coastal motorway.

The reaction to the proposals broke along expected lines. There was a welcome for the employment potential from various trades unions and a furious rejection from the local groups, especially from the Dublin Bay Preservation Society, who had become far more organised in recent years. The matter might have rested there since the Port and Docks Board had made it clear that these were long-term proposals which might take 20 or more years to realise. What injected energy into the debate was a proposal to get the process going by building an oil refinery.

The suggestion that an oil refinery was under consideration had been around for some time and pre-dated the publication of the port studies. It became a firm proposal only in late 1972 when it was announced that an application would be made for a refinery on a 200-acre site to be reclaimed by the Port and Docks Board, east of the Electricity Supply Board (ESB) generating station. While the role of the Port and Docks Board was clear, an added air of mystery was that the identity of those wishing to build the refinery would not be disclosed until permission was given, though it was suggested that Irish interests would be in the minority. This galvanised opposition from local groups and local politicians, and the planning application was rejected by Dublin Corporation in 1973.

The rejection by Dublin Corporation did not end the matter because the Port and Docks Board appealed the decision to the Minister for Local Government. A public inquiry was required and this began in January 1976. The arguments in favour centred on the need for Ireland to have secure supplies of oil products and on the employment potential of the project. The arguments against were very extensive and ranged from the question of amenity to the quality of the information provided by the proposers. Objectors argued that the Port and Docks Board did not own the land on which they were proposing to build and therefore

The modern port on the northside, looking towards the Bull Island (2016) [AUTH]

had no status in the matter. Moreover, they did not have the expertise to manage an oil refinery. There were objections on environmental grounds too, pointing out that environmental protection was weak and no such project should proceed until these had been radically improved. Crucially, it was also clear that the government was lukewarm, at best, about the idea and had other potential locations in mind for a refinery, especially Whiddy Island in Bantry Bay. Following consideration of the inspectors' report, the minister refused permission in May 1976 but the idea of a refinery rumbled on for many years. The application was a salutary lesson to Dubliners who learned that projects with major environmental impacts could be approved for Dublin in the face of opposition from the planning authority, Dublin Corporation. Radical improvements in environmental protection were necessary as a matter of urgency.

Dublin port eventually was able to reclaim some additional land from the bay, parallel to Clontarf, but not to the extent envisaged in this map. Any suggestion of filling in the bay along Sandymount disappeared, as did the proposal to build a motorway there, and it seems very unlikely that any such proposal will be revived in the medium term. This has left Dublin port short of areas for future expansion and there have been suggestions that a radical solution is needed – moving the main business of the port to another location for instance, probably Drogheda.

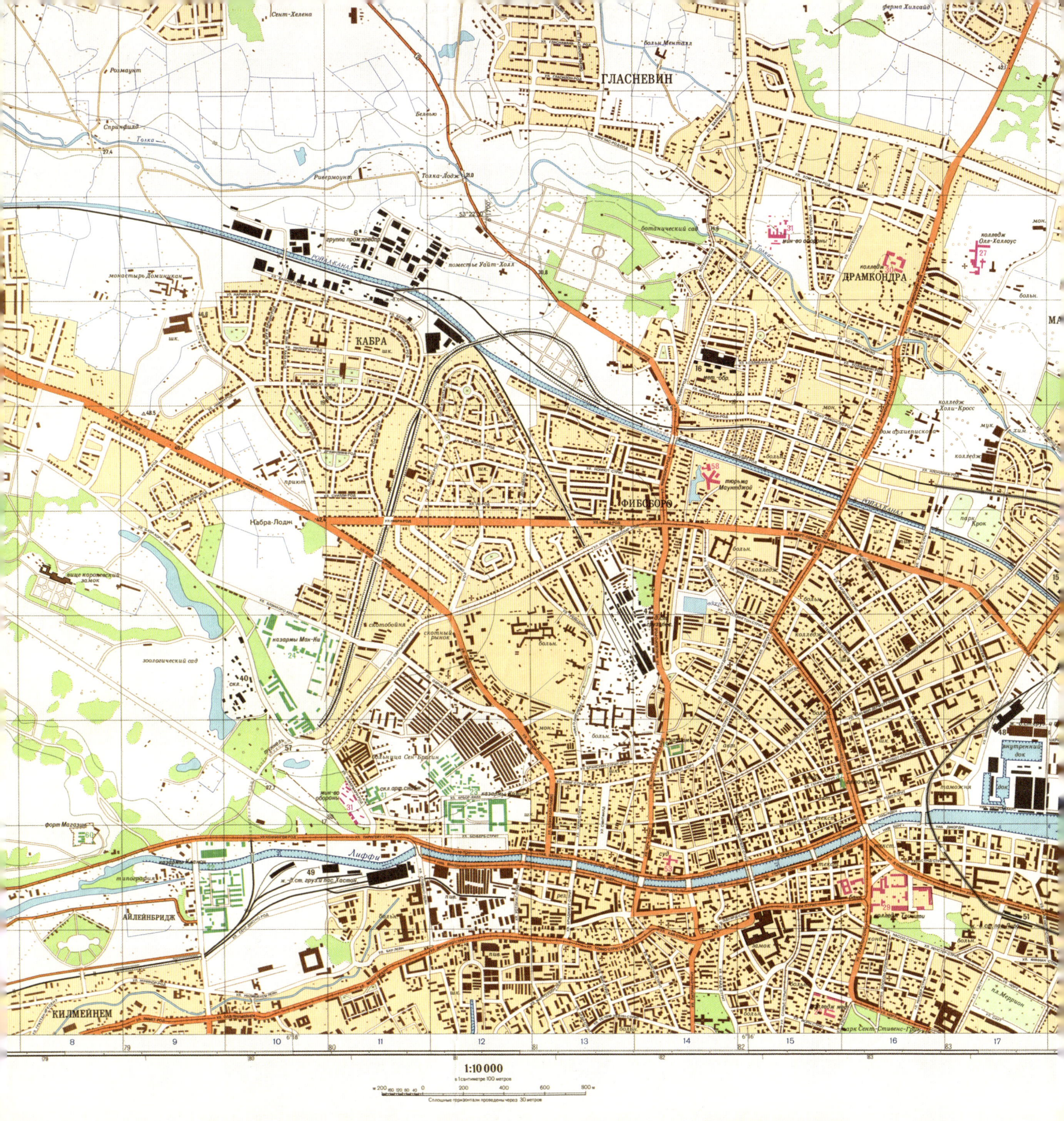

# 1980

## Targeting Dublin: USSR General Staff maps

The January 1960 edition of *Esquire* contained an article by Caroline Bird entitled 'Nine places to hide'. It was an examination of where on the planet would be relatively safe in the event of global nuclear war. As she put it: 'If you really want to be safe from atomic destruction, here is the most up to date of survival surveys – a guide to the few remaining places on this earth where human life would not be destroyed.' There were nine such locations, spread across the world, based on an analysis of likely targets for destruction and the most likely resulting patterns of radioactive fallout. The safest place in Europe was Cork, though it seemed that whoever drew the accompanying map had only a vague idea of where Cork was located. Ireland was also a good location because of low population and the fact that it had most of the resources needed for a modern industrial economy. Indeed, the article suggested that the reason that German companies were investing in Ireland was because of the likelihood of their survival in a post-war world. Not many people in Ireland would have been aware of the advantage held by Cork because *Esquire* was on the list of banned publications, but anecdotally it has long been suggested that this was the catalyst for the 'discovery' of Cork, especially west Cork, by English and Dutch immigrants.

Every household in Ireland received a copy of a booklet entitled *Bás Beatha*. First published in 1965, with a braille edition in 1966, it continued to be distributed up to 1972, though it was not updated. It was offered as a practical manual for survival from radioactive fallout, and the resumption of normal life. There was a short section on how to survive being in the vicinity of a direct hit, but the main focus was on the challenges posed by radioactivity. Advice was given on how to prepare, especially in the creation of a 'refuge' room and the storing of water and necessary items. Under the stairs was a favoured location which could be improved by the placing of dense materials on the stairs above. It was assumed that

*Dublin*, USSR General Staff, 1:10,000, sheet 1 (1980) [AUTH]

there was time to make these preparations and that there would also be time to transmit warnings of imminent threat via the radio.

Viewed from today's perspective and a greater understanding of the effects of a nuclear strike or the persistence of radioactive contamination, it was a hopelessly optimistic publication. In its view, the radioactive threat could be measured in hours or perhaps days and services might be out of action for up to 14 days. On page ten, it was noted that:

> The strange existence in refuge will eventually end. About two days in refuge may be expected but this period may have to be extended somewhat if the authorities think it wise to do so. In any case you will eventually be told when it is safe to go out. It is possible that there may be no conditions attached to your release. In many areas, however, people will be told that they can spend only a specified number of hours per day outside the refuge area.

It was accepted that Ireland, even Cork, could be expected to suffer from the temporary effects of radioactive fallout. However, was Ireland ever under more direct threat? Did anyone bother to have a look at potential targets in Ireland? One of the results of the fall of the Soviet Union was the release of many official publications, and for a brief period there was an abundance of these documents. One such gem was a set of maps of Dublin. Described as being published by the USSR General Staff, the maps were at a scale of 1:10,000 and covered the city in four large sheets, each approximately 110 × 90cm. The standard USSR '1942 co-ordinate system' was used to provide precise locational data, but conventional longitude and latitude was also provided. The maps were dated 1980 but the content seems to relate more to the early 1970s, and they were obviously influenced by the Dublin Popular Edition maps produced by the Ordnance Survey. They were not a direct copy though because the geography of the city is not always totally correct – for example, they got the configuration of College Green incorrect. They were well produced on good paper, with clean lines and a pleasing colour scheme. A short gazetteer identified the location of principal streets and there was a brief geographical description of the city entitled 'Reference'. This was a good introduction to the city dealing with climate, weather, topography, economy and so on. For example, the reader was told:

> The Dublin railway junction includes several stations (incl. about 44–51). The most important of these are freight and passenger stations, stations (ob. 47, 49) with developed track facilities and large stations, as well as the cargo port station North Wall (ob. 50). Nearly all node roads are electrified; stations have a depot, repair shops, warehouses.

So, what was their purpose? They were designed to identify places of 'interest'; not the usual tourist sites but locations that might be important in military terms. Colour coding was added to specific features which were also numbered and named in the gazetteer. Thus, 'government and ad hoc institutions' were shown in pink, and included the universities and main banks, as well as the prisons, but the Oireachtas was ignored. In light green were military and communication sites. The various barracks were shown, as well as a pointer towards the airport. Black was reserved for military and industrial production facil-

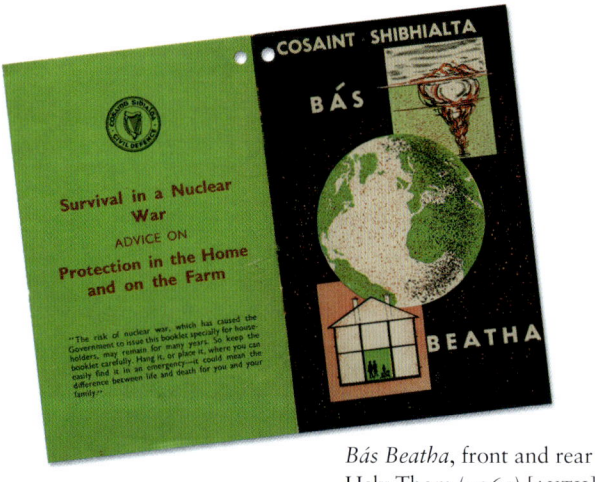

*Bás Beatha*, front and rear covers, Hely Thom (1965) [AUTH]

1980

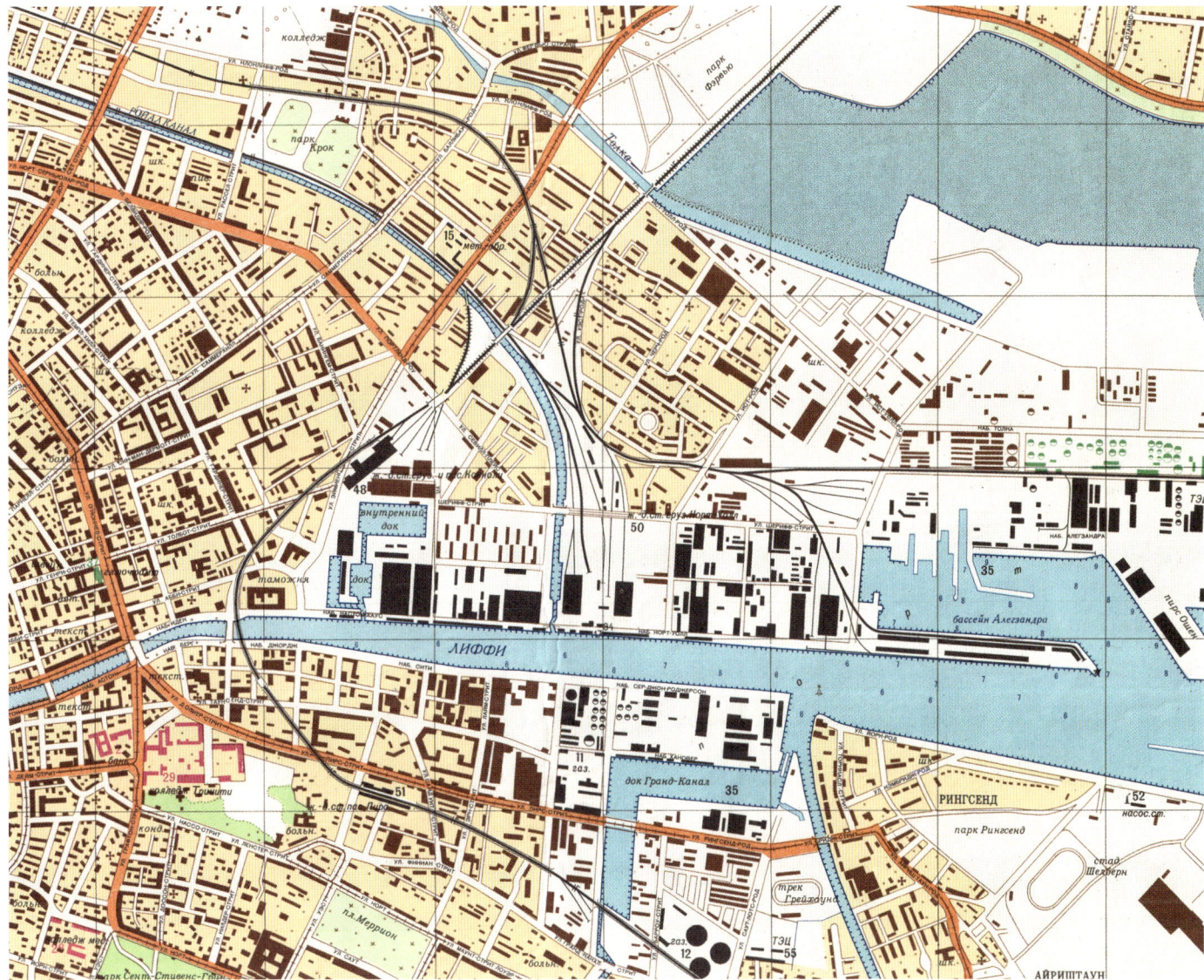

USSR, Dublin Port, extract from *Dublin*, USSR General Staff (1980) [AUTH]. Highlighted are the fuel storage tanks in green, port facilities and gas production in black, Trinity College, Bank of Ireland, College of Surgeons (not the correct location) in red.

ities and here the focus was on the docklands where the map identified the storage tanks and the gas production facilities.

Some comfort might have been taken from the fact that much of the information was out of date. While the map itself was reasonably current, the identification of the Magazine Fort in the Phoenix Park and the Royal Hospital as military instal- lations suggested that their local source was not as good as might be. In a similar vein, University College Dublin might be spared attention because, although the map showed the buildings of the Belfield campus, they were not attributed to the university. Likewise, there was no indication that radio and television now operated in a suburban location at Montrose.

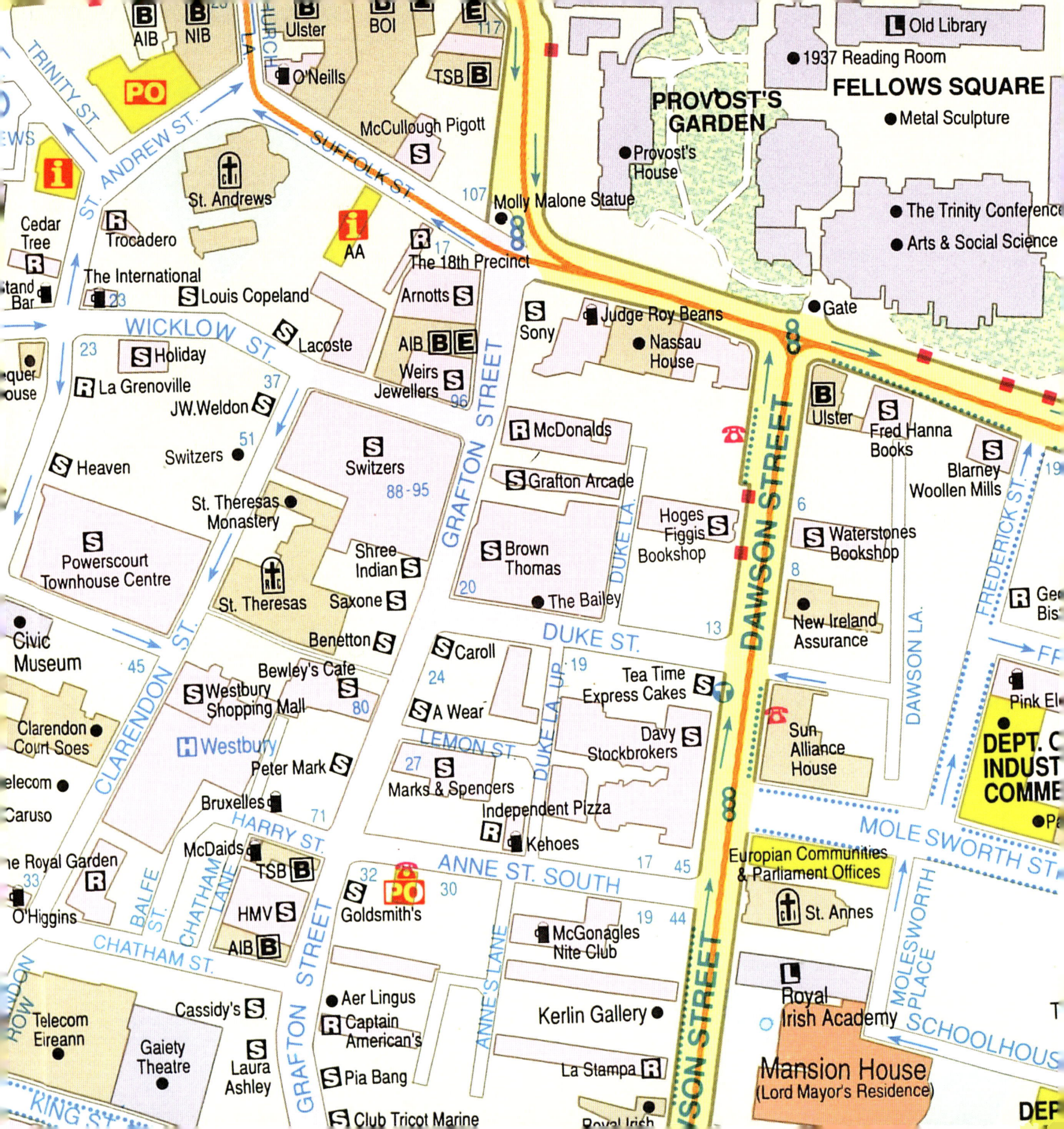

# 1992

## A city centre database

There was a tourism boom in the 1990s and considerable demand for maps of the city that would orient visitors and provide them with essential information. All the major publishing companies produced them either as a stand-alone map or as an insert into a guidebook of the city. One of the more interesting examples was a joint production by ERA-Maptec, based in Dublin, and Shobunsha, then a Japanese publishing company. The map as sold was folded to slightly more than A5 and enclosed within cardboard covers. When opened, there was a map of the entire city on one side at a scale of 1:20,000 and a much more detailed map of the city centre, at a scale of 1:5,000 on the reverse. What was remarkable about the map was the range of information that was shown, far more than on any comparable map. Unfortunately, the 1992 issue proved to be the only edition. ERA-Maptec has now ceased trading and Shobunsha is now Shobunsha Holdings – the company still publishes maps, among other products, but the focus is almost exclusively on Japan.

The maps showed and named both the main and secondary streets, and those with a bus or coach route along them were differentiated by an orange line. Along each of these routes, the stops were shown as red squares. Traffic lights were shown, as was the permitted direction of travel. There was also a symbol for a taxi rank and off-street parking. In the days before GPS systems and digital maps, this would permit a driver to work out the best route in an increasingly complicated city. Granted, it could not be done on the spot (it was very much a desk exercise) but at least the information was readily available. Most roads were shown in yellow, but the pedestrianised streets were in a shade of pink and located on the Earl Street/Henry Street/Mary Street axis, and the environs of Grafton Street on the south side. The necessity to deal with parking meters on other streets was marked by a series of green dots along the building line.

The environs of Grafton Street, extract from *Dublin City Map* (1992) [TCD]

Not all buildings were detailed but those included were differentiated by colour. Yellow was used to identify administrative functions, such as the Department of Education on Marlborough Street, the Inspector of Taxes on O'Connell Street and the offices of Dublin County Council, also on O'Connell Street at that time. Retail businesses were shown in a light pink with a further 'S' symbol to indicate a shop. The retail nature of Henry/Mary Street was very clear with the ILAC centre named (and its internal structure indicated), as was Arnotts with Marks and Spencer and Penny's on Mary Street. The redevelopment of Parnell Street and Jervis Street was just beginning as a consequence of the government-sponsored development incentives, but the plots were still shown as ground-level car parks, a role they had had since the mid-1970s.

Smaller shops were also named, such as Chapters, the bookshop on Henry Street, the Woollen Mills at Merchant's Arch and the Virgin Megastore on Aston Quay. The level of detail was impressive and it was possible to get a good indication of what was available on each street. Earl Street had Dunnes and Boyers while Clerys was nearby on O'Connell Street. Likewise with the Grafton Street area where shops such as A-Wear, Shree Indian, Saxone, Benetton and Teatime Express Cakes were located, as well as the big names such as Switzers and Brown Thomas.

The information did not stop with shops. An 'R' identified restaurants so the visitor could find the Eastern Tandoori, the Cedar Tree or Captain America's and, of course, McDonald's, all in the Grafton Street locale. Also named were the pubs, with most of these further identified by a full beer-tankard symbol. Commercial activity was given brown shading, with the banks differentiated by a 'B' symbol while outlets offering foreign exchange were given an 'E' symbol. There was quite a lot of these in the area between Grafton Street and Henry Street. This was still the era when in-person branch banking was possible and the dense distribution was the legacy of a period with many more individual companies. The same shading was used to identify churches and hospitals. Hospitals were given a green + symbol but churches were more precisely indicated. Each was given a tombstone-like symbol and lettering which indicated the denomination. Thus, St Anne's on Dawson Street was identified as Church of Ireland (CI), while the Church of St Theresa on Clarendon Street was Roman Catholic (RC). Older church sites were also identified. An additional element was that this shading was used to identify residential flat developments. In 1992, most flat developments were social housing – the massive boom in privately owned flats was just beginning – and the map shows both the location and the outlines of the blocks.

There was so much information provided that this was almost a visual Thom's directory. This impression was given greater weight by the inclusion of street numbers on most roads. This very helpfully indicated to any user whether she or he was heading in the wrong direction. Other important items of information included tourist information centres and the location of public phone kiosks. This was not yet the era of the mobile phone – phone cards were a recent innovation – and the location of a public phone was useful information. Of course, finding it in working order was another matter. Another public facility of importance was the location of public toilets. These were a lot less prevalent than they used to be and a map user would have to scour the sheet to find one at Burgh Quay.

All of the statues on public streets were named and among the other cultural elements included were the cinemas, theatres and galleries. There was still a cluster of cinemas in the area around O'Connell Street as well as the Screen on Townsend Street. The Andrews Lane Theatre was mentioned, as was the Olympia, Abbey and Gaiety. The railway stations appeared as bright red and 'Busárus' was helpfully translated as 'bus station'.

Despite all of this information, the map did not seem crowded. The printing was clear and crisp and the colours stood out well. It was not a perfect map – there were small errors – but its imperfections would doubtless have been sorted in subsequent editions. St Mary's Pro-Cathedral was incorrectly named St Mark's. The Irish Life Shopping Centre on Talbot Street had a huge footprint when it was only a minor element of an Irish Life office and flat development. It would

# 1992

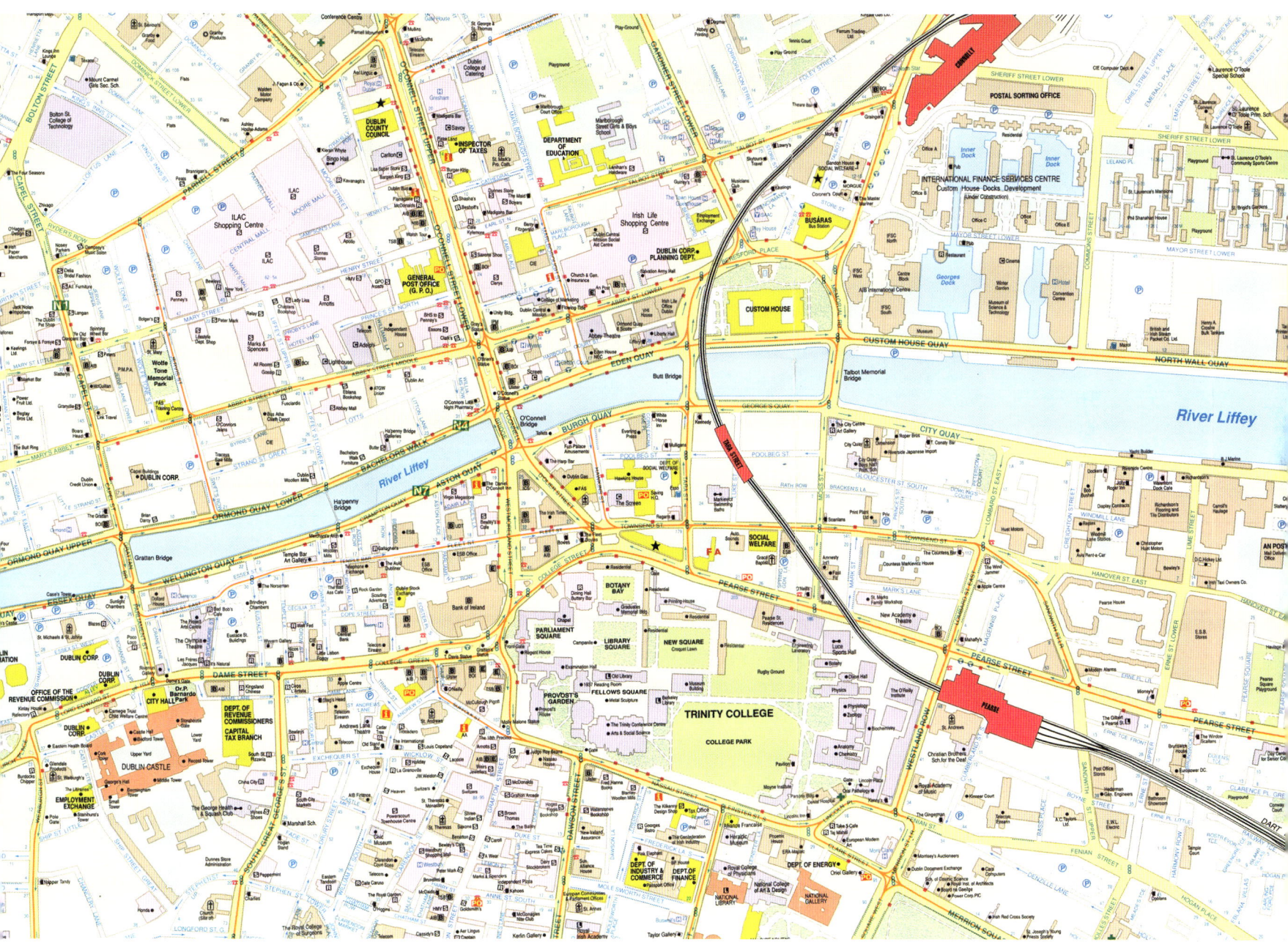

have been a nightmare and a very costly one to keep this map up to date and perhaps that realisation was part of the reason for its disappearance. However, it showed what could be done with a good design and comprehensive information.

ERA-Maptec and Shobunsha Ltd, *Dublin City Map* (1992) [TCD]

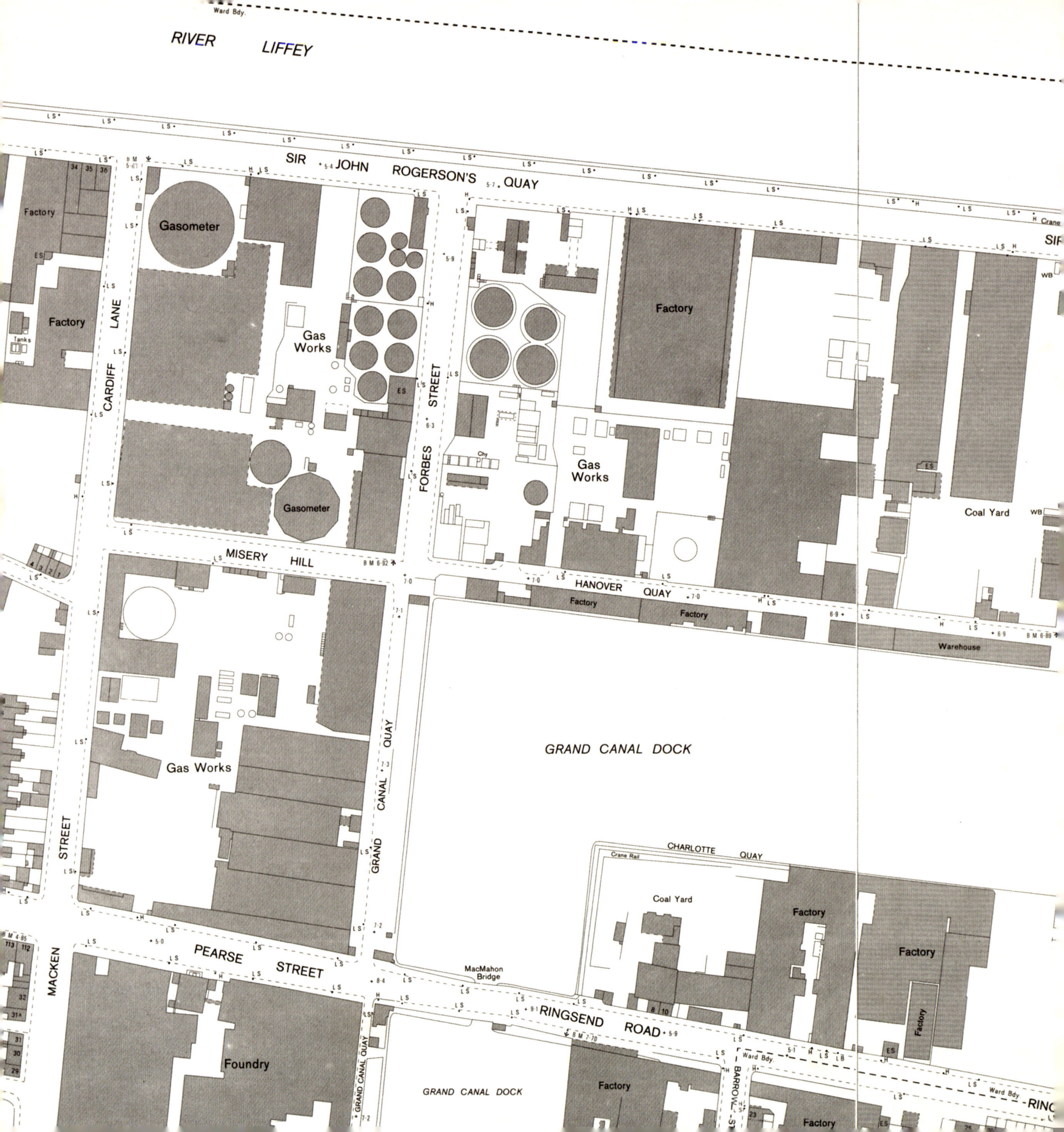
Ward Bdy.
RIVER LIFFEY
SIR JOHN ROGERSON'S QUAY
Crane
Factory
Factory
Tanks
CARDIFF LANE
Gasometer
Gas Works
Gasometer
FORBES STREET
Gas Works
Factory
Coal Yard
MISERY HILL
HANOVER QUAY
Factory
Factory
Warehouse
Gas Works
GRAND CANAL QUAY
GRAND CANAL DOCK
CHARLOTTE QUAY
Crane Rail
Coal Yard
Factory
Factory
Factory
STREET
MACKEN
PEARSE STREET
MacMahon Bridge
RINGSEND ROAD
Foundry
GRAND CANAL QUAY
GRAND CANAL DOCK
Factory
BARROW ST
Ward Bdy.
Factory

# 2000

## Life after town gas: Grand Canal Dock

By the early 1980s, much of Dublin's quays were derelict, the outcome of a combination of factors which included slum clearance and suburbanisation of both population and business. Added to this was the eastward movement of the port, which left behind docklands without an obvious economic purpose. Dublin began to solve these problems by an ambitious programme of incentive-led regeneration which saw block after block of apartments built along the quays up to O'Connell Bridge. In the docklands, an even more ambitious programme of renewal resulted in the area becoming an international financial services district combined with apartment developments of the same kind as further upstream. The market was vibrant and developers were prepared to look at areas where previously they would only have driven through at speed. One such area was where Dublin had produced town gas.

Town gas had been important as a source of lighting, heating and power in Dublin since the early decades of the nineteenth century, but by the 1860s the need to bring order to the market had resulted in the creation of a monopoly supplier – the Alliance and Dublin Gas Consumers' Company (the gas company). They had two major sites: one at Barrow Street was used for storage, and here the distinctive gasometers rose and fell depending on the need for storage, and the other, closer to the river around Grand Canal Dock, was where the gas was produced. This involved heating coal intensively in the absence of air and it was a dirty, smelly and polluting business. Nearby was the 'monster gasometer' on Sir John Rogerson's Quay. This was a fixed structure and its construction in 1933 gave the city a 245-foot high landmark, then the tallest in the city.

The arrival of natural gas in the 1980s left these places without a purpose, until the discovery by the middle classes of 'docklands' as a place to live offered possibilities. The tower

---

Grand Canal Dock, the gasworks site, Ordnance Survey plan, 1:1,000, sheets 3264-7 and 3264-8 (1971) [AUTH]

CLIMB IRELAND FIRST!
The Irish Alpine Club make their assault on the great Gasometer.

The Gasometer, *Dublin Opinion* (December 1933) [AUTH]

was taken down gradually during 1993 but it was close to the end of the century, though, before the future of the storage and production locations became clear. The largest of the storage tanks on Barrow Street was saved and incorporated into a new housing scheme, inevitably called the Gasworks, where construction began in 2003. That left the bigger site from Macken Street eastwards and south to Pearse Street. An early initiative on the part of the gas company (now Bord Gáis) to capitalise on the boom in apartment building along the quays with a £100 million residential development in September 1990 came to nothing. The site was highly toxic

and for some time it was wondered if it would ever be economic to engage in the kind of clean-up that would be necessary. It required designation as part of the remit of the Dublin Docklands Development Authority (DDDA) to get renewal underway. The DDDA acquired the Dublin gas production site in 1998 for a reputed £18 million, regarded as a knockdown price. Then, in March 2000, the Minister for Local Government, Noel Dempsey, approved a proposal to give the DDDA power to develop a planning scheme for the entire Grand Canal Dock area. This plan, covering 29.2ha with an additional 9ha of water, was approved by the minister in January 2001. These interventions, plus a strong continuing demand for apartments in Dublin, ensured the rapid clean-up of the site. The main problem was the soil, which was polluted with a variety of toxins. The agreed plan was to employ a consortium of Pierce Construction and a Belgian company, Soils NV, under the supervision of the Environmental Protection Agency, to deal with the earth. This involved cleaning some soil on site for later use as backfill but the main focus of the operation was to remove the contaminated soil for treatment in Belgium, after which it would be used in road projects across Europe. It was a tricky operation with significant local concern about contamination, and not without controversy. There were complaints about smells and dust flying around, as anyone using the commuter route along Cardiff's Lane and into Macken Street could attest, but the DDDA was adamant that there were no health risks. By early 2002, the DDDA was in a position to invite tenders for sites in the 14 acres where the decontamination was completed. The opportunities included sites for offices, apartments, a hotel, a piazza and underground car parking. There were two big sites for apartments, one at Hanover Quay and the other at the junction of Pearse Street and Macken Street. At the same time, work was beginning on the creation of the internal spaces onto which would face a centre for the performing arts.

The Pearse Street site became Gallery Quay, with 298 apartments together with shops and restaurants available for sale in spring 2003, long before they were available for occupancy. There was a range of types of apartments available, somewhat larger in area than typically available in other quayside locations. Those units facing the water were particularly attractive, and priced accordingly. Forbes Street apartments followed in autumn 2004. This was a development of 124 apartments, some with river views while others faced onto an internal courtyard. There were one-, two- and three-bedroomed units, with the latter having the same area as an equivalent suburban house.

The area was now rapidly moving upmarket, and its commercial character was enhanced when two of the city's major legal firms – Matheson, Ormsby Prentice, and McCann FitzGerald – decided in 2005 to relocate there and to take prominent sites. Work began on the five-star hotel also in 2005 to a design of which some approved and others likened to a 1970s office block.

Character and prestige was provided by the completion of the cultural centre, what became the Bord Gáis Energy Theatre. Having signature structures designed by internationally recognised architects had become a popular device for cities seeking to develop worldwide recognition, and this development soon had two such structures. Linking the Grand Canal Dock project with that of Spenser Dock was the Samuel Beckett Bridge designed by Santiago Calatrava and begun in 2007, though having an uncanny resemblance to his bridge in Buenos Aires. The theatre took a little while to arrive but was available in spring 2010. Designed by Daniel Libeskind, it had his signature discordant lines on the outside, with angular blocks set at varying angles. It looks most impressive from the air but is rather hard to appreciate from ground level. Inside, it had a more conventional look because there is only so much that can work in a theatre. What was more impressive was that it had 2,000 seats, making it twice the size of the Gaiety Theatre and three times the size of the National Opera Theatre in Wexford. Not since Dublin lost the 3,700-seat Theatre Royal in 1962 had the city had such a large venue. Development continued eastwards during the following years and pushed the 'new city' further into the harbour, and there has also been significant development adjacent to the second site, the Gasworks.

The town gas production site (1983) [JB/CPA]

Gallery Quay under construction (November 2004) [AUTH]

The Gasworks development (July 2016) [AV]

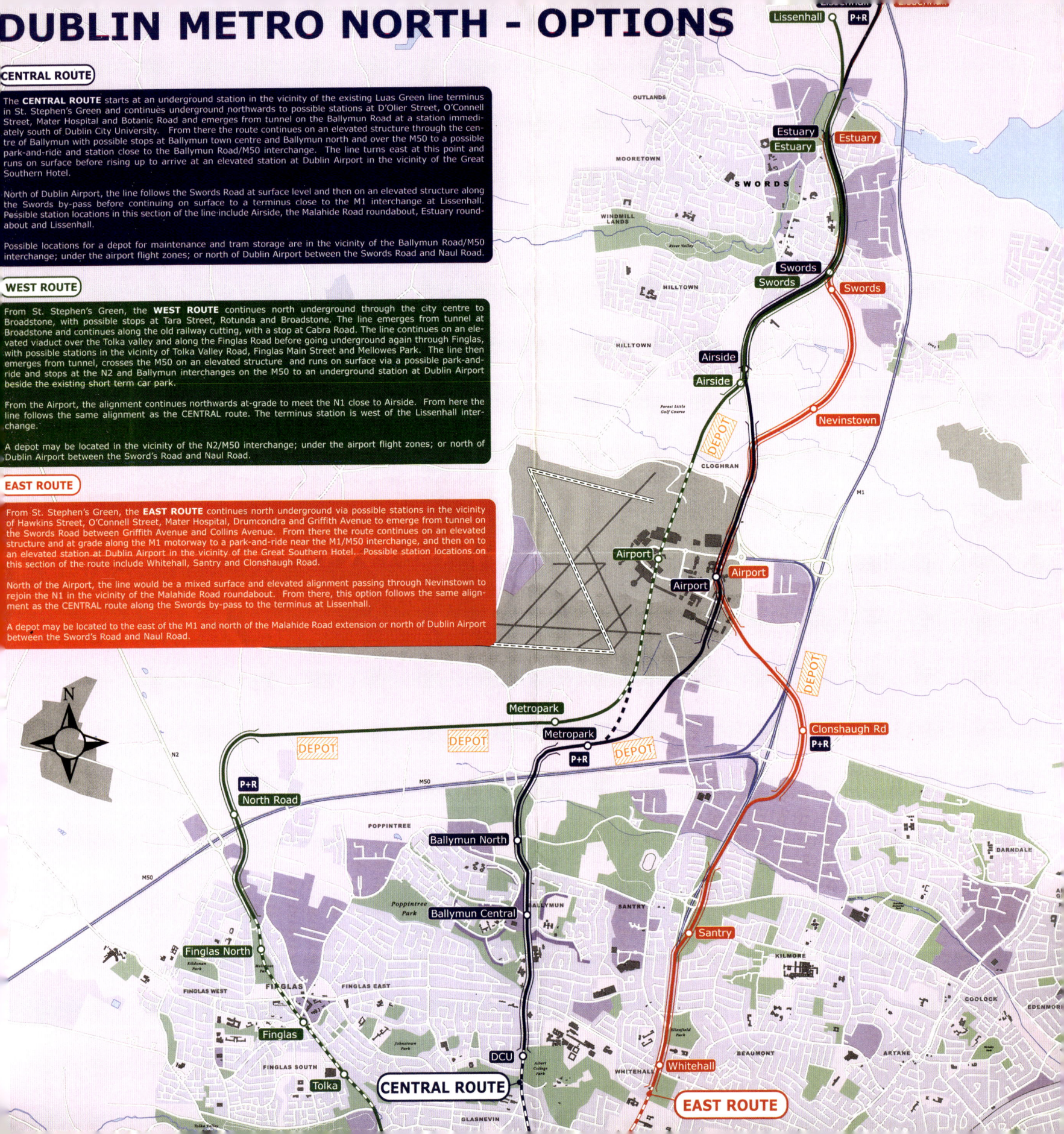

# 2006

## Below ground: an underground system for Dublin

Increasing congestion in London during the second half of the nineteenth century led to the consideration of an innovative solution – an underground railway. Work on the first line, 6km between Farringdon Street and Bishop's Road, Paddington, began in 1860. This first line was made by the cut-and-cover method but subsequent lines were at greater depths using a tunnelling approach. Other European cities were relatively slow to follow but Budapest used cut-and-cover to build a 4km line which opened in 1896. In Paris, the first 10km line for the Métro was opened in 1900. However, the cost and inconvenience of building such infrastructure ensured that underground systems were built in only larger cities, or in cities with particular transport issues.

Despite serious congestion in the city centre, there was never any serious consideration of an extensive underground system. The closest that the city had was the tunnel under the Phoenix Park opened in 1877 by the Great Southern and Western Railway to provide a connection from Kingsbridge Station to the Dublin docklands. A tunnel was also considered to link Amiens Street and Westland Row stations. The absence of detailed knowledge about conditions on the route ensured a long and contested debate, and the idea was dropped in favour of the Loop Line, which was completed by 1891.

The idea of an underground did not figure in any of the traffic solutions proposed for Dublin prior to the 1970s. Railways were out of favour and the trend was towards closing them rather than constructing more expensive versions. The 1970 Dublin Transportation Study (DTS) was not enthusiastic about the need for an underground in the period to 1991, though they did concede that it might be useful to explore it as a means of connecting the various railway stations. As a result, a line appeared on the first Dublin Development Plan map. The DTS report led to the commissioning by Córas Iompair Éireann (CIE) of the Dublin Rail Rapid

---

Railway Procurement Agency, *Dublin Metro North*, public information brochure (2006) [AUTH]

Transit Study (DRRTS) in 1973, and which was completed in 1975. The scope of the study was considerably expanded from a consideration of the underground linkage and it turned into a full-blown assessment of the alternatives for a public transport system. The recommendation was for an electrified rail rapid transit system which would serve the three new towns. This would also serve Ballyfermot and Inchicore, and would terminate with a new underground central station at the Ha'penny Bridge. A new line from Ballymun via Finglas South would connect with the Blanchardstown line and link Cabra with the central area. The existing coastal line from Bray would turn inland at Sandymount and serve the new commercial centres in Ballsbridge and the south-east inner city before joining the other lines at the central station. There would be a new station at Fairview/West Road, and this would connect into the system at Tara Street. The railbed of the Harcourt Street line would become a busway as far as Dundrum (though it had the potential to be extended). It would be a network of 71km, ten of which would be underground, with 47 stations, of which seven would be underground in the city centre. Added to it would be an extensive system of feeder buses, serving areas not on the rail system.

The benefit-cost analysis was not convincing, and nothing happened immediately. However, the mood was swinging back towards the use of rail as part of the solution to Dublin's transport problems. The coastal railway was redeveloped as an electric commuter system – the DART – in 1984. Trams returned to the city centre with the opening of two light rail lines – the Luas – in 2004. These were successful, but a solution was needed for the northern part of the city and a metro system, partially underground, was the preferred option. So began a planning and consultation process for Metro North – ambitiously named as being only the first phase. This would link St Stephen's Green with Swords, via Dublin Airport. The map shown here was that put out for public consultation in 2006. By then, the Railway Procurement Agency had undertaken some preliminary analysis and concluded that there were three routes that would meet a variety of objectives. The western route (22km), would link

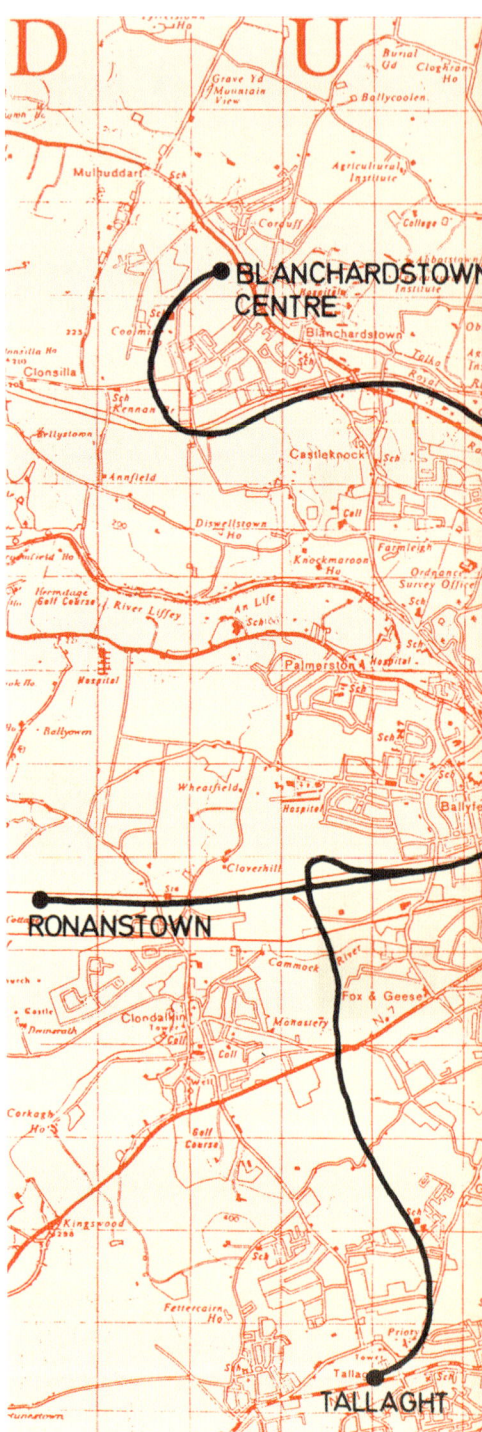

Transport Consultative Commission, DRRTS recommended public transport network, *Report of the Transport Consultative Commission on Public Transport Services in the Dublin Area*, Figure 11.2 (1980) [AUTH]

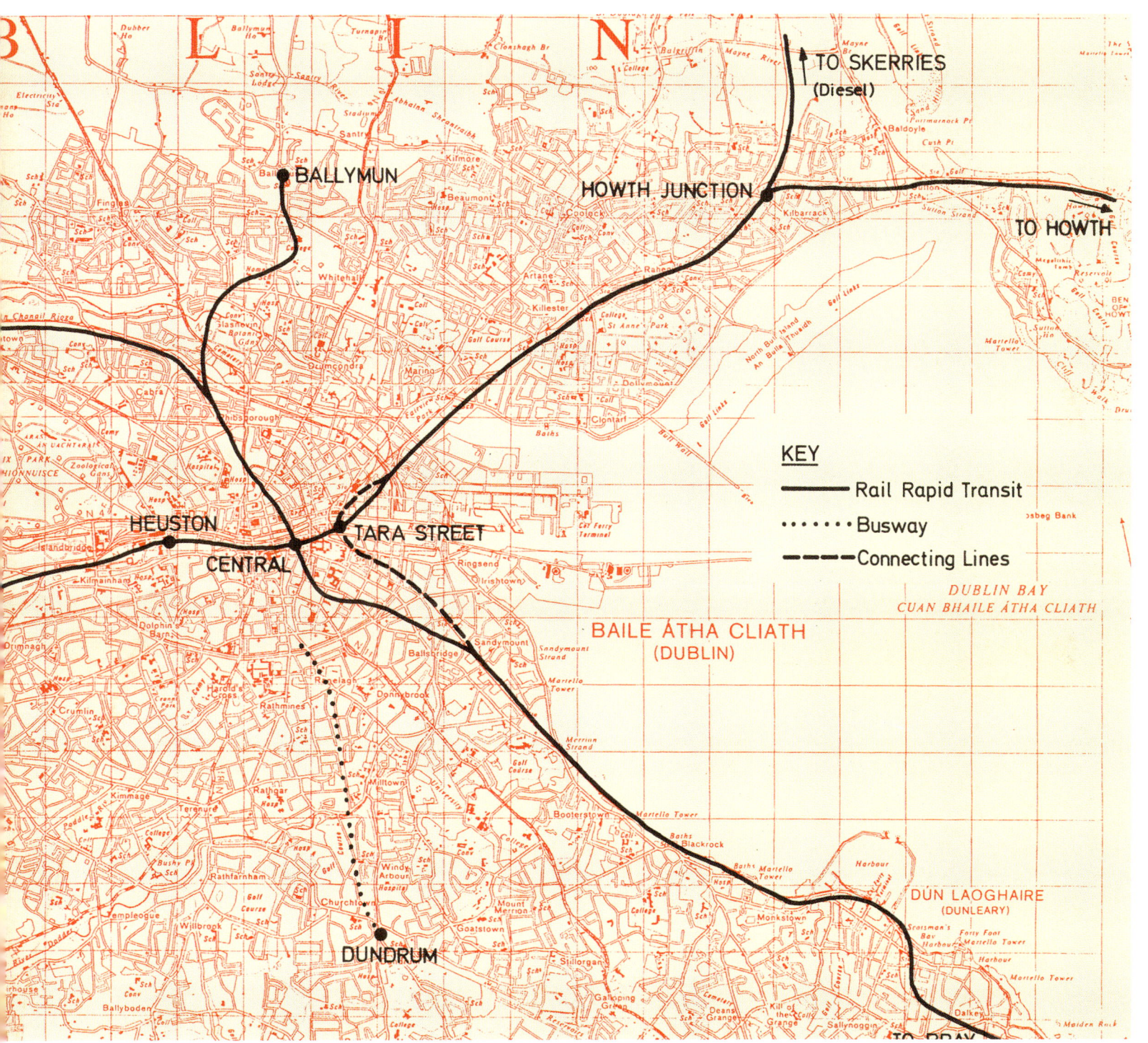

Cabra and Finglas with the city centre and then proceed along largely undeveloped land to the airport. Public transport links to the west of the city were relatively poor and this would meet this need. The central route (17km) had the advantage of directness but it went through long-established and built-up areas and would need a significant underground section as far as Collins Avenue. However, it too would fill gaps in public transport service. The eastern route (17 km) was similar and would come overground only at Whitehall, rather close to the entrance to the Port Tunnel. All three routes would serve the airport, though only the western route would actually link with the terminal buildings; the other two involved a significant walk. There was some variation in the route to Swords and whether the developing Airside business campus was served, but the routes all converged at Swords and onwards then to Lissenhall.

Even at this point in the consultation, the central route had emerged as the preferred option for reasons of construction feasibility, fulfilling transportation needs and the capital cost. Stating this in the consultation document, and noting that this route had been the basis for the RPA's outline business case, diminished somewhat the reality of the consultation process.

The city centre section would be underground no matter which of the three options emerged as the preferred choice. The route would make its way from St Stephen's Green, where it joined with the overground Luas line, to a stop somewhere in the vicinity of Tara Street/Trinity College, and onwards to a station near Upper O'Connell Street.

This was at the height of the economic boom and there was general support for such a large-scale infrastructure project, at least initially. All the main political parties were supportive. However, as the consultation continued, and particularly once it became clear that the decision would be in favour of the central one, the question of the benefit-cost began to become more focused. There was surprise that the costings undertaken had been somewhat vague and the response to the project became less enthusiastic as the reality of the cost began to be understood. There was concern that the benefits did not come close to the costs, even when the RPA proposed that the line would have no more than basic stations. They were contemplating only bare concrete walls, no canopy at street level and only stairs and a lift as access. This led many to wonder what kind of project it really was. The reality of the disruption that would be caused to the city centre also caused people to pause. The initial idea, later scaled back, involved digging up a large segment of St Stephen's Green. While much of the landscape would be restored, there would still be a lot of infrastructure located within the green, thus ruining one of the few squares which the city had. O'Connell Street would have to be essentially rebuilt, with all of the statues having to be removed during construction. The disruption caused by the building of the Luas lines was still fresh in the minds of citizens and they were not entirely convinced by assurances from the RPA that the experience of the Luas construction would ensure that it would be better this time. Along the route, reaction depended on proximity to the line. The experience in Marino with the tunnelling for the Port Tunnel had been generally satisfactory and this did not now raise major concerns. Rather, concern focused on loss of amenity and the location of stations.

The plan was to have the system in operation by 2013 but the economic crash in the years following 2008 put paid to that. While the central route remained the choice, the detail of the route was not finalised and many issues relating to the city centre remained unresolved. The project still remains on the agenda, but it is now accepted that construction will not begin before 2025 at the earliest, with completion by 2035. However, it is far more likely that the western route will become the extension of the Luas Green Line in a shorter timeframe. This already runs along O'Connell Street, via Broadstone, the TU Dublin campus at Grangegorman, to link with mainline and suburban railway line at Broombridge (Liffey Junction).

Dublin Metro North possible routes in the inner suburbs and city centre (2006) [AUTH]

# 2015

## Dublin from space

For most of its history, maps were the only means that Dubliners had to get a sense of what the city looked like from above. The view from high buildings could give a sense of the spatial configuration of nearby places and it is easy to understand why the bird's-eye view of Dublin which the *Illustrated London News* produced in 1846 generated such interest. The satellite era began with Sputnik in 1957, but it was the 1970s before satellite images of the Earth began to become available to the public. It was, of course, rumoured that the military of the superpowers had much more sophisticated imagery available.

As any Dubliner will attest, Dublin is cloudy much of the time. This, combined with the fact that early satellites visited only occasionally, meant that it proved challenging to get good visual images of the city that were free of cloud. Additionally, the resolution level was relatively low, leaving the capacity to read car number plates in the realms of spy fiction.

One of the first detailed images of Dublin came from the Landsat 5 satellite. This was a low-orbit satellite launched in March 1984 and that continued in operation until June 2013. It was jointly managed by the US Geological Survey (USGS) and the National Aeronautics and Space Administration (NASA). During this time it orbited the Earth more than 150,000 times and this image of Dublin comes from 1989 with a resolution of 30 metres. The data was processed by the European Space Agency as part of the Earth Watching project.

The management strategy for Dublin's growth from the 1970s onwards envisaged most growth taking place to the west, with the Dublin mountains preserved for recreation and a flightpath preserved for Dublin Airport. The images show that this is what happened, in broad terms. In the 1989 image, the wide sweep of Dublin bay can be seen together with the Bull Island and the sandbanks around Sandymount. The North and South Bull walls were visible as well as the

---

Dublin, Landsat image, 5TM, 18 July 1989, bands 321 (1989) [ESA]

enclosing walls of Dún Laoghaire harbour. The sinuous line of the River Liffey was visible, as was the Dodder. So too were such distinctive landmarks as Dublin and Baldonnel airports, and the Phoenix Park where the straight line of Chesterfield Avenue stood out. At this time, construction of what became the M50 had begun and the western section was visible as a boundary between the older city and the new suburbs. The extensive nature of development in Tallaght was evident, so too the new housing in Clondalkin, Blanchardstown and Lucan.

Dublin was a quite compact city with relatively little radial development along the main roads out of the city. Even at this scale, it was possible to note a lot of green space within the urban area, particularly in the northern inner suburbs. It was only in the area within the canals that the image suggested dense urban development. July 1989, when this image was taken, was a particularly hot month and parts of the Phoenix Park looked very parched. Elsewhere there was still a lot of green but shades of brown could be seen here and there across the urban landscape. In the countryside, the varied shades indicated the different crops.

The second image is from Landsat 8 and dates from 2015. This satellite was launched in 2013 and is capable of resolutions from 15m to 100m and operates in the visible, near-infrared, short-wave infrared, and thermal infrared spectrums. This image was also processed by ESA and is at 30m resolution. It showed how much the city had grown to the west in the intervening years: the M50 was visible as a complete C-ring around the city and was a dividing line between the older urban landscape and the new suburbs. Similarly, the upgraded M1, M2 and M3 roads cut an identifiable line through the countryside. There had been much infill along the route of the M50 and the brighter objects there, and in adjacent locations, were the various industrial, commercial and retail parks that have been developed: the edge city. A great deal of housing had been added to Tallaght, Clondalkin and Blanchardstown. It was originally intended that there would be distinct green belts between these suburbs and the city but the image shows that while there remained some green space between the western suburbs only the M50 differentiated these from the rest of the city. The image was captured in May so there was no opportunity for the grassy areas in the city to lose their colour and the city appeared much greener as a result. In the countryside, the occasional splodges of yellow indicated the arrival of rapeseed as a crop.

The results from the same kind of development pressure was evident around Dublin Airport. Swords, Malahide and Portmarnock had all grown towards the airport and while there was still a distinct airport 'corridor', the airport was quite circumscribed by development to the north and south. The same was true to a greater extent for Baldonnel Airport to the south-west. Its layout was clear in the 1989 image but by 2015 both housing and commercial developments had appeared around it to the extent that it was quite hemmed in. Though there was some infill along the southern border, the city did not encroach much into the mountains.

There was still a lot of green space visible despite the infill which was going on in the older suburbs. Marino is one of the more distinctive urban landscapes and is readily identified from the air. However, it failed to stand out at this resolution. Instead, it was the regular curves of Cabra and to some extent of Crumlin that stood out. There were clouds over Killiney and Dalkey in 1989, but there was a good view in 2015. River corridors showed up clearly as green borders along the channels and while the southern fringe of the city into Killiney was densely developed, there still was a great deal of green space available.

Dublin, Landsat image, 8OLI, 23 May 2015, bands 432 (2018) [ESA]

# Further reading

Abercrombie, Patrick (1922) *Dublin of the Future* (Dublin, Civic Trust).

Abercrombie, Patrick, Kelly, Sydney and Robertson, Manning (1941) *Dublin Sketch Development Plan* (Dublin, Dublin Corporation).

Abercrombie, Patrick, Gavan Duffy, George and Giron, Louis F. (1942) 'The Dublin Town Plan' [with Comments], *Studies: An Irish Quarterly Review*, 31(122), 155–70.

Abrams, Charles (1961) *Urban renewal project in Ireland (Dublin)*, prepared for the Government of Ireland (New York, United Nations).

Andrews, John H. (1974) *History in the Ordnance Map* (Dublin, The Ordnance Survey Office).

Andrews, John H. (1975) *A Paper Landscape: The Ordnance Survey in Nineteenth-Century Ireland* (Oxford, Oxford University Press). Four Courts Press edition, 2001.

Andrews, John H. (1977) *Two maps of 18th-century Dublin and its surroundings by John Rocque*. Booklet to accompany reprint of Rocque's four-sheet maps of Dublin city 1756 and of County Dublin 1760 (Kent, Harry Margary).

Andrews, John H. (1980) '"Mean pyratical practices": the case of Charles Brooking', *Irish Georgian Society Quarterly Bulletin*, 23, 33–41.

Andrews, John H. (1988) *John Rocque's pocket plan of Dublin, 1757*. Historical note to facsimile reproduction (Dublin, Phoenix Maps).

Andrews, John H. (1989) *John Taylor's map of the environs of Dublin, 1816*. Booklet to accompany the reproduction of the map by Phoenix Maps (Dublin, Phoenix Maps).

Andrews, John H. (1997) *Shapes of Ireland: Maps and their Makers 1564–1839* (Dublin, Geography Publications).

Andrews, John H. (2009) *Maps in Those Days: Cartographic Methods before 1850* (Dublin, Four Courts Press).

An Foras Forbartha (1972) *Dublin Transportation Study. Technical Reports* (Dublin, An Foras Forbartha).

Bannister, David and Morland, Carl (1994) *Antique Maps* (London, Phaidon Press).

Bannon, Michael, J. (ed.) (1985) *The emergence of Irish planning, 1880–1920* (Dublin, Turoe Press).

Bannon, Michael, J. (ed.) (1989) *Planning: the Irish experience, 1920–1988* (Dublin, Wolfhound Press).

Bartlett, William Henry (1846) *The Scenery and Antiquities of Ireland* (London, George Virtue).

Bonar Law, Andrew and Bonar Law, Charlotte (2005) *A Contribution towards a catalogue of the prints and maps of Dublin City and County, Volume 1: prints* (Dublin, The Neptune).

Bonar Law, Andrew and Bonar Law, Charlotte (2005) *A Contribution towards a catalogue of the prints and maps of Dublin City and County, Volume 2: maps* (Dublin, The Neptune).

Boyd, Gary (2005) *1745–1922 – Hospitals, spectacle & vice*. Making of Dublin City series, Volume 3, Joseph Brady and Anngret Simms (eds) (Dublin, Four Courts Press).

Brady, Joseph and Simms, Anngret (eds) (2001) *Dublin through space and time* (Dublin, Four Courts Press).

Brady, Joseph (2004) 'Reconstructing Dublin city centre in the 1920s', in Clarke, H., Prunty, J. and Hennessy, M. (eds) *Surveying Ireland's Past, multidisciplinary essays in honour of Anngret Simms* (Dublin, Geography Publications, 639–64).

Brady, Joseph and Lynch, Patrick (2009) 'The Irish Sailors' and Soldiers' Land Trust and its Killester Nemesis', *Irish Geography*, 42(3), 261–92.

Brady, Joseph (2014) *Dublin 1930–1950 – The Emergence of the Modern City* (Dublin, Four Courts Press).

Brady, Joseph (2014) 'The Liffey and a bridge too far: Bridge-building and governance in Dublin 1870–1960', *Irish Geography*, 47(2), 75–103.

# FURTHER READING

Brady, Joseph (2015) 'Dublin – A City of Contrasts', in Fogarty, Anne and O'Rourke, Fran (eds) *Voices on Joyce*, 77–96 (Dublin, UCD Press).

Brady, Joseph (2016) *Dublin 1950–1970 – Houses, Flats and High Rise* (Dublin, Four Courts Press).

Brady, Joseph (2017) *Dublin in the 1950s and 1960s: cars, shops and suburbs* (Dublin, Four Courts Press).

Brady, Joseph and McManus, Ruth (2018) 'Marino at 100 – a suburb of lasting importance', *Irish Geography*, 51(1) 1–24.

Brady, Joseph and McManus, Ruth (2020) 'Dublin's twentieth-century social housing policies: tenure, "reserved areas" and housing type', *Planning Perspectives*, 35(6), 1005–30.

Brady, Joseph and McManus, Ruth (2021) *Building Healthy Homes* (Dublin, Dublin City Council and Four Courts Press).

Brady, Joseph (2022) *Dublin 1970–1990 – the City Transformed* (Dublin, Four Courts Press).

Buitléir, Muiris de (2013) *A Portrait of Dublin in Maps: History, Geography, People, Society* (Dublin, Gill and Macmillan).

Cameron, Charles (1868) *Lectures on the Preservation of Health* (London and New York, Cassell, Petter and Galpin).

Cameron, Charles (1893) *Report upon the State of Public Health in the City of Dublin for the year 1893* (Dublin, Dublin Corporation).

Clark, Mary (1983) *The book of maps of the Dublin City Surveyors, 1695–1827; an annotated list with biographical notes and an introduction* (Dublin, Public Libraries Department, Dublin Corporation).

Clark, Mary (1985–6) 'Dublin surveyors and their maps', *Dublin Historical Record*, 29, 140–8.

Craig, Maurice, J. (1983) *Brooking's map of Dublin, 1728*. Facsimile reproduction, with architectural notes by Maurice Craig (Dublin, Friends of Trinity College Library and the Irish Architectural Archive).

Clarke, H.B. (2003) *Dublin, part I, to 1610*. No. 11 Irish Historic Towns Atlas (Dublin, Royal Irish Academy).

Clarke, H.B. and Prunty, Jacinta (2011) *Reading the maps: a guide to the Irish Historic Towns Atlas* (Dublin, Royal Irish Academy).

Craig, Maurice (1969) *Dublin 1660–1860* (London, Halcyon Books).

Cullen, Frank (2015) *Dublin 1847: city of the Ordnance Survey* (Dublin, Royal Irish Academy).

Dublin Corporation (1918) *A survey of the north side of the city of Dublin*, Report 13/1918 (Dublin, Dublin Corporation).

Ferguson, Paul (1983) *Dublin – The Castle Sheet 1853*, introduction to reproduction by Alan Godfrey Maps (UK, Alan Godfrey Maps).

Ferguson, Paul (1988) *The A to Z of Georgian Dublin: John Rocque's maps of the City in 1756 and the County in 1760* (Kent, Harry Margary).

Flinn, D. Edgar (1906) *Report of the Sanitary Circumstances and Administration of the City of Dublin with special reference to the causes of the high-death rate* (London, HMSO).

Gilligan, Henry A. (1988) *A history of the port of Dublin* (Dublin, Gill and Macmillan).

Goodbody, Rob (2014) *Dublin, part III, 1756 to 1847*. No. 26 Irish Historic Towns Atlas (Dublin, Royal Irish Academy).

Greater Dublin Commission (1926) *Report of the greater Dublin commission of inquiry* (Dublin, Stationery Office).

Hall, Samuel Carter and Hall, Anna Maria (Mrs S.C.) (1841–3) *Ireland – its scenery and character* (London, How and Parsons).

Herries Davies, Gordon L. (1983) *Sheets of Many Colours: Mapping of Ireland's Rocks, 1750–1890* (Dublin, Royal Dublin Society).

Housing Inquiry (1885) *Report of the Royal commission appointed to inquire into the housing of the working classes. Minutes of evidence etc., Ireland*, British Parliamentary Papers, cd. 4547, London.

Housing Inquiry (1944) *Report of inquiry into the housing of the working classes of the city of Dublin, 1939–43* (Dublin, Stationery Office).

Irish Railway Commission (1838) *Second report of the Commissioners appointed to inquire into the manner in which railway communications can be most advantageously promoted in Ireland* (London, William Clowes).

Joyce, Weston St J. (1921) *The neighbourhood of Dublin* (Dublin, Gill).

Kissane, Noel (1988) *Historic Dublin maps*. National Library of Ireland Historical Documents series (Dublin, National Library of Ireland).

Lennon, Colm (2008) *Dublin, part II, 1610 to 1756: the making of the early modern city*. No. 19 Irish Historic Towns Atlas (Dublin, Royal Irish Academy).

Lennon, Colm (2009) *Dublin 1610 to 1756: the making of the early modern city* (Dublin, Royal Irish Academy).

Lichfield and Associates (1966) *Preliminary appraisal of shopping centre redevelopment in Dublin Centre*. Report No. 1 (London, Nathaniel Lichfield and Associates).

Local Government Board for Ireland (1900) *Report of the Committee appointed by the Local Government Board for Ireland to inquire into the public health of the city of Dublin* (Dublin, HM Stationery Office, Thom and Co.).

Local Government Board for Ireland (1914). *Report of the departmental committee appointed by the Local Government Board for Ireland to inquire into the housing conditions of the working classes in the city of Dublin*. British Parliamentary Papers, 19, cd.7272/7317-xix (Dublin, Local Government Board for Ireland).

Local Government Board UK (1919). *Manual on the preparation of state-aided housing schemes.* Local Government Board (UK, HM Stationery Office).

Local Government (1938) *Report of the Local Government (Dublin) Tribunal* (Dublin, Stationery Office).

McAteer, Desmond (1935) 'Suggested airport for Dublin', *Studies: An Irish Quarterly Review*, 24(93), 73–84.

McAteer, Desmond (1942) 'Merrion reclamation and Dublin town plan', *Studies: An Irish Quarterly Review*, 31 (122), 252–8.

McCullough, Niall (1989, 2007) *Dublin: an urban history: the plan of the city* (Dublin, Anne Street Press).

McDonald, Frank (1985) *The destruction of Dublin* (Dublin, Gill and Macmillan).

Mac Gréil, Micheál (1974) *Educational Opportunity in Dublin* (Dublin, Research and Development Unit, Catholic Communications Institute of Ireland).

McManus, Ruth (1996) 'Public Utility Societies, Dublin Corporation and the development of Dublin, 1920–1940', *Irish Geography*, 29(1), 27–37.

McManus, Ruth (2002, 2021) *Dublin 1910–1940: shaping the city and suburbs* (Dublin, Four Courts Press).

McManus, Ruth (2004) 'The role of public utility societies in Ireland, 1919-40', in Clarke, H., Prunty, J. and Hennessy, M. (eds) *Surveying Ireland's Past, multidisciplinary essays in honour of Anngret Simms* (Dublin, Geography Publications, 613–38).

McManus, Ruth (2019) 'Tackling the urban housing problem in the Irish Free State, 1922–1940', *Urban History*, 46(1), 62–81.

Mapother, Edward Dillon (1866) *Report on the Health of Dublin for the year 1865* (Dublin, Dollard).

Moore, Niamh (2008) *Dublin docklands reinvented – the post-industrial regeneration of a European city quarter* (Dublin, Four Courts Press).

Municipal Corporations (1835) *First Report of the Commissioners appointed to inquire into the Municipal Corporations in Ireland* (London, William Clowes).

Nolan, William and Simms, Anngret (eds) (1998) *Irish Towns: A Guide to Sources* (Dublin, Geography Publications).

O'Brien, Joseph V. (1982) *Dear dirty Dublin, a city in distress, 1899–1916* (Berkeley, University of California Press).

Ó Cionnaith, Finnian (2012) *Mapping, measurement and metropolis: how land surveyors shaped eighteenth-century Dublin* (Dublin, Four Courts Press).

Ó Cionnaith, Finnian (2022) *Land Surveying in Ireland, 1690–1830* (Dublin, Four Courts Press).

Ó Gráda, Diarmuid (2015) *Georgian Dublin – the forces that shaped the city* (Cork, Cork University Press).

Ó'Maitiú, Séamus (2002) *Dublin's Suburban Towns, 1847–1930* (Dublin, Four Courts Press).

Ó'Maitiú, Séamus (2021) *Rathmines. Irish historic towns atlas: Dublin suburbs 1* (Dublin, Royal Irish Academy).

O'Rourke, H.T. (1925) *The Dublin civic survey* (Liverpool, Liverpool University Press).

Osborough, Niall (1996) *Law and the emergence of modern Dublin* (Dublin, Irish Academic Press).

Parker, A.J. (1987) *Dublin Shopping Centres: A Statistical Digest* (Dublin, Centre for Retail Studies).

Prunty, Jacinta (1998) *Dublin slums 1800–1925, a study in urban geography* (Dublin, Irish Academic Press).

R. Travers Morgan and Partners (1973) *Central Dublin Traffic Plan* (London, R. Travers Morgan).

Refausse, Raymond (2000) *A catalogue of the maps of the estates of the Archbishops of Dublin 1654–1850* (Dublin, Four Courts Press).

Rowley, Ellen (ed.) (2016) *More than concrete blocks: Dublin's twentieth-century buildings and their story. Volume 1 – 1900–1940* (Dublin, Four Courts Press).

Rowley, Ellen (ed.) (2019) *More than concrete blocks: Dublin's twentieth-century buildings and their story. Volume 2 – 1940–1972* (Dublin, Four Courts Press).

Schaechterle, Karl (1965) *Dublin traffic plan. Part 1* (Germany, Ulm/Donau).

Schaechterle, Karl (1968) *Dublin traffic plan. Part 2* (Germany, Ulm/Donau).

Shaw, Henry (1850) *The Dublin Pictorial Guide and Directory* (Dublin, Shaw).

Stratten and Stratten (1892) *Dublin, Cork and south of Ireland: A literary, commercial and social review* (London, Stratten and Stratten).

Stationery Office (1980) *Report of the Transport Consultative Commission on Public Transport Services in the Dublin Area* (Dublin, Stationery Office).

Wright, George Newenham (1821) *An Historical Guide to the City of Dublin* (London, Baldwin, Craddock and Joy).

Wright, George Newenham (1831) *Ireland illustrated in a series of views* (London, H. Fisher Son and Jackson.

Wright, L. and Browne, K. (1974) 'A future for Dublin, Special Issue', *Architectural Review*, November, 268–330.

# Index

Abbey Street 193
    improved 121
    Lower 43
    Middle 131, 151, 193
Abercrombie, ideas 120, 123–4, 152, 182, 185
Abercrombie, Patrick xviii, 119–21, 123–5, 139, 171, 179, 180–1, 182–3, 185–9, 203–4
Act of Union 43, 81
Addison, Joseph 49
administration 45, 72, 111–12, 171, 176–7
    local 205
administrative areas 175, 177
advertisements 87–9, 91, 107, 164–5, 167, 191, 196
air, clean 104–5
airport 163, 224, 238, 242
Albert, Prince 201
Alliance and Dublin Gas Consumers' Company 231
alluvium 143
amenities 51, 108, 152, 171, 173, 182, 217, 221, 238
Amiens Street 93, 115, 117, 137, 148
Amory, Jonathan 26
Amory's Ground 26
An Nua Ghaeltacht Teoranta 155
apartments 72, 220, 231–3
archaeology xi, 139–40
architects 30–1, 124, 152, 171, 233
*Architectural Review* xviii, 212, 217
architecture 100, 212
    impressive 77
    regular 35
Arklow 137
Armagh 19, 115
Arnotts 192–3, 228
Arran Quay 9, 217
Artane Castle 25
Ashtown 182
Aston Quay *see* Astons Key
Astons Key 9, 12, 42, 145, 228
atlases xv, xvii–xviii, 5, 15, 64–5, 135, 159, 168
Aungier Street 85, 208
axonometric view 185, 187

Bacon maps 159, 195
Baggot (Baggat) Street 30, 39, 47, 208
Baile Átha Cliath 135–7, 181
Balbriggan 207
Baldonnel Airport 242
Baldoyle 37
Balgriffin 143
Ballast Board 16–17
Ballsbridge 59–60, 69, 195, 212, 236
Ballybough 26, 217
    Bridge 37, 68
Ballybrack 176
Ballyfermot 195, 236
Ballymore Eustace 53
Ballymun 176, 236
Bankes, Thomas 78
banks
    district 88
    in-person branch 228
Bantry Bay 221
Barker, Jonathan 29–31
baronies 135, 137
Barrack Bridge 115
Barrack Street 20, 43
barracks 20, 25–6, 93, 96, 224
    cavalry 96
Barrow Street 231–2
*Bás Beatha* 223–4
Batchelours Walke (Bachelor's Walk) 12, 151
bathing places 27
baths 111, 132
bay xii, xiv, 16–17, 25–7, 34, 37, 55, 57, 124–5, 143, 182, 217, 219–21
    reclamations 179
    shallow xi
Beaux Arts 119
Beggars Bush 47, 93
Beggsboro 161
Belfast 115, 216
Belgrave Square 97
Benediction 164, 165
Beresford Place 121, 208

bird's-eye view xvi, 93, 99–100, 241
bishops 84, 92
Black Pool 6
Black Rock 37, 68
Blackpitts 39, 217
Blackrock 57, 60–1, 65, 69, 107–8, 121, 125, 137, 143, 175–6, 182, 219–20
    Station 172
Blanchardstown 143, 204, 236, 242
Blessington 53, 145
    Basin 39
    Steam Tramway 145
    Street 39
Bligh, William xiv
Blue Coat School 20
Boland's Mill 135
Bolton Street 20, 85, 120
Booterstown 54, 57, 61, 65, 145, 182
Bord Gáis Energy Theatre 232–3
borough 172–3
Botanic Gardens 49–51
Boundary Commission 67–8
Bournville xviii
Bowen, Emanuel 17
Bray 61, 116, 236
    railway line 144
Bride Street 46, 141
bridge 6, 34–5, 55, 61, 147–9, 185, 187–8, 233
    loop line railway 185
    swivel 117, 148
    transporter/lift 148–9
Bridge Street 6
Bridgefoot Street 113, 215
Broadstone 39, 57, 93, 116–17, 238
Brooking, map 12, 19–20
Brooking, Charles xii, 9–10, 12–13, 19, 35, 75, 77
Broombridge 238
Brown Thomas 89, 169, 192, 228
Bull Island xiv, 125, 221, 241
Burgh, Thomas 16
Burgh Quay 12, 208, 228
Busárus 228

buses 109, 145, 205, 236
business districts 116, 128
Butt Bridge 117, 121, 148, 215

Cabinteely 143
Cabra 124, 129, 161, 176, 216, 238, 242
Cameron, report 104–5
Cameron, Charles A. 103, 105, 111
Canal Harbour 39
canals xvii, 17, 39, 47, 55, 64, 69, 136, 145, 212, 216
    motorway 216–17
Capel Street 35, 85, 113, 120–1, 124, 187, 189, 208, 212
Captain America 228
Cardiff's Lane 232
Carlisle Bridge 39, 42, 77, 93, 99, 148
Carmelite Confraternity 84
carriages 20, 30, 61, 85, 89
cartographer 22, 30, 38
    skilled xii
    William Faden 34
cartouche 13, 15, 19–20, 39, 46–7, 53, 57
Cash, Robert xiii–xiv
castle 6, 13, 35, 43, 72–3
Castle Avenue 69
Castle Sheet xv, 71–3
Castle Street 43, 72, 208
Castleknock 182, 216
Castlewood Avenue 97
Cathal Brugha Street 151, 153
cathedral xiii, 21, 83, 120, 185, 187–9
    metropolitan 185, 188–9
Cathedral Street 151, 153
Catholic Emancipation xviii, 81, 163–4
    centenary 164
Cavendish Row 43
central area xvii, 9, 107, 119, 141, 171, 212, 215, 236
Champs-Élysées 121
Chancery Lane 88–9
Chapel Lane 192
Chapelizod 145, 182
Charleville Road 91
chemists 89, 153, 193
Chesterfield Avenue 165, 242
chimneys 75, 100
Christ Church Cathedral xiii, 7, 13, 21, 43, 101, 104, 120, 124, 185, 189
Christ Church Place 88–9
Christian Brothers 84
Church Street 113, 129
churches 6, 13, 20, 82–3, 85, 93, 100–1, 156, 165, 188–9, 228
Churchtown 195
CIE (Córas Iompair Éireann) 235
cinemas 193, 228
circular roads 39, 69, 128, 182, 215
citadel 27
City Architect 139, 152–3
City Basin (Bason) 39, 57, 65, 78–9
city centre xiii–xiv, xvii–xviii, 59–60, 83, 115–17, 120–1, 124, 127–8, 143–5, 151, 163–4, 172, 182, 217, 235–6, 238

city estate 132, 155
City Hall 185, 187–8
city manager 176, 211
City of Dublin Junction 117
Civic Exhibition 139
*Civic Survey* xviii, 139–41, 143–4, 147–9, 171, 173
Civics Institute of Ireland 119–20, 139–40, 144, 149
Civil Service Housing Association 156
Civil War 140, 151, 153
Clanbrassil Bridge 96
Clanbrassil Street 141, 217
Clare Street 30, 208
Clarendon Street 169, 228
clays 105, 112, 143
Clerys 152, 228
Clondalkin 176, 204, 242
Clonliffe Road 69
Clonskeagh 25, 108, 159
Clontarf 16, 26, 37, 69, 108, 116, 125, 132, 136, 141, 175
coal 43, 59, 132, 231
coast, south 57
Cock Lake 17
College (Colledge) Green 6, 12, 39, 42, 71–3, 81–2, 88, 141, 147, 200, 207–8
College, Albert 155–6
colleges 73, 205, 208
Collins Avenue 161, 182, 238
commissioners 35, 41–3, 67, 96–7, 115, 147, 155–6, 211
commons 46, 67, 84, 175
companies
    bus 97, 145
    gas 231–2
    insurance xviii, 88, 167
    rail 107
    shipping 60
Confectioner's Hall 89
congestion 41, 147–8, 173, 235
    points 147
Conyngham Road 43
Córas Iompair Éireann *see* CIE
Cork 3, 5, 19, 115–16, 223–4
Cork Hill 208
Cork Street 93, 161, 215, 217
Corporation 67–9, 127, 129, 131–3, 156, 160–1, 182, 188–9, 200, 203, 211–12, 216
    city estate 6
councils 67, 148, 160, 172, 176–7
    common 67
    elected metropolitan 176
    local 176
county borough 189, 196, 207, 211
county council 53, 175
County Dublin xiv, 27, 53–7, 96, 135, 175–7, 207
Croydon, extension 133
Croydon Park 160
Crumlin 124, 155, 159, 176, 195, 242
Custom House xvi, 9, 13, 43, 65, 93, 99, 117, 121, 124, 185
Cutt Throat Lane 21
cyclopaedia 63, 135

Dalkey 15–16, 37, 61, 108–9, 144, 175–6, 242
Dame Street xiv, 30, 42, 71, 73, 85, 88, 167, 207
Dargan, William 60, 92
Dawson Street 12–13, 169, 228
DDDA *see* Dublin Docklands Development Authority
death rate 105, 111–12
demesnes 49, 53, 57
densities 75, 100, 105, 107, 133, 141, 160, 169
    high 128, 205
    hygienic 141
    lower 96, 161
    population 139–40
department stores 59, 168, 192
design 42, 47, 55, 60, 85, 87, 93, 124, 129, 133, 135, 161
destruction 121, 151, 223
developers 26, 29, 31, 59, 97, 212, 231
development plan 123, 171, 179, 211–12, 220
dignitaries 82, 84
directories xiii–xiv, xviii, 39, 45–6, 64, 87–9, 91
Dirty Lane 21
diseases 103–5, 111–12
distribution 105, 117, 128, 165
district 171–2, 176, 195
    best-preserved Georgian 215, 217
    commercial 208
    dispensary 113
docklands xiii, 65, 117, 129, 225, 231
docks 117, 125, 168
D'Olier Street 39, 42
Dollymount xiv, 55, 57, 107–8, 124
Dollymount Avenue 182
Dolphin's Barn 57
Dominick Street 187
Donnybrook 30, 69, 108
Dorset Street 216
drapers 89, 191
Drumcondra 12, 75, 107–8, 117, 124, 129, 145, 159–61, 175, 216
    stations 117, 144
Dublin Airport 236, 241–2
Dublin and Kingstown Railway Company 60–1, 115
Dublin and Lucan Electric Railway Company 145
Dublin Artisans' Dwelling Company 112–13
Dublin Bakery Company (DBC) 151, 153
Dublin Bar xii, 17, 37, 55
Dublin Bay xi, 15–16, 37, 54, 61, 100, 121, 124, 241
Dublin Bay Preservation Society 221
Dublin Castle 35, 41, 43, 54, 93
Dublin Corporation 12–13, 101, 113, 123–5, 127–9, 131–3, 148–9, 152, 155–6, 175–6, 179, 189, 199–200, 207, 211–12, 216–17, 219–21
Dublin County Council 176, 220, 228
Dublin Docklands Development Authority (DDDA) 232
Dublin Metro North 235, 238
Dublin Metropolitan Constabulary/Police 72, 147
*Dublin of the Future* 119–21, 123–4, 139, 152, 182, 185
*Dublin Opinion* 209, 232
Dublin Port and Docks Board 219–20

# INDEX

Dublin Publishing Company xvii, xix
Dublin Rail Rapid 235
Dublin Sanitary Association 105
Dublin Society 49–51
Dublin Transportation Study (DTS) 215–16, 235
Dublin United Tramways Company (DUTC) 107, 145, 147, 153
Dublin Wicklow & Wexford Railway Company 116
Dún Laoghaire (Dun Leary) 54, 59, 109, 115, 141, 145, 171–3, 176, 182, 207, 211
Dún Laoghaire
    Borough 176, 211, 220
    Civic Survey 145
Duncan, William xiv, 53–7
Dundrum 172, 176, 182, 236
Dunghill Lane 21
Dutch Billy 13, 30, 35, 89
Dyflin 6

Earl Place 152
Earl Street 152, 191, 227–8
Earlsfort Terrace 127
Eblana xi, 2
Eccles Street 39
Eden Quay 145, 151, 208
education xiii, xvii, 63, 113, 139, 196, 228
Eircode 197
Electric Railway Company 108
Electricity Supply Board (ESB) 221
engravings xiii–xiv, 13, 33–4, 37, 47, 50, 54, 65, 88, 96, 99–100
Ensor, John 30
entertainments 72, 193
Essex Bridge 9, 35, 43
Eucharistic Congress 163, 189
Europe xi, xiii, 2, 26, 33, 41, 76, 95, 223, 232
Eustace Street 187
Exchange Building 35
Exchequer Street 73

façades 30, 42, 72, 88–9, 91, 93, 133, 151–2
Faden, William xiv, 37–8
Fairbrothers' Fields 133, 161
Fairview 69, 100, 132, 195
fares 61, 108, 143
Fassaugh Avenue 161
fenestration 30–1
Fianna Fáil 177
Findlater Place 151, 153
Finglas 176, 182, 195, 236, 238
Fitzwilliam (Pembroke) estate 22, 29, 97, 141, 175–6, 188–9
Fitzwilliam Square 39, 208
Fitzwilliams 29–30, 39
flats 129, 172, 228
    tidal 60
Fleet Street 27
Foxrock 144
Francis Street 6, 43, 101, 217
*Freeman's Journal* 61

Gaeltacht Park 155–6
Gaiety Theatre 228, 233

Gandon, James 72
Garden City xviii, 119, 131
Gardiner estate 20–2, 29–30, 39, 47, 64, 120, 128
Gardiner Street 39, 208, 212, 215
Gasometer 232
Gasworks 232–3
Geddes, Patrick xviii, 129, 131
General Post Office (GPO) 87, 151
geography xi–xii, 2, 19, 55, 135, 147, 163, 203, 224
George's Quay 12, 82
Georgian 89
    architecture 212
    character 72, 89, 152, 208
    Dublin 212
Glasnevin 26, 49–51, 85, 117, 124, 132, 136, 155, 160, 164, 175
    cemetery 50, 83–4
    stations 117
Goad, plan 151, 167–8, 191
Goad, Charles Edward xviii, 151–2, 167–9, 191–3
gondola 148–9
governance 95, 175–7
    fractured 175, 211, 219
government 13, 43, 45, 68, 83, 216, 221, 224
Grafton Street 73, 89, 165, 167–9, 191–3, 208, 227–8
Grand Canal xiii, 27, 38–9, 54, 65, 69, 78, 96, 215–16
Grand Canal Dock 78, 125, 141, 231–3
Grand Canal Harbour 39, 79
Grand Jury xiv, 53, 175
*The Graphic* xvi, 99–100, 113, 182
gravels 105, 112, 143
Great Britain Street 84–5
Great Brunswick Street 141
Great Northern Railway 145
Great Southern and Western Railway Company 116–17, 144, 161, 235
green belts 182, 203–5, 242
Gresham, Thomas 92
Gresham Hotel xvii, 92–3, 107, 151
grid 47, 129
Griffith Avenue 132, 136, 156, 182
guidebooks xiii, xvii, 91, 99, 107, 168, 208, 223
Guinness brewery 39, 116

hachuring 37, 65, 136
Hammam Buildings 153
Hampstead Garden Suburb 132
Hanover Quay 232
Ha'penny Bridge 185, 187, 236
harbour xii, 15–17, 55, 57, 61, 65, 78, 100, 233
Harcourt Street 93, 144
    line 116, 236
Harris, Walter xiii, 33
Hawkins Street 141
health xvii, 103, 232
Heffernan, Daniel xvi, 91–3
Hendron Bros 192
Henrietta Street 20, 212
Henry Street 20, 151–3, 191–3, 228
Hibernia 2–3, 201

High Street 101
hospitals xiii, 20–1, 93, 228
hot air balloon 77
hotels xvii, 153, 165, 193, 232–3
house insurance xviii
House of Commons 84, 175
House of Lords 42, 84
housing 35, 95–7, 119, 121, 123–5, 129, 132, 155–6, 171, 173, 176, 182–3, 242
    new 124, 242
    poor 103, 111, 141
    social 123, 161, 173, 228
Housing Inquiry 123, 127
housing problem 140–1, 143
Howth 3, 15–16, 25, 37, 54, 100, 108, 144–5, 176–7, 182
    Castle 16
    Harbour 144
    Head 25, 108
    Ireland's Eye 15
    railway line 109, 144
    tram service 109
    Urban District 176
Hume Street 212
hygiene 105, 139–41

ILAC shopping centre 193, 228
*Illustrated London News (ILN)* xvi, 75–6, 79–82, 84, 92–3, 99–100, 241
*ILN*, panorama 100
Imperial Hotel 89
Industrial Revolution 45, 63
industry 39, 64, 93, 104, 139, 147, 171, 183, 209, 212
    large-scale 78
    linen 93
    polluting 78
infection
    airborne 104
    bacteriological 105
infrastructure 117, 141, 204, 235, 238
inner city 208, 212
    tangent square xviii, 217
institutions 64, 89, 128, 153, 192–3
insurance xviii, 88, 167–8
International Eucharistic Congress xviii, 163–5
international exhibitions 163, 200
*Irish Independent* 156
Irish Life 228
Irish miles 136
*Irish Times* 153, 168–9, 195, 200, 208
Irish Tourist Association xvii
Irishtown 27
Islandbridge xi, 43
    Barracks 93
islands 2–3, 5, 15, 57
Iveagh Trust 113

James' Street 6, 39, 195
Jervis Street 192, 228
    hospital 101
Johnston, Francis 72
Jones' Road 69, 216

249

Kelly, Sydney 171, 182, 187
Kevin Street 217
Kilbarrack (Killbarrack) 37, 176
Kildare House 20, 30
Killester 16, 176, 195
Killiney 176, 242
Kilmainham 161, 175, 201
Kimmage 57, 176
Kingsbridge (King's Bridge) xvi, 85, 93, 115–16, 235
Kingstown 57, 59–60, 65, 82, 100, 108, 115, 117, 175
    railway 60–1, 115

land, undeveloped 60, 64, 97, 172, 216, 238
Landsat 241–2
landscape xiii, xvi, xix, 16, 25, 29, 39, 55, 64, 163, 165
lanes 20–1, 71, 96, 101, 111, 148, 192, 215–17
    narrow 43, 192
    stable 20, 30
Larcom, Thomas 68, 95
layout 25, 31, 71–2, 133, 161, 167, 201, 204–5, 242
Léar Sgáil 135–7
Leeson Bridge 96
Leeson Street 30, 39, 77, 93, 208
Leinster House 29, 77, 200–1
Leinster Lawn 201
Leinster Square 97
Leinster Street 208
Liffey Street 187
lighthouse xiv, xvii, 37
lightship 17, 54
Lincoln Place 208
linear towns 204
linen xiv, 53, 159
    bleaching fields 39
Linnaean system 50
Local Government Board 111, 129, 132, 152
    inquiry 119
Local Government Tribunal 175, 177
Lord Aberdeen 119, 200
Lord Lieutenant 16, 34–5, 43, 68, 84, 91, 119, 200
Lord Mayor 12, 34, 67, 83
lotteries 35
Lucan Electric Railway Company 145

Mabbot Street 12
Mac Dubhghaill, Cathal 135–7
Macken Street 232
Magazine Fort 26, 75, 225
Malahide 176, 242
Mallet, Alain Manesson xi–xii
Malton, engravings 33, 37
Malton, James xiii, 13, 33–5, 37–8
manufacturing 112, 169
Mapother, Edward 103, 113
Marino 25, 55, 124, 129, 131–2, 155, 159–61, 238, 242
    housing scheme 131–3
market xiv, xvii, 43, 113, 115, 168, 231
Marlborough (Marleborough) Street 12, 141, 151, 208, 228
Martello Towers 16, 53–4, 57, 61

Mary Street 192, 228
Mecklenberg Street 84
merchants 6, 73, 87–8
Merchant's Arch 228
Merrion 29, 61, 79, 215
Merrion Gates 60, 183
Merrion Row 208
Merrion Square 22, 27, 29, 31, 39, 77, 80, 83, 85, 188–9, 201, 208
Merrion Street 20, 29–30, 46, 215
Metropole Hotel 89
Midland Great Western Railway Company 65, 116–17, 144
military 45, 224, 241
Minister for Local Government 155, 203, 221, 232
Monkstown 59–60, 172
Montrose 225
monuments xiii, 43, 85, 91, 121, 199–201
Moss Street 207
Mosse, Bartholomew 20
Mountjoy Square 29–30, 39, 128, 212
Municipal Boundaries Commission 175

Nassau Street 83, 89, 168–9, 208
Nelson, Admiral Horatio 199
Nelson Pillar 65, 78, 100, 108–9, 165, 199–200
network 107, 116–17, 141, 143, 145, 216, 236
New City Pictorial Directory 87–9
New Gardens 46
Newfoundland 129
Newgate prison xiii
North Dublin Union 93, 116
North King Street 113, 215, 217
North Lotts 13, 26–7, 129
North Strand, the 26, 69
North Wall 55
    cargo port station 224

O'Connell, Daniel xviii, 80–5, 116, 199
    funeral procession 81–2
O'Connell Bridge 147–8, 164–5, 185, 231
O'Connell Street 120, 141, 147, 191–3, 199, 207–8, 217, 228, 238
offices, civic 187–9
oil refinery 221
Oireachtas 148, 176, 199, 201, 224
Old Cabra Road 156
open spaces 21, 39, 65, 124, 132–3, 160, 173, 205
Ordnance Survey 19, 68, 71, 95–7, 139–40, 143, 155–6, 164, 191, 195, 199–200, 224, 231
    5-foot plan 71
    6-inch plan 143, 156
    25-inch plan 139
    plan 95–7, 155–6, 199–200, 231
Ormond Market 113
Ortelius, Abraham 3
Owen, Robert 63
Oxmantown Green 20
Oxmantown Road 145

Palmerston Park 108
panorama 75–6, 78
parcels 61, 108

Parkgate Street 43, 93
parking xviii–xix, 164–5, 173, 201, 205, 207–9, 227, 232
parks xviii, 27, 65, 123, 133, 160, 163–5, 173, 183, 185, 208–9
    community 160
    linear 124, 133
    public 173, 182, 189
    residential 30
    retail 242
parliament 42–3, 49, 201
parliament building 72–3, 201
Parliament Street xiv, 35, 41, 71, 187–8, 207
Parnell, Charles Stuart 199
Parnell monument 217
Parnell Square 208, 212
Parnell Street 192, 208, 215, 217, 228
passages xii, 21, 41, 124, 192, 211
Patrick Street 113, 212, 217
Pearse Street 141, 195, 207, 232
pedestrian route 165, 205
perspective xii, xvi–xvii, 10, 13, 75, 78, 92, 100, 140, 145, 224
Petrie, George xv–xvi, 75, 78
Phibsborough 195
Phoenix Park 26–7, 41, 46, 49, 75, 78, 92, 117, 121, 163–5, 235, 242
Pidgeon House 26, 37
piles 16–17
Pimlico 215
Pinkeen River 137
planning authority 21, 189, 221
planning scheme 179, 232
Pontifical Mass 164
Poolbeg lighthouse xiv, xvi, 55, 220
Poolbegg 17
Poole, John xiii–xiv
population 35, 96, 104, 107, 141, 171, 205, 231
port 13, 16, 26, 57, 61, 64, 93, 100, 147–8, 219, 221
    asylum 57, 59
Port and Docks Board 148–9, 175, 219–21
Port Sunlight xviii, 132–3
Portland Row 128
Portmarnock 176, 242
Portobello Barracks 96
Portolan charts 1
Post Office Dublin Directory and Calendar 87
Post Office Public Utility Society 156
postal districts 195–7
    new 195, 197
postcards 85, 108, 153, 169, 201
postcodes 197
Poulaphouca 145
poverty xviii, 29, 77, 96–7, 101, 103, 111–12
power 117, 152–3, 199, 203, 207, 231
power station 125
Powerscourt xiii
Price, Charles 15–17
Princes Street 193
printers 81, 131, 169
printing 5, 9, 64, 72, 128, 164, 195, 228
prisons 15, 93, 224

# INDEX

Pro-Cathedral 82–4, 164
promenade xvii, 35, 168
Ptolemaic, geography xi, 2–3
public buildings xiii, 6, 13, 20, 33–4, 71, 91, 93, 123
public health 103–5, 111, 171, 176
Public Health Committee 104
public utility societies 155, 161
purchase
    compulsory 153
    tenant 133, 160–1

quays 9, 13, 35, 43, 77, 85, 141, 148, 152, 188–9, 231–2

radials
    routes 39, 65, 124, 182
    super-normal 124
radio 193, 224–5
Radio Éireann 193
radioactivity 223
Railway Clearing Company 115–17
Railway Procurement Agency 235–6
railways 57, 59, 61, 65, 95, 107, 115–17, 136, 235
    atmospheric 61
    coastal 236
    stations 93, 116, 136–7, 147, 228, 235
Ranelagh 54, 69, 145
Ranelagh Gardens 20
Rathdown 137
Ratheny 37
Rathfarnham 107–8, 176, 182
Rathgar 107
    Improvement Act 96
Rathlin Road 161
Rathmines 55, 69, 91, 95–7, 108, 165, 175–6, 195
    Improvement Act 96
    township 96, 107
Rathmines Road 96
RC *see* Roman Catholic
reclamation 125, 183, 219–20
reconstruction xiii, 151–2, 182
Redmond's Hill 85
reform 63, 68, 177
Regional Planning Act 171, 176, 203
regular streets 13, 37, 77, 88
Renaissance 29, 41, 49, 120
reserved area policy 161
residential areas 20, 29, 88, 107, 145, 182, 195–6, 204–5
residents 97, 129, 156, 197, 208, 221
restaurants 169, 193, 232
Richmond, Bridewell 93
Richmond Barracks 161
Richmond Hill 97
Ringsend 16–17, 26–7
River Dodder 12, 27, 39, 124, 141, 242
River Liffey xi, 6, 9, 12, 34–5, 39, 42, 64–5, 195–6, 209, 212, 217, 219
River Tolka 17, 37, 50, 69, 124, 137
road system xviii, 173, 182, 185, 187, 189

roads 30, 57, 120, 124–5, 156, 159, 161, 163, 165, 182, 203, 205, 211–12, 215–17, 227–8
    100-foot 132, 136
    circumferential 124, 187
    new 30, 54, 132, 143
Robertson Manning 171–3, 181–2, 187
Roches Stores 192
Rocque John xii–xiv, 19–23, 25–7, 30, 35, 37, 39, 41, 47, 57, 71
Roman Catholic (RC) 12, 20, 124, 187–8, 228
roof lines 30, 96, 133
rooms 13, 54, 129, 133, 156
Rotunda 35, 85, 141
    Hospital 46, 121
routes 42–3, 59, 61, 65, 107–9, 116–17, 121, 145, 148, 212, 215–17, 235–6, 238
    bus 165, 195, 205
    central 238
    circuitous 83
    coastal 65, 115
    eastern 238
    numbers 145, 159
Royal Arcade 73
Royal Barracks 20, 25, 64, 78, 92–3, 116
Royal Canal 39, 47, 57, 69, 116–17, 216
Royal Circus 39, 47
Royal Commission 67
Royal Dublin Society (RDS) 200
Royal Exchange 15, 65, 71
Royal Hospital 25, 78, 92–3, 201, 225
Royal Infirmary 64
rubbish removal 97, 105
ruinous 101, 120, 133, 167
Rutland Square 43, 121, 129

Sackville Mall xiii, 20–1, 29, 35, 42, 116
Sackville Street 39, 78, 92, 99, 116, 120–1, 128–9, 151–3, 199
    Lower 87, 89, 151
    Upper 84
Safety First Association 208–9
Sallynoggin 172
Saltaire xviii
Samuel Beckett Bridge 233
sandbanks 17, 26, 37, 55, 241
    significant xii
    south Bull 17
Sandyford 216
Sanitary 104, 111–12
sanitation 105, 111
    public 112
Saxone 228
Scalé, Bernard 22, 27
Schaechterle Karl xviii, 215–16
SDUK (Society for the Diffusion of Useful Knowledge) xv–xvi, 63–5
SDUK
    plan of Dublin xv, 64–5
    publications 64
sea 2, 15, 17, 37, 60, 68, 220
    bathing 27
    charts 15
self-service 191

setbacks 133, 160–1
settlement xi–xii, 2, 26, 57, 65, 69
sewers 97, 105, 141, 216
Shaw, Henry 87, 89
Shelbourne Hotel xvii, 93
Sherrard, Thomas 42
Ship Street 43
shopping centres 168, 182–3, 201
shops 35, 88–9, 165, 168, 191–3, 205, 228, 232
*Sketch Development Plan* xviii, 171, 179, 181–3, 185, 187, 189
Skinners Row 6, 21, 43
slum clearance 101, 124, 187, 231
Smithfield 12, 113, 163, 217
society xiv–xv, 45, 63, 84, 91, 124, 156
soil 105, 112, 232
South Bull 37, 241
South Circular Road 47, 78, 161, 216
South Dublin Union Workhouse 93
South Great George's Street 43, 59, 73, 77, 85, 189, 192, 208
South Wall xiv, 16–17, 37, 54–5, 92, 125, 183, 219–20
souvenirs xviii, 81, 163–4
Soviet Union 204, 224
speculative builders 96, 160
Speed, map xi, 6, 25, 34, 37
Speed, John 1, 5, 7, 9, 19
Spenser Dock 233
spot heights 136
spring tide 16–17, 55
squares 30–1, 39, 77, 79, 91, 93, 97, 129, 133, 189, 215–16
    residential 30, 97
St Andrew Street 73
St Ann's Church 13
St Audeon 101
St Mary's Pro-Cathedral 82, 228
St Patrick 21, 163
    Cathedral xiii, xvi, 99, 101
St Stephen's Green 13, 20, 22, 30, 33, 35, 47, 85, 91, 93, 238
    East 215
    North and Merrion Row 208
    South 215
    West and Grafton Street 208
stations 117, 224, 236, 238
    generating 221
    passenger 116–17, 224
statue 20, 93, 199–201, 228, 238
    equestrian 20, 35, 65, 72, 77
    Victoria 201
storage 43, 165, 201, 231–2
    tanks 225, 232
street directories xiv, xvi, 45, 78
street line 129, 160
streetscape 20, 22, 30, 33, 35, 39, 47, 88–9, 91, 185
subscribers 64, 76, 87, 91, 168
subscriptions 76, 87, 93, 200
suburban
    developments 30, 69, 127, 143
    housing areas 131, 133
    locations 37, 123, 155, 209, 225

suburbs xii, xiv, xvii, 6, 9, 16, 27, 54–5, 95–6, 127,
    159–61
  coastal 143
  developing 65, 69
  garden 119, 131
  inner 238, 242
  model 131, 133
  new xviii, 159, 242
Summerhill 128, 141, 217
supermarkets 191–2
supplements 63, 76, 99, 164
Sutton 17, 108, 145
Swan River 55, 96
Switzers 89, 168–9, 192, 228
Swords 155, 236, 238, 242
Sydney Parade 61
Synod Hall 100–1

Talbot Street 228
Tallaght 176, 182, 204–5, 242
Tara xi
Tara Street 236
Templeogue 57, 145, 182
tenements 89, 101, 103–4, 111, 127–8, 168, 173,
    192
  first-class 129
  multi-family 95
Terenure 108, 145, 176
terminus 60, 108, 115–16, 145
terraces 31, 55, 96
  short 160–1
  small 97
theatre 1, 5, 228, 233
  new national 121
Thom
  directory 87, 167, 196
  map 113, 228
Thom, Alex 87, 107
Thomas Street 43, 212
threat 197, 212, 224

tides xii, 17, 55
  ebbing 17
  high 55
  low xii, 17, 68
  neap 17, 55
toilets, public 141, 165, 228
topography 6, 60, 65, 224
towers 13, 54, 75, 77, 85, 93, 120, 231
town planning xviii, 119, 139, 179, 203
Townsend Street 42, 228
townships 96, 116, 175
  coastal 175–6
  independent 65, 69, 96
traffic 120, 124, 139, 143–5, 147–9, 171, 173, 205,
    207, 212, 215
  congestion 120, 143, 182, 207
  lights xviii, 156, 227
trains 61, 82, 144–5
tram, lines 57, 148, 155
trams 95, 107–9, 123, 144–5, 147–8, 163, 165, 236
Travers Morgan xviii, 215–17
trees 20, 30, 132
  sycamore 13, 20, 35
  yew 50
Trinity College 12, 34, 42, 72–3, 77, 82–3, 88, 91,
    93, 99, 225
tuberculosis 111, 137
Tudor Walters Report 132
Turkish Baths 153
typhoid fever xvii, 103–5

underground 232, 235–6, 238
University College Dublin 127, 225
Unwin, Raymond xviii, 124, 129, 131, 152
urban landscape xvii–xviii, 39, 151, 212, 242
urban renewal 151, 163, 193, 211–12, 231–2

vandalism 200, 208
Vartry water system 105
Viceregal Lodge 26, 41

Viceroy 6, 35, 41, 43, 72
Victoria, Queen 200–1
vignettes xii, xvi, 25, 91
Vikings xi–xii
villages 49, 54, 69, 156
  model 60, 63, 131
vista 35, 77, 121, 185, 187–8

Walsh, Revd David 49
Warburton, John 49
wards xvii, 47, 69, 85, 141
warren 20, 43, 101, 192
water 6, 17, 39, 55, 63, 65, 100, 105, 111, 223,
    232–3
  low 13, 16–17, 55
water supply 39, 97, 141
Waterford Street 208
weather 35, 104, 112, 141, 224
Wellington Quay 187
Wellington Testimonial 65, 93, 121
Westland Row 59–60, 117, 208, 215
  station 93, 137, 148, 235
Westmoreland Street 39, 42, 71
Whitehall 195, 216, 238
Whitelaw, James 49
Whitworth Hospital 64
Wide Streets Commission 20, 22, 27, 29–30, 33, 35,
    39, 41–2, 47, 71, 73, 88, 182
Wilson's Dublin Directory 45–7
Wood Quay 141
workhouses 64, 93
workrooms 104, 169, 193
Wright, Myles 203–5, 216

York Street 208–9

zonation, land-use 123, 182